6th International Conference on High Performance Marine Vessels 2006

(HPMV China 2006)

In Association with 2006 China International Boat Show

Shanghai, China
7-8 April 2006

ISBN: 978-1-61567-851-8

Printed from e-media with permission by:

Curran Associates, Inc.
57 Morehouse Lane
Red Hook, NY 12571
www.proceedings.com

Some format issues inherent in the e-media version may also appear in this print version.

Printed by Curran Associates, Inc. (2010)

The Royal Institution of Naval Architects
10 Upper Belgrave Street
London SW1X 8BQ UK

Phone: 44 207 235 4622
Fax: 44 207 259 5912

publications@rina.org.uk

TABLE OF CONTENTS

A Gas Turbine Modular Model for Ship Propulsion Studies

G. Benvenuto, University of Genova, Dipartimento di Ingeneria Navale e Tecnologie Marine, Italy
U. Campora, University of Genova, Dipartimento di Macchine, Sistemi Energetici e Trasporti, Italy

SUMMARY

The paper presents an original modular mathematical model, based on physical laws, for the dynamic simulation of open cycle gas turbine (GT) plants, specifically developed for ship propulsion plants studies.
Each gas turbine component (i.e. compressor, turbine, combustor) is modelled in a specific module, by means of time dependent momentum, energy and mass equations and non linear algebraic equations. The great application flexibility of the simulator modules gives the possibility of modelling different gas turbine plant configurations.
In the paper the General Electric LM 2500 gas turbine is simulated as a significant application example. The simulation results are presented and compared with experimental data provided by the gas turbine manufacturer.
As far as the ship propulsion application is concerned, the described gas turbine model may be used and has been already used as part of a more complex, previously developed, propulsion plant simulator. In particular the propulsion system of a fast trimaran ferry, consisting of two steering waterjets driven, through reduction gears, by two gas turbines, has been considered and its dynamic behaviour analysed with reference to deceleration-acceleration manoeuvres, whose results are presented in the paper.

NOMENCLATURE

afr air fuel ratio
c_p constant pressure specific heat
c_v constant volume specific heat
f, f' functions
g, g' functions
HP high pressure
h specific enthalpy
K constant
J rotor inertia
LHV fuel lower heating
BFSC brake specific fuel consumption
M mass flow rate
Ma Mach number
n shaft speed
P power
p pressure
Q' heat flow
T temperature
t time
u specific internal energy
V GT component volume
Y momentum-loss parameter
β compressor pressure ratio

Δ difference

γ ratio of specific heat

η efficiency

ε turbine pressure ratio

ω angular velocity

ρ density

σ coolant mass flow partition coefficient

Ω cross-sectional area

1 compressor inlet

2 combustor inlet

3 1st stage *HP* turbine inlet

4 1st-2nd stage *HP* turbine intermediate volume inlet

5 2nd stage *HP* turbine inlet

6 *HP*-power turbines intermediate volume inlet

7 power turbine inlet

8 power turbine outlet

Subscripts

a ambient condition

av average value

b burning

bl air bleed

bloss bearing loss

br brake

C compressor

co coolant

cb combustor

f fuel

g gas

i inlet

in GT inlet duct

loss friction loss

m motor

o outlet

out GT outlet duct

nd non-dimensional

T turbine, turbine stage

1. INTRODUCTION

In the latest years the demand for higher speeds in the passenger ferry market has stimulated the research of new ship types and suitable propulsion systems able to satisfy the required performance in both steady state and transient conditions.

In this context the Department of Naval Architecture and Marine Engineering (DINAV) of the University of Genova, in a joint research carried out in collaboration with other Universities, has developed a concept design of a high speed ro/ro vessel based on the trimaran configuration. Due to the required high speed, the propulsion solution adopted, among other options, was a gas turbine/waterjet combination.

In previous papers the authors presented methodologies and models for the performance evaluation and the dynamic simulation of waterjet propulsion systems.

In this paper the attention is focused in particular on the gas turbine simulation model. Indeed, the performance optimisation of the gas turbine plants used for the propulsion of ships requires the availability of efficient and reliable models to simulate the behaviour of these engines in both steady state and transient conditions.

For this reason, GT simulation codes were developed starting from about 1950 [1], as soon as analog or digital computers were available.

At the beginning, for the GT dynamic simulation, linear [2] or iterative [1,3,4] methods were employed. Afterwards, thanks to the computer technology developments, only the inter-component volumes method [1,5,6,7,8] was utilised. This last approach guarantees in fact a best accordance between simulation and physical phenomena [1].

In the last years, the availability of computer software [9,10] specifically introduced for the transient simulation and analysis of physical phenomena, has allowed the development of GT dynamic simulation models [11,12] organised in modular structure.

The paper presents a modular computer code for the transient simulation of open cycle gas turbine plants, developed in a MATLAB-SIMULINK® environment [10], specifically oriented to dynamics or control studies.

The main targets pursued in developing the model are the following:

- To generate a 'rational' set of modules, each one simulating a GT component by means of physically based mathematical tools;
- To develop a flexible code, able to simulate different GT plant configurations with different degrees of accuracy;
- To achieve low calculation times, also when using a personal computer;
- The possibility of using the model also as part of a more general code for the simulation of more complex systems (i.e. gas-steam combined plants, ship or aircraft propulsion plants);
- To generate a multipurpose GT dynamic model to be used not only in the selection of the best plant configuration, but also in the optimisation of the governor apparatus and of single plant components.

In order to reduce the calculation time, the inter-component volumes method is preferred here to more sophisticated approaches for the GT dynamic simulation [13,14], especially as regards the compressor-turbine calculation.

Nevertheless, since the blade cooling flow in the modern GT high pressure turbine can reach also the 30% of the compressor mass flow, it is decided that this fact cannot be neglected in the turbine expansion calculation. A stage-by-stage expansion calculation model is adopted for the cooled turbine performance determination.

In the paper, after a brief description of the simulator modules, the procedure is applied, as an example, to the simulation of the General Electric LM 2500 gas turbine. Steady state and transient results obtained by the simulator are presented and compared with experimental data provided by the gas turbine manufacturer.

In order to experiment the model validity also in transient conditions, a ship propulsion example was considered, by combining the gas turbine model with a previously developed waterjet system code. The propulsion system consists of two steering waterjets driven, through reduction gears, by two LM 2500 gas turbines. This plant configuration has been proposed for a fast trimaran ferry, whose main characteristics and propulsive performance were presented in previous papers. In this paper the dynamic behaviour of the propulsion system is analysed with particular reference to acceleration-deceleration manoeuvres whose simulated results are presented and discussed.

2. MATHEMATICAL MODEL

The basic features of the model are assumed as follows:

- one-dimensional flow;
- time dependent dynamic and thermodynamic parameters;
- the ideal gas equation is used for the fluid model, internal energy and enthalpy depending (non linearly) from temperature and (for burned fuel gas) also from gas composition [15];
- a distributed parameters process can be simulated by referring to a lumped parameters scheme, with control volumes suitably defined;
- energy and mass accumulation are neglected in the compressor and turbine modules;
- thermal capacity of the metal walls is not considered.

2.1 FUNDAMENTAL EQUATIONS

The working fluid processes are governed by the continuity equation:

$$\frac{d(\rho V)}{dt} = M_i - M_o \qquad (1)$$

and by the energy equation:

$$\frac{d(\rho V u)}{dt} = (M_i h_i - M_o h_o) + M_f LHV + Q' - P \qquad (2)$$

The dynamics of the rotating shafts is determined by the momentum equation:

$$\frac{d\omega}{dt} = \frac{1}{J\omega}(P_m - P_{br}) \qquad (3)$$

2.2 MODULE ORGANISATION

The simulator is structured in a modular arrangement. Each module is pertinent to a GT plant component. A generic module (fig. 1) is characterised by state and global variables, as well as by input and output.

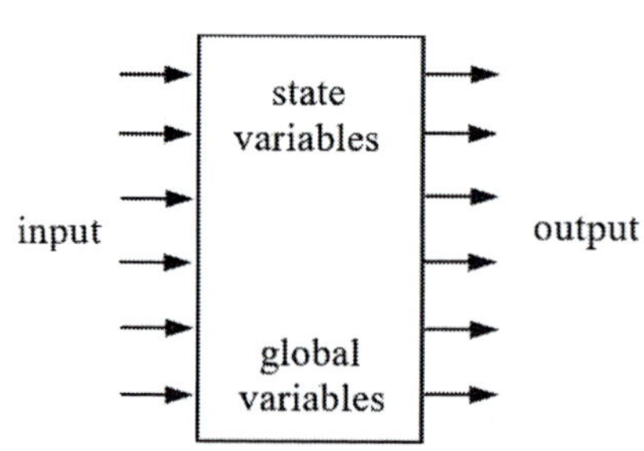

Figure 1: Generic component module.

The mechanical and thermodynamic conditions of a GT component (module) are characterised by its 'state variables' (fig. 1), which are influenced by both 'input' and 'global variables'.

The 'input' are transmitted through input ports of the module, coming from output ports of other modules of the system, or input ports of the simulator. These parameters may be, for instance, fluid related quantities (mass flow, pressure, temperature) or mechanical quantities (shaft speed, power).
The 'output' are transmitted through output ports of the module (see fig. 1) to other blocks of the system.
The 'global variables' are a set of data (i.e. geometrical data, constants of equations) stored in the "workspace" of the simulator and always available to all of the modules.

3. MODULES DECRIPTION

3.1 COMPRESSOR MODULE

Input: stagnation inlet and outlet pressure (p_i and p_o respectively), stagnation inlet temperature (T_i) and shaft speed (n).
Output: stagnation outlet temperature (T_o), outlet air mass flow (M_o) and outlet air bleed (M_{bl}), shaft power (P_C).
For the compressor simulation, steady state performance maps [16] are used (see fig. 2; in this figure and in the following ones the data are normalised by referring the quantities to the Maximum Continuous Rating (MCR) values).
In the compressor module the non-dimensional mass flow rate ($M_{C\,nd}$):

$$M_{C\,nd} = M_C \frac{\sqrt{T_i}}{p_i} \tag{4}$$

and the isentropic efficiency (η_C) are expressed, in 2D matrix form, as functions of the compressor pressure ratio ($\beta = p_o/p_i$) and of the non-dimensional shaft speed ($n_{nd} = n/\sqrt{T_i}$).
The compressor behaviour is determined according to the following calculation scheme.
The effective outlet temperature (T_o) is given by:

$$T_o = T_i \left[1 + \frac{1}{\eta_C} \left(\beta^{\frac{\gamma-1}{\gamma}} - 1 \right) \right] \tag{5}$$

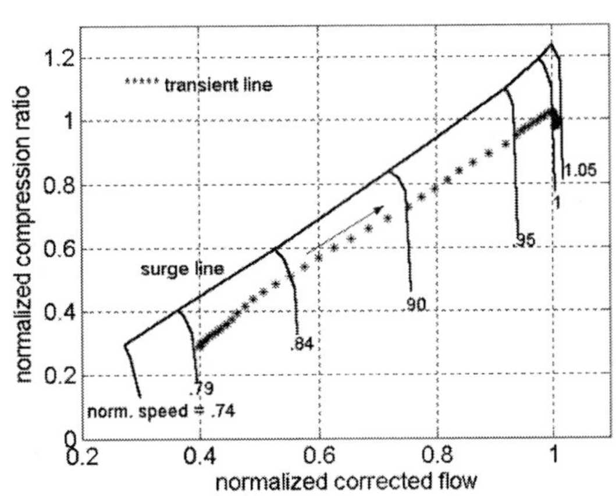

Figure 2:Compressor map and acceleration transient line.

By using the compressor map of figure 2, the non-dimensional mass flow rate ($M_{C\,nd}$) may be determined. Then the compressor mass flow rate (M_C) is obtained from equation (4).
So it is possible to calculate the mechanical power input of the compressor (P_C):

$$P_C = M_C \, c_p \, (T_o - T_i) + P_{C\,bloss} \tag{6}$$

where the term $P_{C\,bloss}$ represents the bearings losses, evaluated as function of the compressor shaft speed (n):

$$P_{C\,bloss} = K_{C\,bloss} \, n^3 \tag{7}$$

As regards the outlet ports of the module, the compressor mass flow rate (M_C) is subdivided into two distinct ports: M_o and M_{bl}.

The mass flow rate M_o represents only the air sent to the GT combustor, which is part of the total mass flow rate of the compressor (M_C), according to the relation:

$$M_o = M_C \, \sigma_C \tag{8}$$

σ_C being a coefficient less than 1 describing the compressor mass flow partition.

The M_o port of the compressor module must be linked with the corresponding inlet port (M_i) of the combustor module.

The mass flow rate M_{bl} represents the air bleeding used to cool the turbine blades, whose value is given obviously by:

$$M_{bl} = M_C \, (1 - \sigma_C) \tag{9}$$

The M_{bl} port of the compressor module must be connected with the corresponding inlet ports of the turbine blocks.

3.2 TURBINE MODULE

This module is able to simulate the turbine expansion in case of both cooled and non cooled turbine blades.

In case of cooled turbine blades, a stage-by-stage expansion scheme is adopted and the turbine module is used for the stage expansion determination. A multistage turbine can be modelled by connecting in series more turbine modules.

Similarly to the compressor model, steady state characteristic maps are used for the stage performance determination. However the possibility of computing stage-by-stage the turbine expansion together with the choice of taking into account also the blade cooling effect, has implied a rather different approach in the turbine calculation procedure, as compared with that adopted for the compressor.

The input/output of this module are:

Input: stagnation inlet and outlet pressure (p_i and p_o respectively), stagnation inlet temperature (T_i) of the main flow, inlet air bleeding stagnation temperature (T_{bl}) and mass flow rate (M_{bl}), shaft speed (n).

Output: stagnation outlet gas temperature (T_o), inlet and outlet mass flow rate (M_i and M_o respectively) of the working gas, shaft power (P_T).

In the turbine stage module, the steady state performance maps [16] are given in 2D matrix form. More in detail, the non-dimensional mass flow rate ($M_{T\,nd}$):

$$M_{Tnd} = M_T \frac{\sqrt{T_i}}{p_i} \tag{10}$$

and the isentropic efficiency (η_T) are both functions of the stage pressure ratio ($\varepsilon = p_i/p_o$) and of the non-dimensional shaft speed ($n_{nd} = n/\sqrt{T_i}$):

In case of cooled turbine stage, the calculation scheme proposed by El-Masri [17] (see fig. 3) is adopted, with some modifications introduced in order to reduce the calculation time.

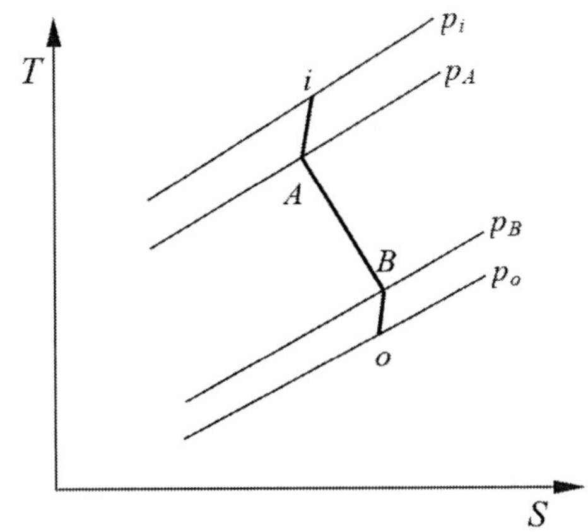

Figure 3: Cooled turbine stage calculation model.

The overall temperature and pressure drops in the stage (*i-o* in fig. 3) are subdivided in three parts:

(*i-A*) stator row-cooling flows mixing;

(*A-B*) adiabatic work-extraction expansion;

(*B-o*) rotor row-cooling flows mixing.

The gas stagnation temperature at the adiabatic expansion starting point (T_A in fig. 3) is determined by applying the energy balance equation to the stator row-cooling flows mixing (*i-A*):

$$T_A = \frac{M_i\, c_{pi}\, T_i + M_{co\,iA}\, c_{p\,co}\, T_{bl}}{\left(M_i + M_{co\,iA}\right) c_{pA}} \tag{11}$$

where the coolant mass flow ($M_{co\,iA}$) is a fraction of the total air bleeding coming from the compressor (M_{bl}), its value being determined by the coolant mass flow partition coefficient (σ_{iA}):

$$M_{co\,iA} = M_{bl}\, \sigma_{iA} \tag{12}$$

where the σ_{iA} parameter is specified for each row. The summation of these parameters, along the overall cooled expansion, must be equal to one.

In the turbine stage calculation, the stator row-cooling flow mixing losses are expressed as functions of the momentum-loss parameter (Y) [17]. So the total pressure at the mixing outlet (p_A in fig. 3) is given by:

$$p_A = p_i \left\{ 1 - 2\,Y \frac{M_{co\,iA}}{M_i} \left[1 + \left(1 - \frac{\gamma - 1}{2} Ma^2 \right)^{\frac{-\gamma}{\gamma - 1}} \right] \right\} \quad (13)$$

where Ma is the absolute Mach number in the mixing zone.

An equation similar to equation (13) may be used in order to calculate the stagnation pressure (p_B in fig. 3) in the inlet section of the rotor row-cooling flows mixing zone (B-o in fig. 3).

It is then possible to determine the expansion ratio (ε) of the stage work-extraction part (A-B in fig. 3):

$$\varepsilon = \frac{p_A}{p_B} \quad (14)$$

So the effective outlet temperature (T_B) of the adiabatic expansion phase may be obtained as:

$$T_B = T_A \left[1 - \eta_T \left(1 - \varepsilon^{\frac{\gamma - 1}{\gamma}} \right) \right] \quad (15)$$

where the turbine efficiency (η_T) is evaluated from the turbine performance map.

The mass flow rate relative to the adiabatic work-extraction phase (M_{AB}) is expressed by the re-arranged equation (10):

$$M_{AB} = M_{T\,nd} \frac{p_{Ai}}{\sqrt{T_A}} \quad (16)$$

where the non-dimensional mass flow rate $M_{T\,nd}$ is derived from the turbine performance map as function of the expansion ratio ε (eq. 14), and from the non-dimensional shaft speed ($n_{nd} = n/\sqrt{T_A}$).

The stage mechanical power (P_T) is then calculated as:

$$P_T = M_{AB}\, c_p \left(T_A - T_B \right) - P_{T\,bloss} \quad (17)$$

taking into account the mechanical bearings losses as previously done for the compressor (see eq. 7).

The turbine module outlet ports: stage outlet stagnation temperature (T_o) and mass flow rate (M_o), are calculated by using respectively the energy balance equation (11), applied to the rotor row-cooling flows mixing phase (B-o in fig. 3), and the continuity equation:

$$M_o = M_{AB} + M_{co\,Bo} \quad (18)$$

where the coolant mass flow ($M_{co\,Bo}$) is determined with an equation similar to equation (18).

The block outlet port: inlet mass flow rate (M_i) is calculated with an equation similar to equation (18).

The procedure above described may be applied also to a non-cooled turbine stage, by using the same turbine module.

In this case, referring to the stator row-cooling flows mixing part (i-A in fig.3), the coolant mass flow ($M_{co\,iA}$), determined in equation (12), is put to zero by assigning a zero value to the coolant mass flow partition coefficient ($\sigma_{i,A}$), or alternatively, because the total air bleeding from the compressor (M_{bl}) is already null (GT without cooled turbine(s)).

This simplifies automatically the equations (11) and (13).

In case of non-cooled turbine stages, the stator row-cooling flows mixing part (*i-A* in fig. 3) has no effect on the stage expansion calculation. The same conclusion may be applied to the rotor row-cooling flows mixing part (*B-o* in fig. 3).

In this case, it has to be noticed also that a multistage turbine can be simulated by means of a single turbine module, by extending to the overall turbine expansion the same performance map approach used the for the stage calculation [16].

The model arrangement adopted for the simulator permits this option without the need of modifying the turbine module.

3.3 COMBUSTOR & INTERMEDIATE VOLUME MODULE

The GT combustor is modelled as an adiabatic capacity, taking into account the time-dependent accumulation of mass and energy.

The variables and parameters considered in this module are as follows:

State variables: stagnation pressure (p) and total internal energy (u) of the plenum.

Input: fuel mass flow rate (Mf), stagnation inlet gas temperature (T_i), inlet and outlet mass flow rates of the working gas (M_i and M_o respectively).

Output: stagnation inlet and outlet pressures (p_i and p_o respectively), stagnation outlet temperature (T_o).

The combustor simulation is based on the following calculation scheme.

Inside the plenum the gas composition and thermodynamic state are considered uniform, moreover the speed variations are neglected. The fluid properties at the combustor outlet section are considered equals to those existing inside the volume.

By using the re-arranged continuity equation (1):

$$\frac{dp}{dt} = \frac{\gamma RT}{V}\left(M_i + M_f - M_o\right) \tag{19}$$

and the energy equation (2):

$$\frac{du}{dt} = \frac{1}{\rho V}\left(M_i h_i - M_o h_o\right) + M_f LHV \eta_b \tag{20}$$

it is possible to determine the combustor plenum state variables: stagnation pressure (p) and total specific internal energy (u).

In the energy equation above the combustion efficiency (η_b) is introduced in order to compensate the adiabatic wall hypothesis assumed for this component.

The stagnation outlet pressure and temperature (p_o and T_o respectively) are then calculated by means of the simple equations:

$$p_o = p \tag{21}$$

$$T_o = \frac{u}{c_v} \tag{22}$$

By evaluating the total pressure loss across the combustor (Δp_{cb}) as function of the quadratic flow velocity in the plenum:

$$\Delta p_{cb} = K_{cb\,loss}\left(\frac{M_o}{\rho\Omega_{cb}}\right)^2 \tag{23}$$

the inlet combustor pressure may be determined:

$$p_i = p_o + \Delta p_{co} \tag{24}$$

The combustor module is naturally allocated between the compressor and turbine modules of the simulator, as required by the input/output parameters organisation. At the input flow sections of the compressor and turbine modules, the pressure values are requested as input parameters, while at the outlet flow sections the mass flow rates are calculated as output parameters. The contrary happens in the combustor module where the pressures and the mass flow rate parameters assume an opposite role.
For this reason, in order to perform a stage-by-stage turbine calculation, it is not possible to connect in series two turbine modules, without interconnecting them to a combustor module (named more correctly in this case 'intermediate volume' module).
The calculation scheme of the intermediate volume module is near the same used for the combustor module, but in this case a zero value is imposed to the fuel mass flow rate (*Mf*) and, in order to reduce the calculation effort, the energy equation (20) may be substituted by the more simple relation:

$$T_o = T_i \tag{25}$$

The same considerations apply also to the connection in series of two compressor modules. Also in this case, in fact, use is made of an intermediate volume module.

3.4 SHAFT DYNAMICS MODULE

This block is used to determine the time variation of the shaft rotational speed.
The variables and input/output considered are as follows:
State variables: previous value of the shaft angular velocity (ω).
Input: motor power (P_m), brake power (P_{br}) of the considered shaft.
Output: new value of the shaft speed (n).
The shaft dynamic balance is expressed by the dynamic shaft equation (3). In this equation the motor power (P_m) is given by the sum of the outlet power parameters (P_T) coming from the turbine modules, pertinent to the analysed shaft. In the same equation, the brake power (P_{br}) is the sum of the shaft compressor powers and/or of the load powers.
Once determined the new angular velocity, the module output parameter (n) is obviously given by:

$$n = \frac{\omega}{2\pi} \tag{26}$$

3.5 INLET & OUTLET MODULES

In order to simulate the GT inlet and outlet flow ducts, these two blocks, where the ambient conditions: stagnation ambient pressure (p_a) and temperature (T_a) are provided as global variables, are used to calculate the pressure losses.

As regards the **inlet module**, its input and output parameters are as follows:

Input: compressor mass flow rate (M).

Output: stagnation pressure and temperature (p_o and T_o respectively) at the outlet section.

In this module, the stagnation outlet pressure is determined by the equation:

$$p_o = p_a - \Delta p_{im} \tag{27}$$

where Δp_{in} is the total pressure loss across the GT inlet duct, evaluated with an equation similar to (23).

The outlet section stagnation temperature is calculated with the equation:

$$T_o = T_a \tag{28}$$

The input and output of the **outlet module** are as follows:

Input: inlet section stagnation temperature (Ti) and mass flow rate (M).

Output: inlet section stagnation pressure (p_i).

The inlet section stagnation pressure is determined with the equation:

$$p_i = p_a + \Delta p_{out} \tag{29}$$

where the total pressure loss across the GT outlet duct (Δp_{out}) is evaluated in analogy with the corresponding term of equation (27).

4. MODEL VALIDATION

The presented gas turbine model has been applied to the simulation of the General Electric LM 2500 gas turbine (fig. 4), consisting, of a two shafts arrangement (gas generator and power turbine) widely used also in the ship propulsion plants.

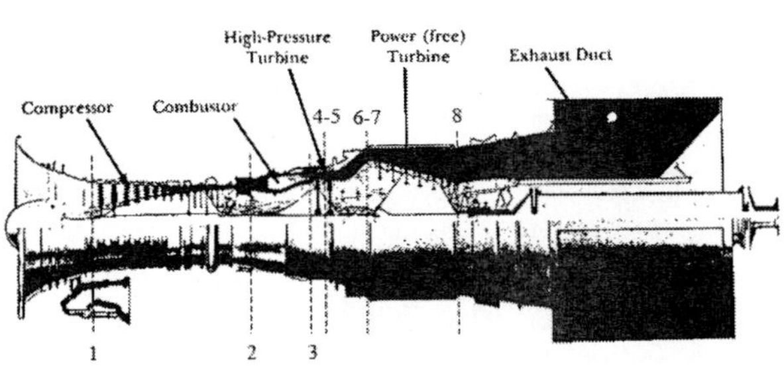

Figure 4: LM 2500 gas turbine and calculation stations (from: FIAT AVIAZIONE, LM2500 handbook).

The gas generator consists of a sixteen stage axial compressor, an annular combustor and a cooled, two stages, high pressure turbine. The gas coming from the high pressure turbine expands in a six stages power turbine, which is only aerodynamically coupled with the gas generator. In this application the power turbine is connected with an electric generator. The MCR (maximum continuous rating) power delivered by the power turbine considered in this application is 26 MW.

In figure 5 the overall scheme adopted to simulate the considered gas turbine is shown.

The graphical SIMULINK representation puts into evidence all the connections between the input/output ports of the simulator modules.

The GT plant stations indicated in fig. 4 correspond to the calculation stations of the simulator scheme visualised in fig. 5.

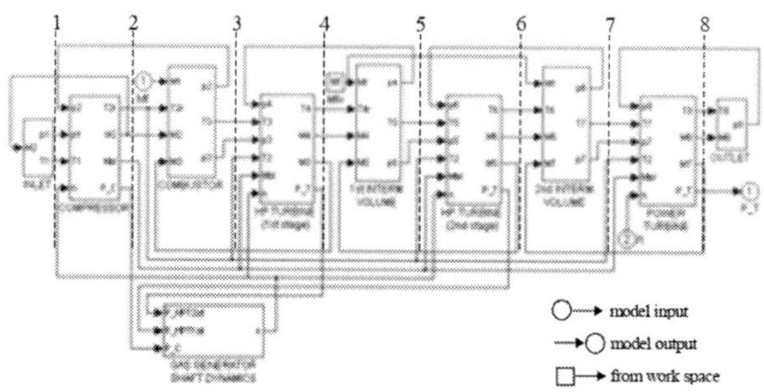

Figure 5: LM 2500 gas turbine overall simulator scheme.

The high pressure turbine has two stages, both cooled, consequently, two turbine modules are required for the simulation of this turbine (fig. 5). Instead, only one turbine module is used for the six stages power turbine.

The GT plant variables are characterised by very different time variations (stiff system). SIMULINK [10] provides several algorithms for the integration of the differential equations set typical of a stiff system. In the presented test case, the variable-step Rosembrock algorithm was adopted, by selecting in a suitable way the integration step range. By using this equation solver on a 800 MHz PC, the calculation time of a transient load variation is about the 10% of the real time.

The validation of the gas turbine simulator is based on a comparison with experimental results in case of both steady state and transient situations.

plant component	Process variable mismatches [%]
compressor	outlet total pressure -0.68
	outlet total temperature 1.29
	outlet mass flow rate -0.10
combustor	outlet total temperature -1.73
high pressure turbine	outlet total temperature -1.97
	outlet mass flow rate -0.10
	shaft speed 0.23
power	outlet total temperature

turbine	-1.86 shaft power -1.32

Table 1: Comparison between experimental and simulation results at gas turbine MCR conditions.

Table 1 shows this comparison referred to the GT plant design conditions (Maximum Continuous Rating). As shown, the mismatches turn out to be less than 2%.

In the gas turbine performance map of figure 6, the experimental (dashdot) lines of constant Brake Specific Fuel Consumption (*BFSC*) are compared with the calculated (solid) lines, showing a substantial good agreement in the entire GT working area.

The same agreement is shown also in figure 7, which reports the experimental (dashdot) and calculated (solid) curves of constant temperature at the power turbine inlet.

The availability of experimental data relative to a transient load variation (acceleration) of the LM 2500 gas turbine has allowed the model validation also in case of dynamic simulation of transient situations.

The acceleration is determined by imposing a time variation law for the fuel mass flow rate. The simulated acceleration transient line (asterisks) is reported on the compressor map of figure 2.

In fig.8 the simulated time variation of the stagnation temperature at the power turbine inlet is compared with the experimental values. The same comparison is shown in fig. 9 for the shaft speed of the gas generator.

On the base of the analysis reported, it may be deduced that the developed model presents a good reliability for the simulation of gas turbine plants in both steady state and transient conditions.

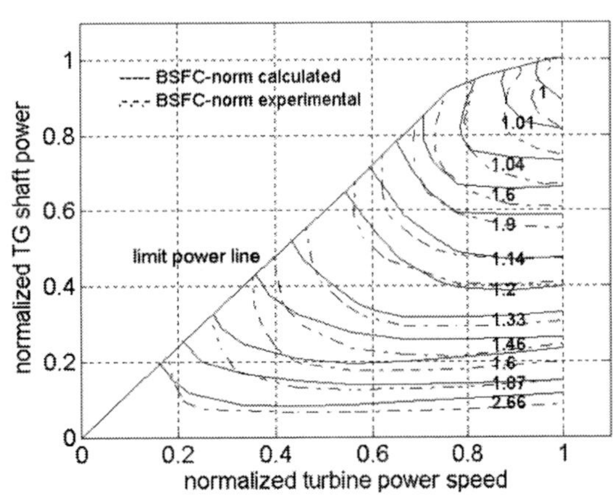

Figure 6: LM 2500 break specific fuel consumption map.

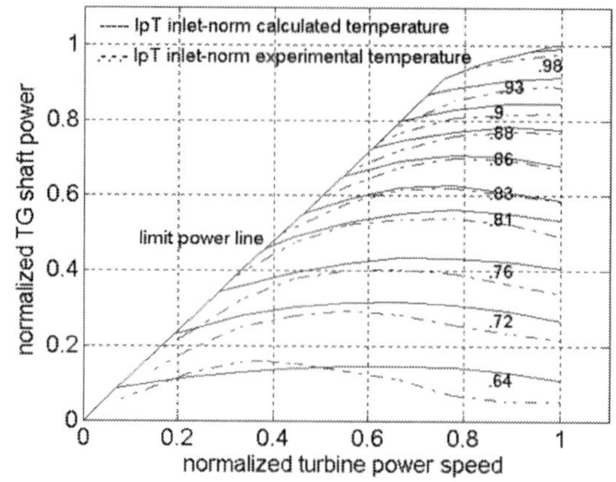

Figure 7: LM 2500 power turbine inlet temperature map.

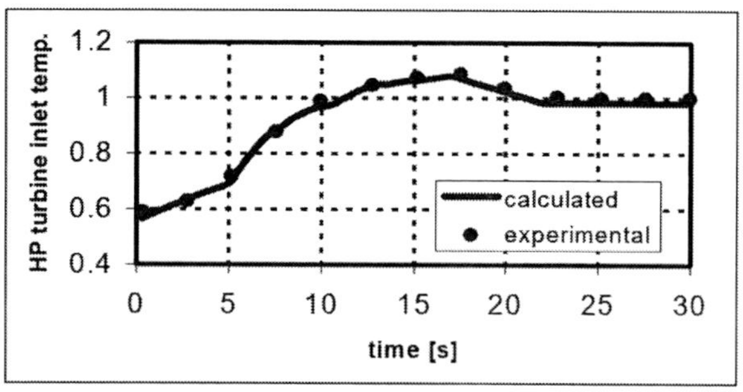

Figure 8: Power turbine inlet temperature transient.

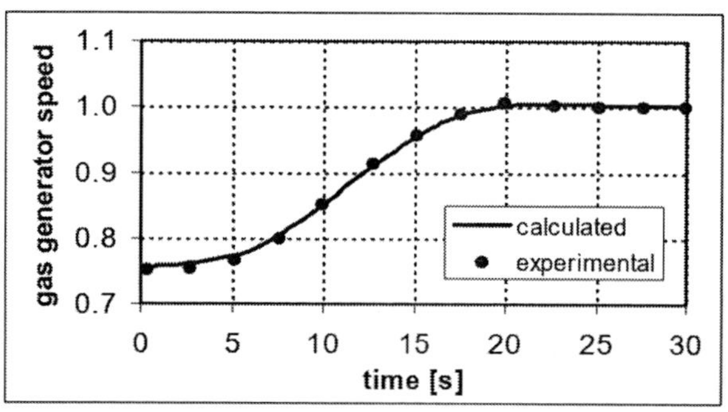

Figure 9: Gas generator shaft speed transient.

5. SHIP PROPULSION PLANT SIMULATION

As application example, the above discussed and validated gas turbine model has been used in connection with the simulation model of a waterjet system already presented and validated [18], suitably designed for the propulsion of a high speed trimaran ro/ro vessel (figure 10). The main characteristics of the considered ship are reported below:

- B_{CH}=10 m, chosen in such a way to arrange four rows of cars in the main hull
- L/B_{CH}=12 for resistance and seakeeping performances
- L=120 m, compatible with the value of the length to beam ratio assumed
- L_{SH}=45 m, chosen according to hydrodynamic reasons
- B_{SH}=3 m, in order to arrange propulsion devices
- B=28 m, maximum total breadth

- T=3.2 m resulting by weight estimation
- Δ= 2000 t

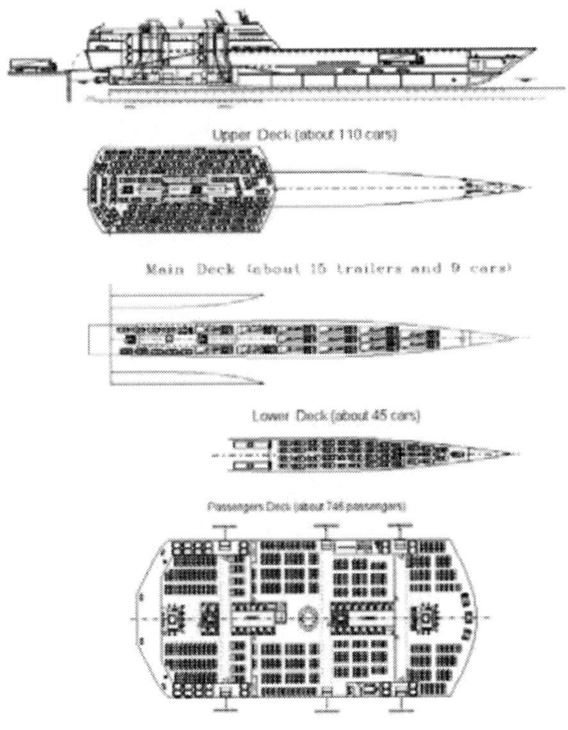

Figure 10 – Trimaran general arrangement.

In figure 11 the MATLAB-SIMULINK® simulator scheme of the considered waterjet propulsion plant is presented. The 'gas turbine' block of figure 11 has been discussed in detail before.

A brief description of the others sub-models (waterjet, hull resistance, ship dynamic, shaft dynamic, telegraph, engine governor), used for the simulation of the considered trimaran propulsion system is reported below, while the reader is referred to [18] for a more detailed description.

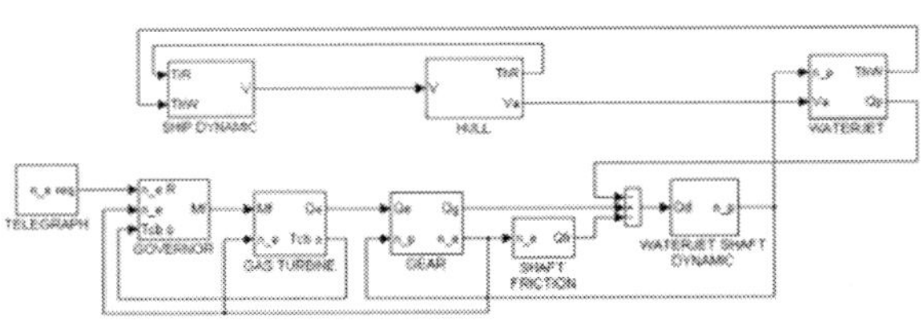

Figure 11 - Waterjet propulsion plant simulator scheme.

The waterjet model is based on the classical momentum equation for what concerns the prediction of the waterjet thrust curves and on available literature data for what concerns pump specific speed and cavitation characteristics. The developed procedure is able to obtain the waterjet performance map including the cavitation limits in a wide range of operating conditions.

In the simulator scheme of figure 11, relative to a ship traveling on a straight line, the hull model receives as input the ship velocity and gives as output the hull resistance. This is done by means of a regression formula based on the results of systematic towing tank tests on the trimaran hull performed at DINAV laboratory.

The ship dynamics is determined by the Newton law equation of a ship traveling on a straight line.

This waterjet pump shaft speed is obtained as solution of the momentum shaft dynamic equation (3).

The 'telegraph' block in figure 11 converts the telegraph position into the required speed for the GT power turbine ($n_{e\,R}$). This output, obtained by a simple SIMULINK table, is then transmitted to the engine governor.

As shown in figure 11, the 'governor' block senses the required and the effective GT power turbine speed and determines the fuel mass flow in order to obtain the required speed. A PID controller scheme was adopted for the power turbine speed governor. An over-limit control is used in order to avoid GT engine torque overloading.

A proportional over-limit control is also used in order to prevent over temperature in the gas generator (GG) turbine inlet section.

5. TRIMARAN SHIP TRANSIENT RESULTS

The analysis presented in the next section is focused to investigate the dynamic response of the presented GT model when used in a waterjet/gas turbine propulsion system plant simulator.

To this purpose, a vessel manoeuvre consisting in a deceleration followed by an acceleration is considered. This transient situation is determined, starting from a steady state ship speed of 35 knots, by the imposition of a decrease of the telegraph output value from 75 to 43 % in 120 seconds (see figure 12), followed afterwards by 130 seconds in which the ship speed is stabilised at 20 knots, then by a telegraph increase from 43 to 89 % in 100 seconds, determining, at the transient end, a trimaran speed of 40 knots (see figure 12).

The same figure 12 reports also the time variation of the GT combustor air/fuel ratio and the waterjet thrust during the considered transient manoeuvre.

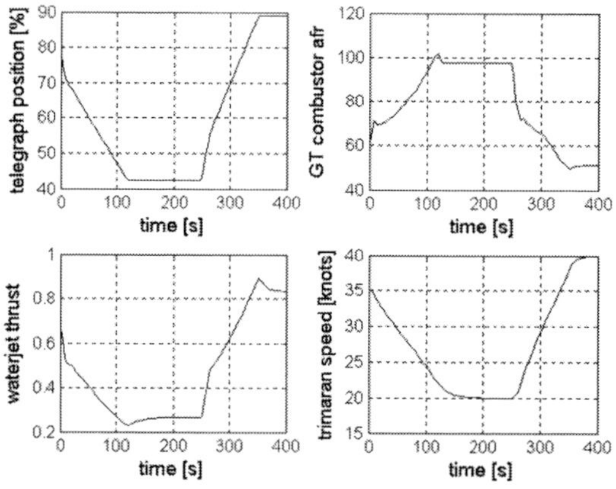

Figure 12 – Main ship and propulsion parameters variation during the manoeuvre.

In the following figures 13, 14 and 15 the working point trajectories (asterisk curves) during the considered transient manoeuvre are reported on the GT map, GT compressor map and waterjet map respectively.

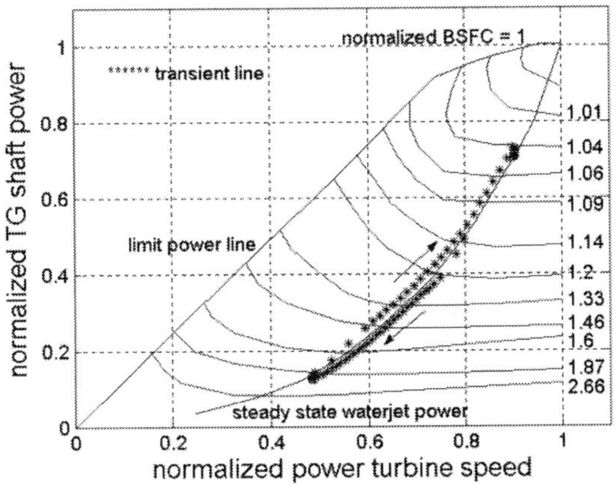

Figure 13 – Gas turbine map and transient working point trajectory.

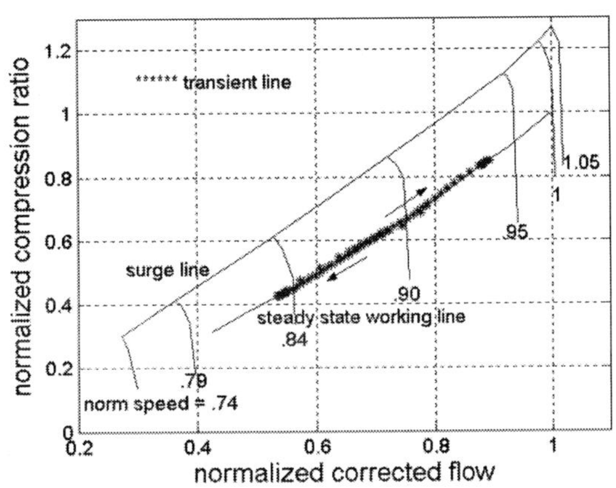

Figure 14 – Gas turbine compressor map and transient working point trajectory.

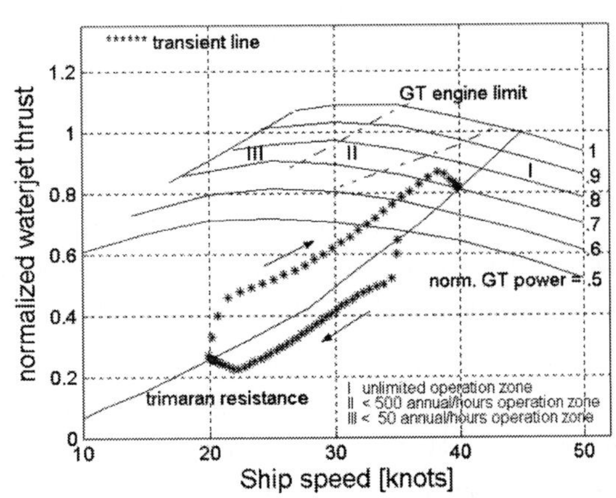

Figure 15 – Waterjet map and transient working point trajectory.

Finally, the following figures 16 and 17 shows the time variation of the most important GT parameters during the considered transient manoeuvre.

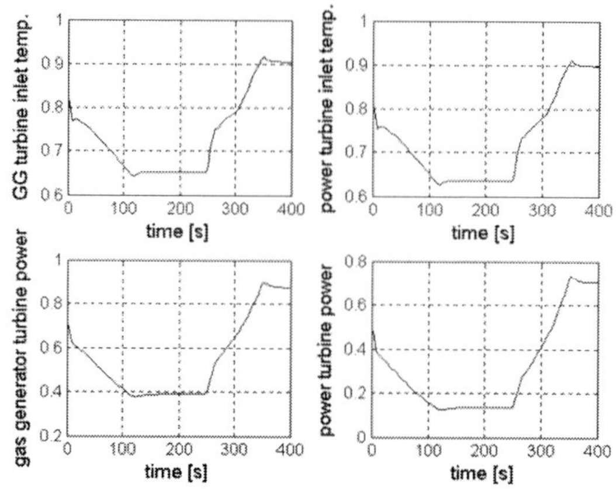

Figure 16 – Time variation of important gas turbine size during the transient.

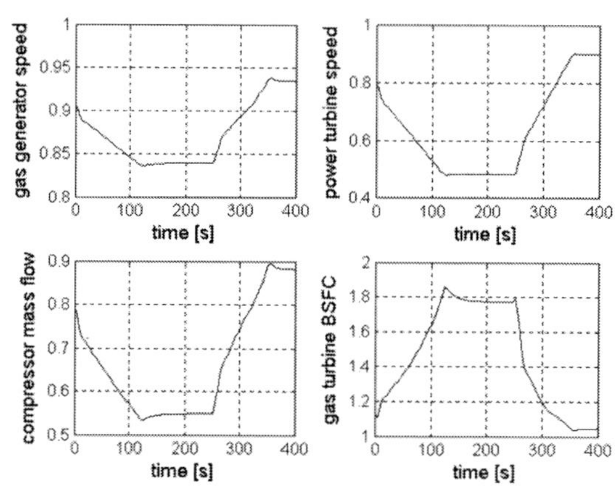

Figure 17 – Time variation of important gas turbine size during the transient.

6. CONCLUSIONS

A modular model for the dynamic simulation of open cycle gas turbine plants, specifically developed to be used in ship propulsion plants studies, has been presented and described in detail.

The modular approach used to generate the GT components offers the possibility of simulating a number of different GT plant configurations.

The test case considered for the experimental validation of the model has demonstrated that the proposed simulation technique seems to be a good compromise between model simplicity, required to reduce the computer time, and results accuracy.

The presented application example, illustrating the use of the GT model in a more complex waterjet/gas turbine ship propulsion plant, has confirmed the validity of the approach and its conformity to the prefixed targets.

The proposed simulation model, allowing to analyse a great number of GT parameters without penalising the calculation stability and the computer time, represents a useful tool for the performance evaluation and optimisation of ship propulsion plants characterised by the use of gas turbine plants as prime mover.

7. REFERENCES

[1] Saravanamuttoo, H. I. H., and Fawke, A. J, "Digital Computer Methods for the Prediction of Gas Turbine Response," SAE Paper No. 710550, 1971.

[2] Taylor, B. L., and Hoppenheimer, F. L., "Investigation of Frequency-Response Characteristics of the Engine Speed for a Typical Turbine-Propeller Engine," National Advisory Committee for Aeronautics, report No. 1017, 1951.

[3] Torella, G., Colantonio, A., and Riviello, P., "Dynamic Simulation of a Turbojet: Theoretical and Experimental Data," L'aeronautica missili e spazio, december, in Italian, 1981.

[4] Torella, G., and Liotti, G., "Numerical and Experimental Analysis of the Transient Behaviour of a Turboprop," 42th National Congress of the Italian Thermo-technical Association, Genova, Italy, September 8-11, 1987.

[5] Sellers, J., and Teren, F., "Generalised Dynamic Engine Simulation Techniques for the Digital Computer," AGARD-CP-151, 1974.

[6] Saravanamuttoo, H. I. H., and Mac Isaac, B. D., "Aerothermodynamic Factors Governing the Response Rate of a Gas Turbine," AGARD-CP-151, 1974.

[7] Cottingon, R. V., "Total Powerplant Simulation," AGARD-CP-151, 1974.

[8] Campora, U., and Gioia. F., "Dynamic Performance Prediction of an Aeronautical Gas Turbine," 44th National Congress of the Italian Thermo-technical Association, Cosenza, Italy, September 12-15, in Italian, 1989.

[9] Spelta, S., "Integration Method of the LEGO's Code,"ENEL-CRA, Internal Report No. 341/82, Milano, Italy, in Italian, 1982.

[10] Anon., "SIMULINK: "User's Guide," The MathWorks, Inc., 1993.

[11] Crosa, G., Pittaluga, F., Trucco, A., Beltrami, F., Torelli, A., and Traverso, F., "Heavy-Duty Gas Turbine Plant Aerothermodynamic Simulation Using Simulink," Transaction of the ASME, Journal of Engineering for Gas Turbine and Power, Vol. 120, pp. 550-556, 1988.

[12] Agresti, M., Camporeale, S. M., and Fortunato, B., "An Object-Oriented Program for the Dynamic Simulation of Gas Turbines," International Gas Turbine & Aeroengine Congerss & Exhibition, Munich, Germany, May 8-11, 2000.

[13] Campora, U., 1990, "Dynamic Performance Estimation of Axial-Flow Gas Turbines," 45th National Congress of the Italian Thermo-technical Association, Cagliari, Italy, September 18-21, in Italian, 1990.

[14] Shobeiri, M. T., Attia, M., and Lippke, C., "GETRAN: A Generic, Modularly Strusctured Computer Code for Simulation of Dynamic Behavior of Aero- and Power Generation Gas Turbine Engines," Transaction of the ASME, Journal of Engineering for Gas Turbine and Power, Vol. 116, pp. 483-494, 1994.

[15] Benson, R. S., and Whiteouse, N. D., 1979, "Internal Combustion Engines," Pregamon Press International Library, Oxford, England.

[16] Cohen, H., Rogers, G. F. C., and Saravanamuttoo, H. I. H., "Gas Turbine Theory," (Third Edition), Longman Scientific & Technical, Harlow, Essex, England, 1987.

[17] El-Masri, M. A., 'GASCAN – an Interactive Code for Thermal Analysis of Gas Turbine Systems' - Transaction of the ASME, Journal of Engineering for Gas Turbine and Power, Vol. 110, pp. 201-209, 1988.

[18] G. Benvenuto, U. Campora: "Dynamic Performance Analysis of a Gas Turbine/Waterjet Propulsion System for a Fast Trimaran Ferry" – 7th International Conference on Fast Sea Transportation FAST 2003, Ischia, Italy, October 7-10, 2003.

8. AUTHORS BIOGRAPHY

Giovanni Benvenuto: Full Professor, University of Genova, Faculty of Engineering, Dipartimento di Ingegneria Navale e Tecnologie Marine, Italy

Director of the Department, Professor of Marines Propulsion Plants at the Genova University.

Research topics: dynamic simulation of marine propulsion plants, mathematical modelization of internal combustion engines, environmental impact assessment of marine transport systems, low emission energetic systems.

Ugo Campora: Associated Professor, University of Genova, Faculty of Engineering, Dipartimento di Macchine, Sistemi Energetici e Trasporti, Italy.

Professor of "Thermal Machinery" in the degree courses of Naval Architecture and Marine Engineering of Genova and La Spezia University.

Research topics: Internal Combustion Engines, Power Plants, Fluid Machinery, Marine Plants Simulation.

Performance Prediction of a Planing Craft by Dynamic Numerical Simulation

M. Altosole, G. Benvenuto, M. Figari

Dipartimento di Ingegneria Navale e Tecnologie Marine, Università di Genova - Italy

ABSTRACT

This paper presents mathematical models for the dynamic simulation of a planning craft and a waterjet propulsion system. The models, developed in a Matlab-Simulink software environment, are structured in modular form, in order to describe the various elements of the system as individuals blocks (hull, prime mover, gear, waterjet, etc.), taking into account their interactions. In this way it is possible to characterise the dynamic behaviour of both the single components and of the whole propulsion plant.

The module of the planning behaviour of a craft is a new feature of the general dynamic simulation model already developed by the authors. With this new module it is possible to account for the resistance and trim variation during boat acceleration in order to properly catch the interaction between the planing hull and the propulsion system.

The waterjet model, already presented in previous papers, has been recently improved in order to take into account the waterjet pump efficiency variation according to the different operating conditions. The developed model is able also to take into account the effects of the cavitation on the waterjet performance. The waterjet propulsion model has been validated both in steady state and transient situations through comparisons with data provided by waterjet manufacturers.

An application has been made for the prediction, at a design stage, of the acceleration performance of a waterjet propelled small craft.

NOMENCLATURE

β = deadrise angle, [deg]

ε = inclination of thrust line relative to keel, [deg]

τ = trim angle of keel, [deg]

Δ= weight of boat, [kg]

a = distance between D_f and CG (measured normal to D_f), [m]

A_j = waterjet area [m^2]

b = beam, [m]

CG = center of gravity

d = draft of keel at transom, [m]

D_f= viscous component of hull drag, (assumed as acting parallel to keel line, midway between keel and chine lines), [N]

e = distance between N and CG (measured normal to N), [m]

f = distance between T and CG (measured normal to shaft line), [m]

F_z = propulsor force normal to shaft [N]

H = pump head [m]

h_2 = height of the exit section referred to the undisturbed water surface [m]

I_{yy} = mass moment of inertia [kg m^2]

J_p = rotational inertia moment [kg m^2]

L_c = wetted chine length (from transom to spray root interception with chine), [m]

LCG = longitudinal distance of center of gravity from transom, measure along keel, [m]

L_k = wetted keel length, [m]

M_m = engine torque [Nm]

M_r = propulsor torque [Nm]

M_s, $M_{s\,ad}$ = mass and added mass of the hull [kg]

N = pressure and buoyant forces on the hull [N]

n = pump rotational speed [rpm]

P_B = pump power demand at the prime mover [kW]

Q = pump flow rate [m^3/s]

R = ship resistance [kN]

R_t = thrust required by the hull

t = thrust deduction factor

T= propulsor thrust [kN]

V= ship velocity [m/s]

V_a = mean velocity of approach, evaluated at a station in front of the intake ramp tangency point [m/s]

w = wake fraction

z_2 = distance between the nozzle and the bottom plating

η_m = mechanical efficiency

η_N = nozzle efficiency

η_P = pump efficiency

η_r = relative rotative pump efficiency

ζ = inlet duct loss coefficient

1. INTRODUCTION

The application of simulation techniques to marine propulsion systems represents a research field developed at the University of Genoa for some years.

The increasing interests in waterjet propelled craft (patrol boats, fast pleasure craft) and ships (fast ferries) stimulated the research activity towards the application of simulation techniques also to this kind of propulsion systems driven by both high speed diesel engines and gas turbines [10-16].

Because of the complexity of the theoretical treatment of the waterjet systems and in sight of the different aims to which simulation studies may be developed for, different approaches have been used during the years. In particular the authors have developed two types of models with different degrees of accuracy. Steady state models, able to compute the waterjet performance maps, including cavitation limits, in a wide range of operating conditions, and dynamic models able to describe the transient behaviour of waterjet propulsion systems.

Waterjet plants are nowadays more and more used for the propulsion of fast ships, ferries and pleasure craft of different sizes. Although several design methods have been published over the years [17] there is not yet a consolidated theory or methodology able to help the naval architect or the marine engineer to select the proper waterjet for a defined application. The common practice of ship designers is to refer to the waterjet manufacturers for the selection, with the obvious advantage of the accuracy of results but with the drawback of the lack of information about the used procedures.

If the dynamic performance (i.e. accelerations, manoeuvring capabilities) of the boat represents a specific requirement for the owner, its assessment may represent a further problem for the boat designer because of the lack of pertinent literature.

The present study comes from the need of the designer to have a tool to evaluate, in a preliminary way but with sufficient accuracy, the type of waterjet to be used and its performance, taking into account also dynamic conditions, i.e. accelerations.

In this paper a mathematical model for the dynamic simulation of a waterjet propulsion system is presented.

The transient performance of a craft is an important parameter for both design and operational points of view. In particular the motivation of this work was the assessment of the acceleration behaviour of a planing boat and the evaluation of the elapsed time from rest to planing condition.

The numerical model, developed in a Matlab-Simulink software environment, is structured in a modular form. The various elements of the propulsion chain (hull, prime mover, gear, waterjet, etc.) have been modelled as individuals blocks and are connected each other in a way to take into account the interactions. In this way it is possible to characterise the dynamic behaviour of both the single components and the whole propulsion plant.

The new feature of this simulator, compared to the propulsion dynamic model presented in previous papers, is the possibility to account for the resistance and trim variation during the simulation in order to properly catch the interaction between the hull and the propulsion system.

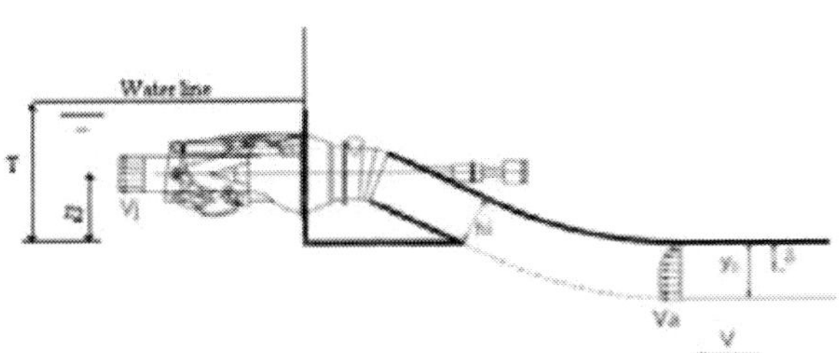

Fig. 1 - Waterjet scheme

This last target has been achieved by the realisation of an other mathematical model, called "Planing Hull Simulator", developed in a Matlab-Simulink software environment.

2. "WATERJET PROPULSION SIMULATOR"

To assess the dynamic performance of a craft, a comprehensive approach, able to represent the whole propulsion system (prime mover, power transmission, propulsor), is required.

The mathematical model is based on the Newton's law applied to the longitudinal motion of the boat and to the rotation of the shaftline.

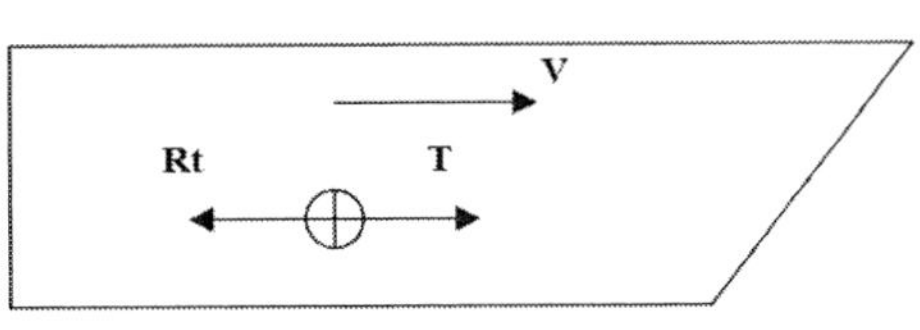

Fig. 2 - Ship longitudinal motion

With reference to Fig. 2 the equation of motion results:

$$(M_s + M_{ad})\frac{dV(t)}{dt} = T(t) - R_t(t)$$

Eq. 1 - Longitudinal motion

The added mass is considered as a constant.

The resistance of the hull is modelled with a 1-D table as a function of the speed: the towing tank resistance curve R for V>0, a function of the forward resistance for V<0.

This way of ship resistance modelling is valid for displacement crafts, but is less valid for planing craft.

The thrust required by the hull is :

$$R_t = \frac{R}{1-t}$$

Eq. 2 - Required thrust

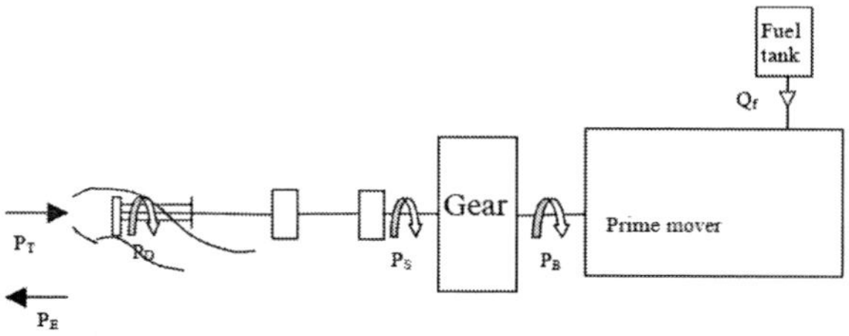

Fig. 3 - Propulsion system

With reference to **Fig. 3** the equation of motion results:

$$2\pi J_p \frac{dn(t)}{dt} = M_m(t) - M_r(t)$$

Eq. 3 - Shaft dynamics

The forces and torques required by the model are computed by several sub-models representing the different elements of the propulsion chain (mainly: prime mover, waterjet, reduction gear, hull resistance, telegraph, engine governor).

The dynamic performance of the boat is obtained from the simultaneous solution of the coupled Eq. 1 and Eq. 2 in the time domain.

The implementation of the numerical simulator is done in MATLAB-SIMULINK® software environment, a wide used platform for the simulation of dynamic systems. The main model is presented in Fig. 4.

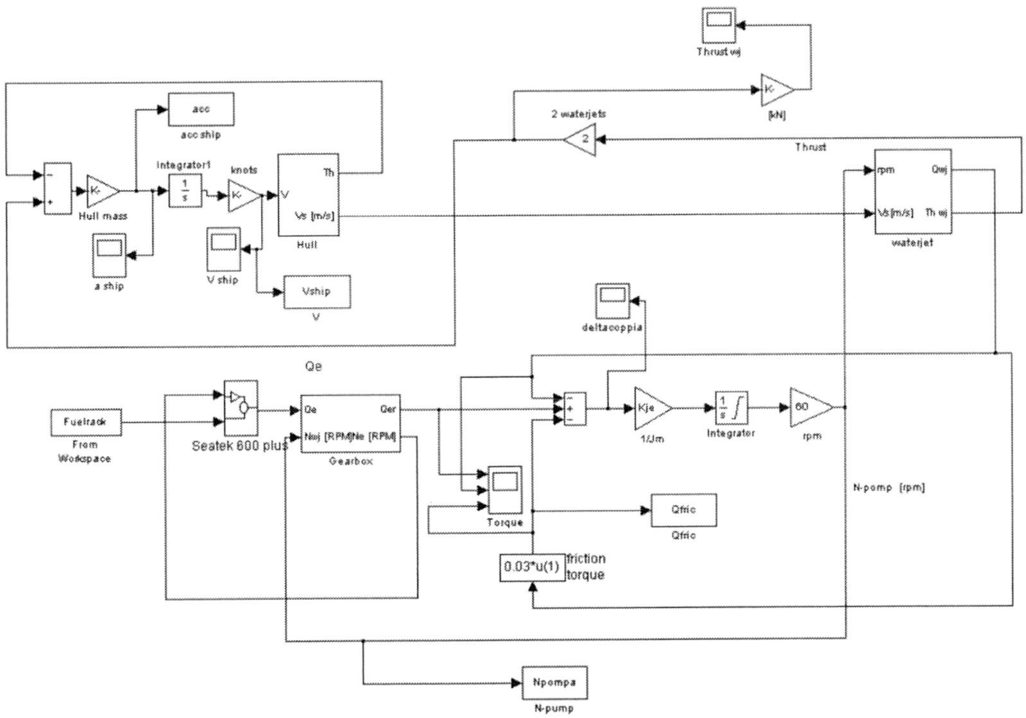

Fig. 4 – Numerical model

A brief description of the main sub-models used for the simulation of the considered craft propulsion system is reported in the following sections.

2.1 THRUST

In the proposed waterjet model the pump flow rate Q is calculated from the equilibrium of the required head pump, obtained by the application of Bernoulli theorem to the waterjet ducting system, with the head pump offered by the pump shaft power P_B:

$$\begin{cases} H = \dfrac{Q^2}{2g\eta_N A_j^{\,2}} - (1-\zeta)\dfrac{V_a^{\,2}}{2g} + h_2 \\ H = \dfrac{\eta_p \eta_r \eta_m}{\rho g Q} P_B \end{cases}$$

The following equation is obtained:

$$\frac{Q^3}{2g\eta_N A_j^{\,2}} - (1-\zeta)\frac{V_a^{\,2}}{2g} Q - \frac{\eta_p \eta_r \eta_m}{\rho g} P_B = 0$$

Eq. 4 – Pump flow rate cubic equation

Whose analytical solution gives the pump flow rate Q, depending on the following coefficients:

$$\alpha = \frac{1}{2g\eta_N A_j^2}$$

$$\beta = -(1-\zeta)\frac{V_a^2}{2g}$$

$$\gamma = -\frac{\eta_p \eta_r \eta_m}{\rho g} P_B$$

Therefore the pump flow rate is varying with time because of β and γ coefficients; indeed these two coefficients are depending respectively on velocity V_a and shaft speed n ($P_B = Cn^3$).

Once calculated the pump flow rate Q, it is possible to obtain from this model the thrust produced by the waterjet system as:

$$T = \rho Q\left(\frac{Q}{A_j} - V_a\right)$$

2.2 ENGINE MODEL

In literature, various approaches are proposed for the dynamic simulation of marine diesel engines; an extensive bibliography on this topic is reported in [3, 4].

In the most simple schemes, the engine performance is calculated by means of diagrams or empirically derived equations (i.e. engine torque or power as functions of rotational speed and fuel consumption).

This approach is quite simple and is suitable for a rough simulation of the engine dynamic behaviour in a short time. It does not allow a careful description of thermal and mechanical stresses induced in critical elements of the power plant.

3. "PLANING HULL SIMULATOR"

The purpose of this part of the work is to develop a simulator able to predict the resistance and the equilibrium running trim of a planing boat.

The following sketch (Fig. 5) shows the forces and moments acting on a planing hull in a more general way with respect to the previous approach of

Fig. **2**.

The first part of Fig. 5 shows the hydrodynamic forces on the planing surface, the second part shows the hydrodynamic forces on the appendages, the third part shows the propulsor forces (waterjet or propeller).

In this more general approach three equations of motion are simultaneously solved: the vertical and horizontal equilibrium and the pitch equilibrium.

In this way it is possible to calculate velocity, trim and position of the planing hull.

$$\begin{cases} \sum F_x = M\dfrac{dv_x}{dt} \\ \sum F_z = M\dfrac{dv_z}{dt} \\ \sum M = I_{yy}\dfrac{d^2\tau}{dt} \end{cases}$$

Eq. 6 – Motion equations

The equations of equilibrium to assess the performance of planing craft require the knowledge of the forces and moments acting on the system at each time frame.

The starting point for the evaluation of the forces is the Savitsky method [20], rearranged by Hadler [21], as presented in Eq. 7, Eq. 8, Eq. 9.

For the vertical forces:

$$\begin{aligned} \Delta = &(N + p\cdot N_a + r\cdot N_r + p\cdot N_p - p\cdot N_s)\cos\tau + \\ &+ (p\cdot T + \Delta D_f)\sin(\tau+\varepsilon) + \\ &- (D_f + p\cdot D_a + r\cdot \Delta D_r + D_k)\sin\tau + \\ &+ p\cdot F_z\cos(\tau+\varepsilon) \end{aligned}$$

Eq. 7 – Vertical equilibrium

For the horizontal forces:

$$\begin{aligned} &(p\cdot T + \Delta D_f)\cos(\tau+\varepsilon) = (D_f + p\cdot D_a + r\cdot \Delta D_r + \\ &+ D_k)\cos\tau + D_w + (N + p\cdot N_a + r\cdot N_r + p\cdot N_p + \\ &- p\cdot N_s)\sin\tau + p\cdot F_z\sin(\tau+\varepsilon) \end{aligned}$$

Eq. 8 – Horizontal equilibrium

For the pitching moments:

$$\begin{aligned} &N\cdot e + p\cdot N_a\cdot e_a + r\cdot N_r\cdot e_r + p\cdot N_p\cdot e_p + \\ &- p\cdot N_s\cdot e_s + D_f\cdot f_f + p\cdot D_a\cdot f_a + \Delta D_r\cdot f_r + \\ &+ D_k\cdot f_k + (p\cdot T + \Delta D_f)f + p\cdot F_z\cdot e_f = 0 \end{aligned}$$

Eq. 9 – Pitch equilibrium

The unknowns in the foregoing equations of equilibrium are evaluated by a simultaneous solution of these equations together with the planing formulas for lift, drag, and center of pressure.

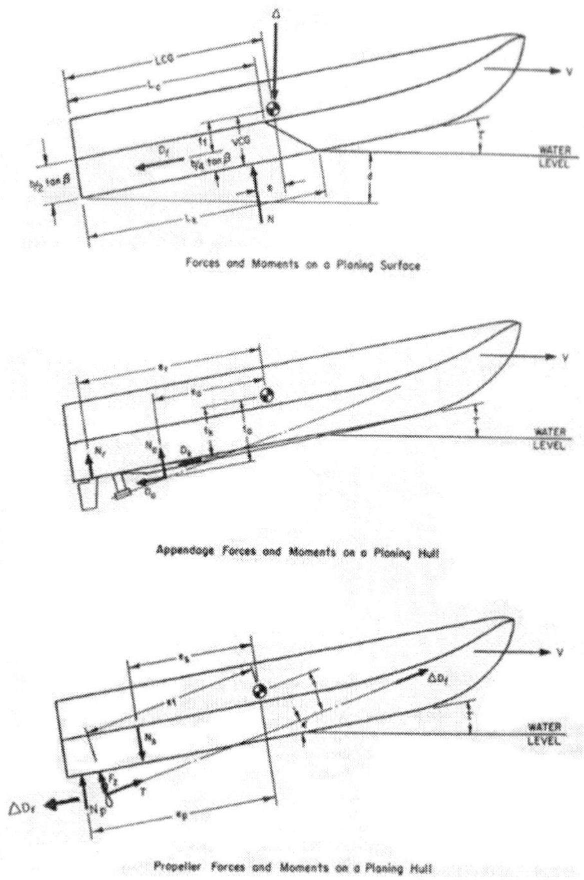

Fig. 5 - Forces and moments on a planing hull

Both Savitsky and Hadler formulations are based on a known boat velocity V. In a time domain simulation the velocity of the boat represents one of the results of the calculation, this requires a re-arrangement of the equations in a more suitable form as presented in

Eq. 10,

Eq. 1, Eq. 2.

$$(N+N_r+N_a+N_p-N_s)\cos\tau+(-D_f-D_a-D_k-\Delta D_r)$$
$$\sin\tau+F_z\cos(\tau+\varepsilon)+(T+\Delta D_f)\sin(\tau+\varepsilon)-\Delta$$

Eq. 10 – Vertical Forces Fz

$$-(N+N_r+N_a+N_p-N_s)\sin\tau+(-D_f-D_a-D_k-\Delta D_r)$$
$$\cos\tau-F_z\sin(\tau+\varepsilon)+(T+\Delta D_f)\cos(\tau+\varepsilon)-Dw$$

Eq. 11 – Horizontal Forces Fx

$$-(N\cdot e+N_r\cdot e_r+N_a\cdot e_a+N_p\cdot e_p-N_s\cdot e_s)+$$
$$-(D_f\cdot f_f+D_a\cdot f_a+D_k\cdot f_k+\Delta D_r\cdot f_r)+$$
$$-F_z\cdot e_f+(T+\Delta D_f)f-Dw\cdot fw$$

Eq. 12 – Pitch Moments Mz

The N forces represent forces acting normal to the hull, the D forces represent different drag components acting along the keel. All of the forces are evaluated with different formulations described in [22].

The simulation starts with the main dimensions of the boat, displacement, arms distances, deadrise angles, initial conditions and the actual thrust T as the principal input.

The main program, "Planing Hull Simulator", Fig. 6, contains two principal blocks, "equilibrium" and "equations of motion", where Eqs. 10,11,12 and Eqs. 6 are solved respectively.

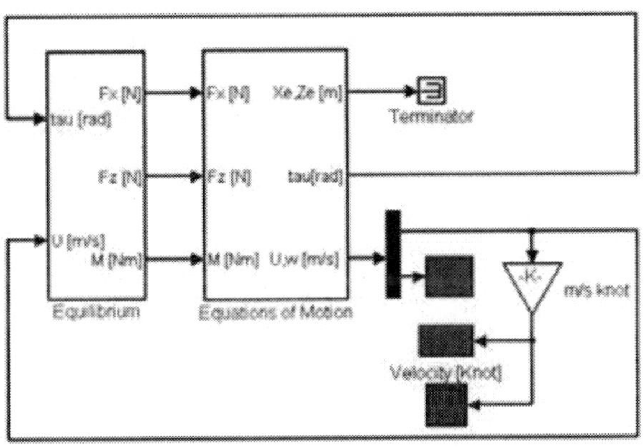

Fig. 6 – Planing hull simulation

The output of the code is represented by the time histories of the significant variables of the system.

In this version of the code the simulation is obtained by imposing a pre-determined time history T(t) to the propulsor thrust T.

4. RESULTS AND CONCLUDING REMARKS

A comprehensive approach, able to represent the whole propulsion system (prime mover, power transmission, propulsor and hull), is required to assess the complete dynamic performance of a planing craft. However, the technical difficulty to reach this final target suggested firstly to develop two different simulators, one for the propulsion system and another for the dynamic equilibrium of the planing hull.

The two developed simulators have been validated in both steady state and transient situations through comparisons with data provided by waterjet and boat manufacturers.

An application of the "Waterjet Propulsion Simulator" has been made for the prediction, at a design stage, of the acceleration performance of a waterjet propelled small craft, while another application for the "Planning Hull Simulator" has been made to assess the resistance and trim variation during the acceleration boat.

The following pictures show the main variables provided by the two simulators during different transient manoeuvres. More precisely Fig. 7 shows the varying with time waterjet thrust, craft acceleration and speed,

computed by the "Waterjet Propulsion Simulator" through the application to the model of a defined law for the engine fuel rack, without taking into account the boat equilibrium running trim.
On the contrary, Fig. 8 and Fig. 9 show the planing boat speed and trim computed by the "Planing Hull Simulator" during a different transient manoeuvre, where the principal characteristics of the boat, initial conditions and working propulsor thrust are the principal inputs of the mathematical model.
However the research is in progress and the next step will be to combine together the described calculation procedures in order to catch the effect of the interaction between the boat trim and the propulsion plant thrust.

5. FUTURE WORK

As pointed out before the next stage of the work will be to link together the two developed models, in order to properly catch the interaction between the planning hull and the waterjet propulsion system.
The final target of the research is the realisation of a complete propulsion simulator to be used in the initial phase of the design process and for training purposes. This last objective needs a proper visualization of the dynamic system in a three-dimensional virtual reality environment, as shown in the simple image of Fig. 10, in order to extend the capability of the used software (Matlab and Simulink) to the world of virtual reality; to this aim it will be also necessary to provide real-time virtual interaction with the dynamic models, indispensable feature for a proper training simulator.

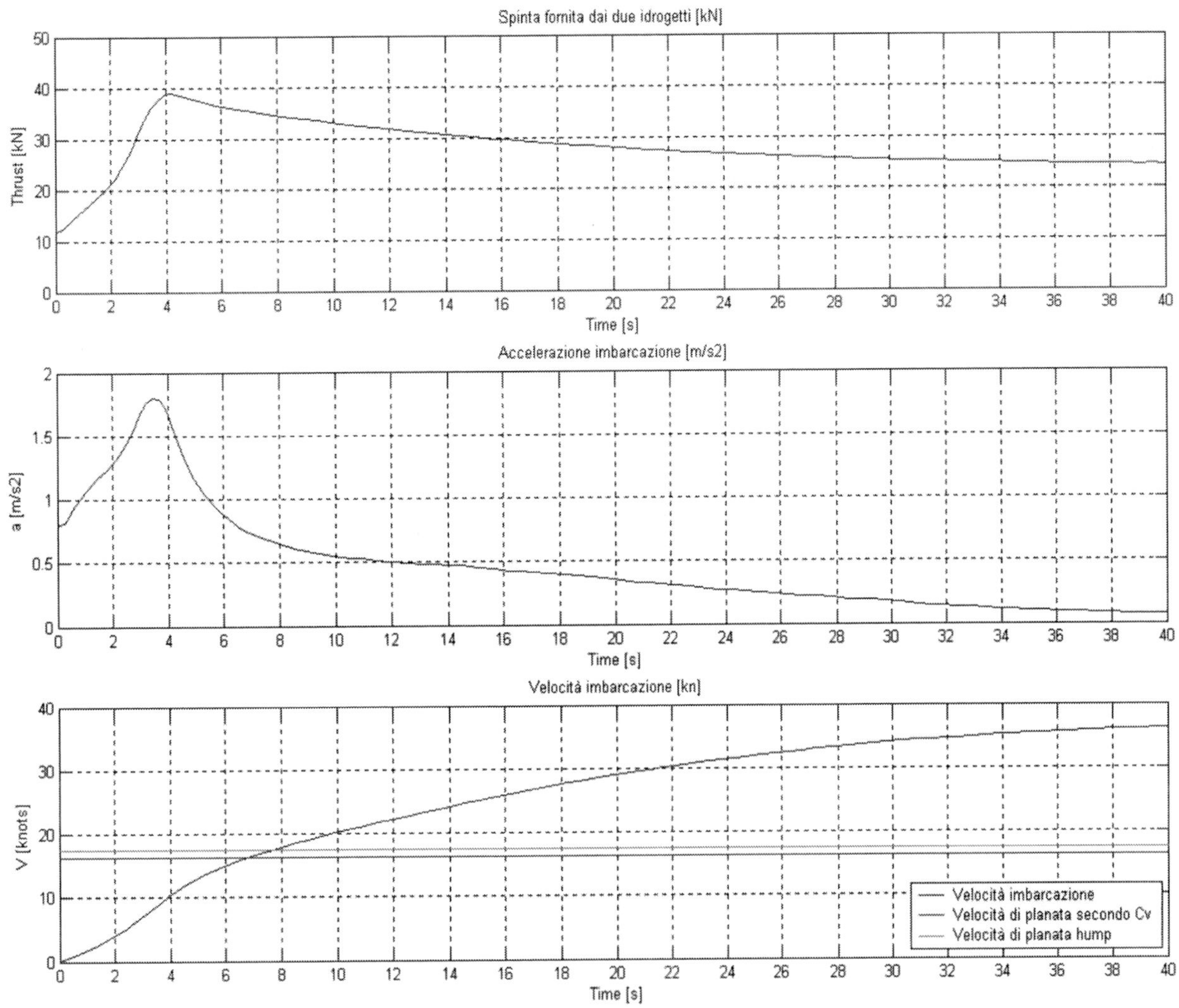

Fig. 7 – Waterjet thrust, craft acceleration and speed

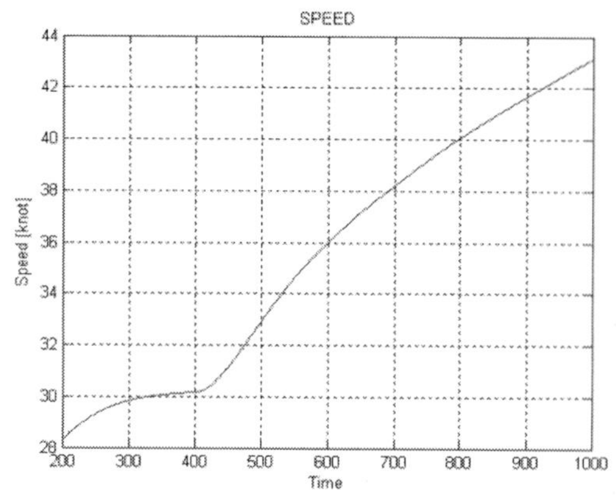

Fig. 8 – Boat speed

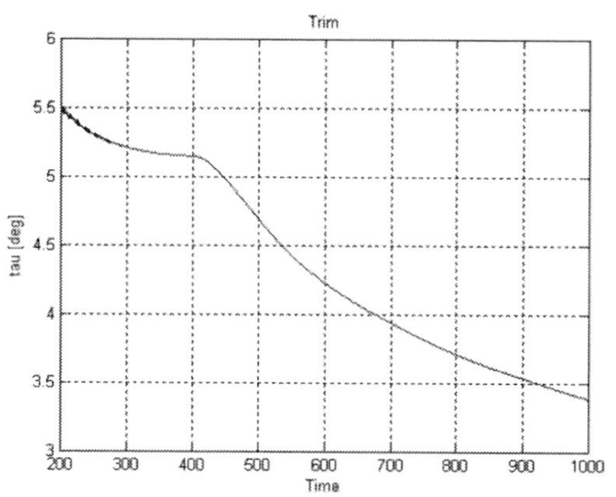

Fig. 9 – Boat trim

Fig. 10 – Virtual reality applied to the mathematical model

6. REFERENCES

1. R.V. Thompson, "Marine Propulsion Machinery: the System Approach", ICMES 84
2. A. Fowler, "Microcomputer-based Simulation of Marine Propulsion Systems", TransIMarE, Vol. 100, 1988
3. G. Benvenuto, U. Campora, G. Carrera, E. Rizzuto, "A Dynamic Diesel Engine Model for Ship Manouvrability Prediction Studies", MARIND 96, First International Conference on Marine Industry, Varna, Bulgaria, June 2-7, 1996.
4. G. Benvenuto, U. Campora, G. Carrera, P. Casoli: "A Two-Zone Diesel Engine Model for the Simulation of Marine Propulsion Plant Transients", MARIND 98, Second International Conference on Marine Industry, Varna, Bulgaria, September 28, October 2, 1998.
5. G. Benvenuto, U. Campora, G. Carrera, M. Figari: "Simulation of Ship Propulsion Plant Dynamics in Rough Sea", 8° International Conference on Marine Engineering Systems (ICMES 2000), New York, USA, May 22-23, 2000.
6. G. Benvenuto, U. Campora, G. Carrera, M. Figari. Ship Stopping Manoeuvre: Simulation of the Propulsion System Behaviour. International Conference on Ship and Shipping research, NAV 2000, September 20-22, Venice
7. G. Benvenuto, S. Brizzolara, M. Figari, "Simulation Of The Propulsion System Behaviour During Ship Standard Manoeuvres", PRADS 2001.
8. M. Altosole, M. Figari, S. D'Arco, Design And Optimisation Of Propulsion Systems By Dynamic Numerical Simulation, NAV 2003, Palermo, Italy
9. U. Campora, M. Figari, Numerical Simulation of Ship Propulsion Transients and Full Scale Validation, Journal of Engineering for the Maritime Environment, Vol. 217, 2003
10. Benvenuto, G., Campora, U., Figari, M., Ruggiero, V. jr, A Dynamic Model for the Performance Prediction of the Waterjet Propulsion System, PRADS'98 – Elsevier Science B.V., 1998
11. Benvenuto, G., Figari, M., Ruggiero, V. jr "Waterjet Propulsion System Performance Prediction by Dynamic Numerical Simulation"– 5st Symposium "High Speed Marine Vehicles", Capri, March 25 – 26, 1999.
12. Benvenuto, G., Figari, M., Moreira, L., Guedes Soares, C., "Dynamic Modeling of Waterjet Propulsion Plants" - Proceedings of IMAM 2000, International Maritime Association of Mediterranean, IX Congress, Ischia (Italy), April 2 – 6, 2000.
13. Becchi, P., Benvenuto, G., Figari, M., "Procedura di progetto di apparati di propulsione a idrogetto", DINAV, Ottobre 2000.
14. Figari, M.: "Mathematical Modelling and Simulation of a Waterjet Propulsion Plant" - 33 WEGEMT School, 2000.
15. M. Altosole, G. Benvenuto, M. Figari, Waterjet Performance Prediction, NAV 2003, Palermo, Italy
16. Benvenuto, G., Campora, Dynamic Performance Analysis of a Gas Turbine/Waterjet Propulsion System for a Fast Trimaran Ferry, FAST 2003, Ischia (Napoli), Italy
17. Allison, J. L., Marine Waterjet Propulsion, Transactions of SNAME, Vol. 101, 1993.

18. ITTC (1987) Report of the High Speed Marine Vehicle Committee, International Towing Tank Conference 1987, pp. 304-313.

19. Alexander Keith, Waterjet versus propeller engine matching characteristics, Naval Engineers Journal, May 1995.

20. D. Savitsky, "Hydrodynamic Design of Planing Hulls".

21. J. B. Hadler, "The Prediction of Power Performance on Planing Craft".

22. M. Figari, A. Ferreira, Time domain performance simulation of a planing craft, Report DINAV n. R5MF04, 2004

DAMAGE IDENTIFICATION AND HEALTH MONITORING OF COMPOSITE STRUCTURES USING VIBRATIONS MEASUREMENTS AND PIEZOELECTRIC DEVICES

I Bovio and **L Lecce**, Department of Aeronautical Engineering – University of Naples "Federico II", Italy

SUMMARY

The development in high speed marine vehicles field needs progress in material technologies. Composite is largely used, but not all its properties are well-known. To improve safety and performances the capability to identify structural damage, in real time, is of primary importance. Purpose of the paper is to present a new application of a Non Destructive Test based on vibrations measurements. The proposed new method is based on the acquisition and comparison of Frequency Response Functions (FRFs) of the monitored structure before and after an occurred damage. Structural damage modify the dynamical behaviour of the structure such as mass, stiffened and damping, and consequently its FRFs, making possible to identify and quantify a structural damage. The activities, presented in the paper, mostly focused on a determining of a representative "Damage Index" for identifying and analysing damage on composite structures. Besides it has been carried out a dedicated neural network algorithm.

1. INTRODUCTION

Purpose of the paper is to present an innovative application inside the Non Destructive Testing field based on vibrations measurements, developed, at the Department of Aeronautical Engineering of the University of Naples "Federico II" (Italy), by the authors during the last three years, and already tested for analysing damage of many structural elements. The aim has been the development of a Non Destructive Test (NDT) which meet to most of the mandatory requirements for effective health monitoring systems, simultaneously reducing as much as possible the complexity of the data analysis algorithm and of the experimental acquisition instrumentation; these peculiarities may, in fact, not be neglected for an operative implementation of such a system. The proposed new method is based on the acquisition and comparison of Frequency Response Functions (FRFs) of the monitored structure before and after an occurred damage. Structural damage modify the dynamical behaviour of the structure such as mass, stiffened and damping, and consequently the FRFs of the damaged structure in comparison with the FRFs of the sound structure, making possible to identify, to localize and quantify a structural damage. The activities, presented in the paper, mostly focused on a new FRFs processing technique based on the determining of a representative "Damage Index" for identifying and analysing damage on aeronautical composite panels. Besides it has been carried out a dedicated neural network algorithm aiming at obtaining a "recognition-based learning"; this kind of learning methodology permits to train the neural network in order to let it recognises only "positive" examples discarding as a consequence the "negative" ones. Within the structural NDT a "positive" example means "healthy" state of the analysed structural component and, obviously, a "negative" one means a "damaged" or perturbed state. With this object in view the neural network has been trained making use of the same FRFs of the healthy structure used for the determining of the Damage Index, as positive examples. From an architectural point of view magnetostrictive devices have been tested as actuators, and piezoceramic patches as actuators and sensors [1,2]. These techniques promise to bring a step forward for the implementation of an automatic "health monitoring" system which will be able to identify a structural damage in real time, improving the safety and reducing maintenance costs.

2. DAMAGE INDEX METHOD

It is common knowledge that structural damage modify the dynamical behaviour of the structure such as mass, stiffened and damping. It has evaluated the variations occurred by means of FRF, which is the ratio between the Fourier transform of the signal used to excite the structure in a point and the Fourier transform of the signal response acquired by a sensor in another point. In fact the FRFs of the sound structure are different from FRFs of the damaged structure. It has determined a damage index to evaluate the variations of FRFs of the monitored structure owing to an occurred damage. The index gives, directly, the measurements of the damage.

$$Index = \frac{\sum_{i=1}^{n} \left| FI_i - FD_i \right|}{\sum_{i=1}^{n} FI_i}$$

FI_i and FD_i are the amplitude of the FRFs of the sound and damaged structure at the "i" frequency. The index is the ratio between the absolute value of the arithmetic mean of the deviation between FRFs of the sound and damaged structure and the arithmetic mean of the FRFs of the sound structure.

3. NEURAL NETWORKS FOR HEALTH MONITORING

A critical point for the validation of the developed damage analysis approach, based on the comparison of the amplitudes of (FRFs), consists in a "statistical test" assessing the confidence level relatively to the mentioned amplitude differences in order to verify that FRFs variations were effectively due to structural perturbations instead of environmental influences (noise, temperature, humidity, vibrations, etc...) which excite not linear behaviour of the experimental set-up. That statistical test permits to identify the threshold dividing healthy configurations from damaged ones. Furthermore, to quantify the amplitudes of FRFs differences, it has been developed a dedicated neural network algorithm.

The damage identification problem can be classified as a typical example of binary learning ("healthy" or "damaged"). A neural network able to implement binary learning can be modelled following two approaches: the "discrimination-based learning" and the "recognition-based learning". In the first approach the network is trained using both "positive" and "negative" samples in order to learn how to discriminate among them; in the second one the network is trained using only "positive" samples and it is able to recognise only these. Within this work has been implemented a system based on the latter approach.

The engine of the damage analysis system is represented by an "autoassociative" neural network (or "auto-encoder") made of three layers: an input one, an hidden one and obviously the output one, fully connected one each other [3].

An auto-associative neural network is a feed forward type network trained only by positive samples in order to rebuild the input on the output layer. If the training phase is successful the network is able to find the common features that samples present in order to extract few general laws permitting to recognise positive unknown examples. The positive samples are represented by the "healthy" configuration' FRFs. Following the training phase, the auto-encoder will be able to reconstruct more or less accurately on the output layer the positive samples, while it will always reconstruct wrongly the negative samples. So, a bad reconstruction of the input layer on the output one is a clear symptom of an anomalous dynamic behaviour of the monitored structure.

Once the auto-encoder has been implemented and trained, in order to use it as a "classifier" for the health status of a structural component, it is necessary to complete the system using a threshold to divide operatively the two samples classes (positive or "healthy" and negative or "damaged"). To do so it will be preliminary necessary to create an index quantifying the reconstruction level of the input layer on the output one; it is obviously directly related to the reconstruction error that the network presents when trying to recreate the input on its output layer. The error, on its side, will be bigger about negative samples then positive ones [4].

4. THE EXPERIMENTAL SET-UP

A typical aeronautical composite panel made by means of RFI technology available in the labs of the Department of Aeronautical Engineering was chosen as test-article:

- 14 plies made of multi-axial HTA 3/6 K (520mm x 520mm x 2mm);
- epoxy resin 977-2 by Cytec;
- panel lay-up: [(90/0/+/–)(+/0/–)]s.

The panel has been constrained to a frame by means of four iron springs (figure 1).

Figure 1: Test-article

Employing alternately the piezoceramic patches as sensors and actuators it has been possible to acquire 72 Frequency Response Functions (9 actuators x 8 sensors). Moreover the frequency range of the acquisitions was fixed by taking in account the "coherence function" behaviour: this function, if evaluated between the response and the actuation patch signal presents values very close to 1 if the input and output signal are correlated; those values become much lower at frequencies where the sensor measurements are not correlated with the excitation, or, in other words, where the input mechanical energy does not reach anymore the sensors. Following that analysis the operative frequency range was fixed from 0,5 to 9,0 kHz measured through 1361 spectral lines (this last number is related to the signal acquisition device).

5. IMPLEMENTATION OF THE NEURAL NETWORK

As mentioned in the introduction the neural network has been trained making use of the same FRFs of the healthy structure used for the determining of the Damage Index, as positive examples.
Owing to reciprocity, since the system is linear, we have now used 36 FRFs only, reducing the redundancy of the information, so the network was lighter.
It had avoided to implement a single neural network including all the FRFs due to the high number of neurones needed (48996 = 36 x 1361). Also considering to implement a single network where neurones were associated with FRFs, some problems were forecasted: in such case in fact it could have been possible that for distinct experimental campaigns two FRFs related to different sensor-actuator couples resulted very similar; this would be a symptom of structural modifications but the network could deduce the opposite situation if the actuator-sensor couple is not explicitly declared. In the end it has employed only one kind of network which is trained separately for each actuator-sensor couple. That has resulted in 36 neural networks equal from an architectural point of view but each one trained by FRFs referred to the healthy configuration and measured by a specific actuator-sensor couple.
The neural network algorithm was implemented in MATLAB environment by using Neural Networks Toolbox. The number of "training" FRFs groups were, for the test-articles, respectively five and thirteen.

5.1 RECONSTRUCTION INDEX AND "THRESHOLD" EVALUATION

Once completed the training phase it raised up the problem on how to concretely employ the networks for identifying damage. The idea was to develop a "reconstruction" index representing the ability of the networks in rebuilding (and as a consequence, recognise) the input FRFs received. A good recognition level would have been connected with a "healthy" status of our structure. It has been defined for each network the reconstruction error between input I and output O:

$$E_n = (O_n - I_n)^2 \quad n=1,..,N$$

where n represents the number of neurones (or acquired FRFs spectral lines). Following it has been defined an overall index of reconstruction of the input vector as:

$$R = mean(E_n) + iqr(E_n)$$

where "mean" is the individual error mean and "iqr" represents a measurement of their dispersion in a statistic sense. It is possible to plot a graph S(R) of the output of the neural network obtained. That graph shows, for each value of R, the ratio between the FRFs (used for the post process), which have a reconstruction R value higher then the fixed one, and the total number of FRFs acquired during a single test campaign. The range of S is from 0 to 1. Practically, the curve S(R) is an estimator of the panel healthy status (figure 3).

Even if no damage has occurred, FRFs, which have been acquired for each actuator-sensor couple on different times, could be not equal between them because of environmental disturbance. Once the FRFs of the sound structure are settled, the more different FRFs are the higher reconstruction error is, even if the damage is not present. For that reason is very important to consider the FRF maximum discrepancy of an actuator-sensor couple such that the neural network is able to reconstruct the input correctly. To do this the FRFs of the sound structure have been perturbed by means of increasing disturbance so much that it has been possible to specify the limit. In order to perturb the reference FRFs it has summed to each FRF a vector having the same dimension. Random values, whose probability distribution is normal with a variable mean and standard deviation, form the vector. Those random values have the same order of magnitude as FRF; to make the number of generated negative values negligible, since FRFs magnitudes are positive quantities, the mean has been restricted to be greater or equal to the standard deviation.

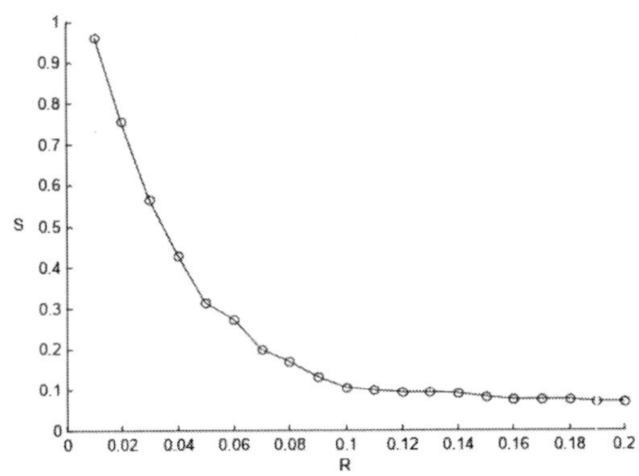

Figure 3: Structure status estimator curve

Such vectors have been added to the reference FRFs so as to generate two various families of perturbed FRFs. First family (fig. 4) has been determined assuming the same disturbance, having a mean greater or, at least, equal to the standard deviation, for all reference FRFs. The relevant curves are characterised by a quick descent of the parameter S from 1 to 0. Second family (fig. 5) has arisen from considering that a real disturb modify the FRFs acquired during a campaign, which represent the healthy status of the panel in a particular instant, with several quantities; with this object in view perturbations having several intensities have been imposed randomly to the reference FRFs. If we take in account the output of the neural network whose input data are the gradually increasing perturbations of the second family FRFs, we note that the reconstruction error increase and the estimator curves shift coherently toward the right.

After having trained a neural network it is possible to estimate the error which is in its output by means of the linear regression between the network output and the corresponding target (during the training we assume that target coincides with input data). By means of that estimation it is possible to obtain delimiting the "healthy configuration" and the "damaged configuration" zones. In between there is a "doubt zone", a family of curves that can give good or bad reconstruction (figure 4).

Figure 6 represents the evaluation of the threshold: the curve of the second family intercepting the last blue curve of the first family (limit of good reconstruction) is chosen. The interception between the "variable" perturbation curve and the lower limit of the "doubt zone" is fixed as the healthy status threshold.

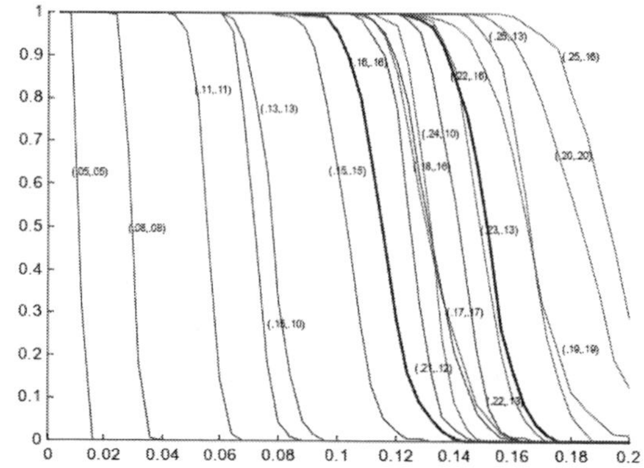

Figure 4: Analytical "constant" perturbations

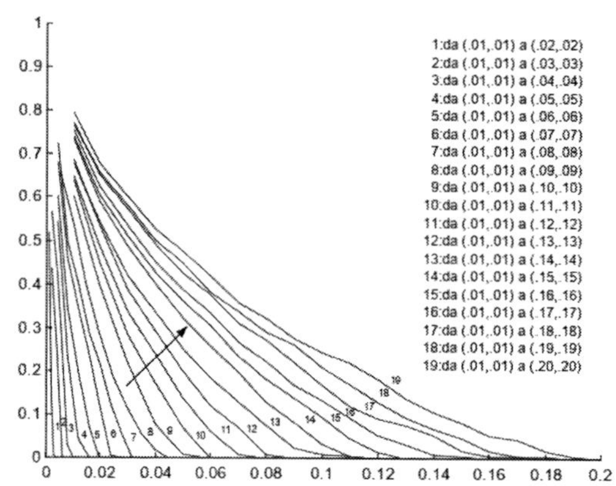

Figure 5: Analytical "variable" perturbations

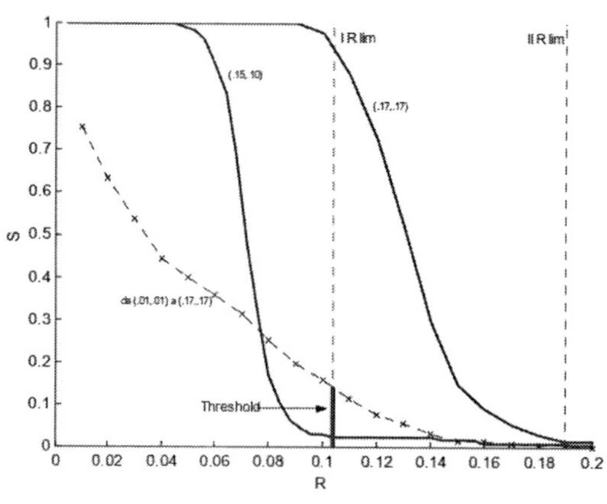

Figure 6: Threshold

6. EXPERIMENTAL RESULTS

Once the threshold representing the health status of the structure was identified, two experimental tests have been carried out in order to test the "Damage Index" method, the neural network and the discrimination approach.

The first experimental test has consisted of an addition of a mass of 4 grams close to the piezoceramic patch #1 and, then, close to the patch #6. Imposing that kind of simulated perturbation it has been possible to test the capability of the neural network in identifying structural changes independently by their location. Besides, at the end of the tests, it has been possible to restore the composite panel to the initial configuration.

The second experimental test has consisted in an impact test close to the piezoceramic patch #2. A sphere made by steel, whose mass is 146,4 grams and the diameter is equal to 4 centimetres, has been used for the impact. The edges of the composite panel have been stiffened, and the panel has been placed horizontally. That sphere has been dropped from a height of 2 metres, so the impact energy has been equal to 2,87 joule.

About the Damage Index method, it is possible to show a graph for each couple of actuator-sensor. To sum up the results, they have been gathered in a graph which contains nine groups of bars. Every group represents the sum of indexes of the couple actuator-sensor, in which the sensor is fixed.

As for the first experimental test (figure 7) each group consists of three bars: the first represents the sensitivity index, which is determined using FRFs of the healthy structure acquired at different times to measure the experimental error and the environmental noise and vibrations which can influence the FRFs; the second and third bar are the indexes obtained after the addition of the masses.

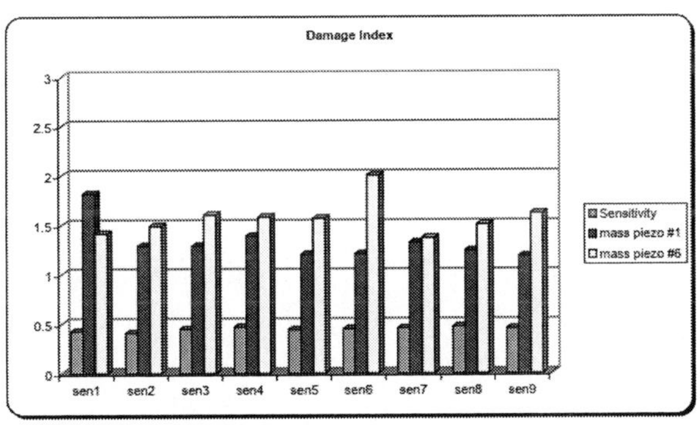

Figure 7: Addition masses

It can be noted that the index has recognized the reference structure and identified and localized the perturbation. All patches have given higher DIs than the sensitivity indexes and it can be noted that the piezoceramic patches which are close to the added mass have the highest indexes, so it is possible to assert that the damage has been localised. There is not the quantification because of the masses were no added in the same time.

As for the second experimental test (figure 8) it is possible to show the results as same as the first test. Also in that case the Damage Index method has identified and localized the damage, in fact all damage indexes are higher than the sensitivity ones, especially the second, very close to the impact point.

In both cases the localisation is not so evident because of the system is very sensitive, in fact also the sensors far from the area have felt the presence of the damage.

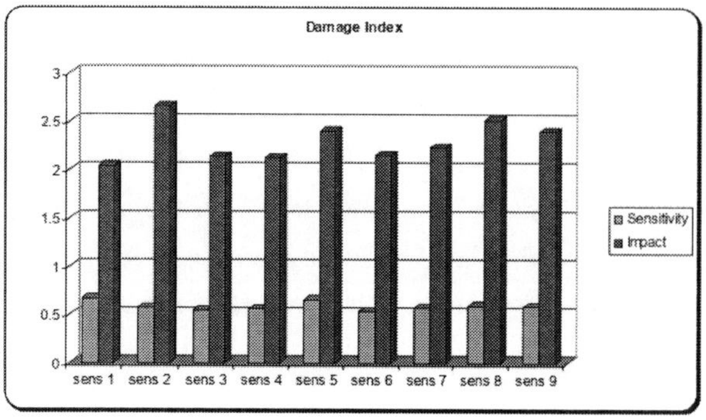

Figure 8: Impact test

About the neural network, once the threshold representing the health status of the structure was identified, thirteen campaigns of acquisition, each of them consists of 36 FRFs, have been acquired for the training.
As for the first experimental test, it is possible to assert that the neural network has recognized the reference structure and has identified the perturbation. It is possible to notice from the output graph reported in figure 9 that the reference structure curve, which has been obtained by means of the acquisition of FRFs of the sound structure, is very close the x-axes, while the addition masses curves are over the threshold.
Furthermore the position of the mass is irrelevant for the identification, in fact the curves are practically super imposable.

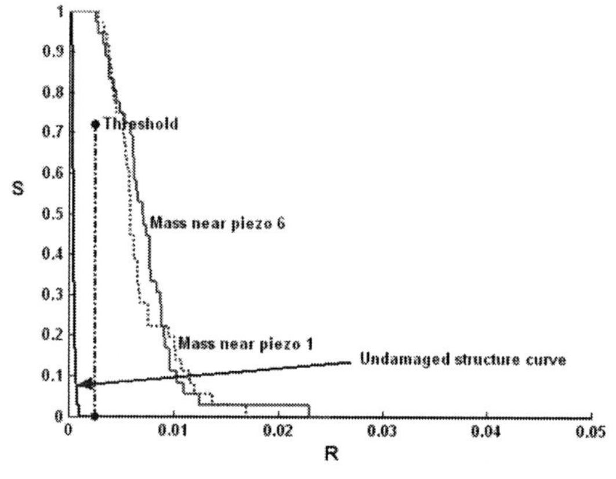

Figure 9: Addition masses

As for the second experimental test, the impact by means of the iron sphere, probably a damage has occurred. The neural network has confirmed it, in fact its output (figure 10) puts the impact curve far from the threshold, and very far from the undamaged structure curve. The FRFs, which have been acquired after the impact test, are definitely changed. That damage could be a delamination.

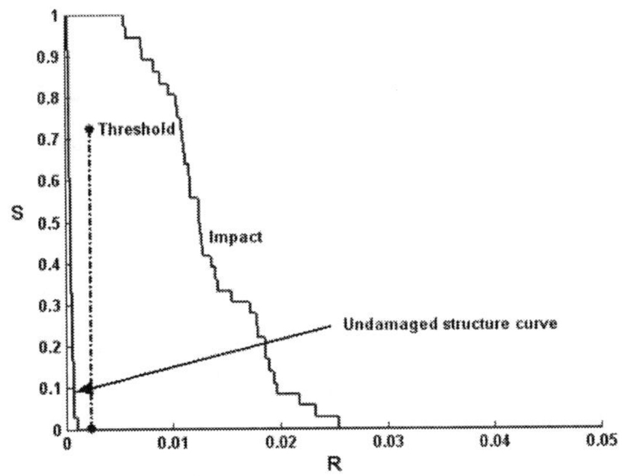

Figure 10: Impact test

7. CONCLUSIONS

The aim of that work has been the development of Non Destructive Tests (NDT) which meet to most of the mandatory requirements for effective health monitoring systems, simultaneously reducing as much as possible the complexity of the data analysis algorithm and of the experimental acquisition instrumentation; these peculiarities may, in fact, not be neglected for an operative implementation of such a system. Those techniques promise to bring a step forward for the implementation of an automatic "health monitoring" system which will be able to identify a structural damage in real time, improving the safety and reducing maintenance costs.

8. REFERENCES

[1] Chen J. Y., Sharpe Lonnie Jr., Jonkowsky Amy L. - *"An overview of vibrational-based nondestructive evaluation techniques"* - Proceedings of SPIE (Society of Photo-Optical Instrumentation Engineers), Volume 3397 (1998).
Maouk M., Zimmerman D.C. - *"Structural Damage Assessment using a Generalised Minimun Rank Perturbation Theory"* - AIAA Journal, vol.32, No.4, pp. 836-842, April 1994

[2] S. K. Thyagarayan, M. J. Shulz, P. F. Pai, J. Chung: *"Detecting structural* damage *using Frequency Response Functions*" – Journal of Sound and Vibration 1998 N. 210.

[3] Fedele P., Cafasso G., Lecce L. - *"A new technique for damage identification and health monitoring of structures using piezoelectric sensors and actuators"* - Proceedings of Seventh International Conference on Adaptive Structures (ICAST), September 1996, Roma (ITALY).

[4] F. P. Sun, Z. Chaundry, C. Liang, C. A. Rogers: *"Truss Structure Integrity* Identification *Using PZT Sensor-Actuator*" – Journal of Intelligent Material Systems and Structures, Vol. 6, Gennaio 1995.

[5] N. Japkowicz: "*Concept-Learning in the Absence of Counter-Examples: An Autoassociation-Based Approach to Classification*" – Dissertation submitted to the Graduate School-New Brunswick Rutgers, The State University of New Jersey in partial full fillment of the requirements for the degree of Doctor of Philosophy Graduate Program in Computer Science – Ottobre 1999.

[6] T. Petsche, A. Marcantonio, C. *Darken*, S. J. Hanson, G. M. Kuhn, I. Santoso: "*A neural network autoassociator for induction motor failure prediction*" – "*Advances in Neural Information Processing Systems 8.*", MIT, Cambridge, 1996.

[7] D. E. Rumelhart, G. E. Hinton, R. J. Williams: "*Learning internal representations by error propagation*" – "*Parallel Distributed Processing*" MIT Press, Cambridge, 1986.

9. BIOGRAPHIES

Igor Bovio is an Aeronautical Engineer. Born July 18, 1973 in Naples, Italy. He got the PhD title in 2006, after having graduated with full marks in 2002. He published his degree thesis regarding the Health Monitoring field on the "Italian Journal of Engineers and Architects Association". He has been a member of the local committee of the "International Workshop on Corrosion in Aircraft Structures and Ground Vehicles" in Ravello, Italy, and the "First World Congress on Corrosion in the Military" in Sorrento, Italy. He carried out a stage at "Officine Aeronavali Venezia", an ALENIA company. He attended several technical courses about "ATR 42-300 and 72-200 Maintenance General Familiarisation", "Smart Material and Structures", "Beamforming and Acoustic Holography", "Scanning Laser Vibrometry". He published some papers with regard to health monitoring and neural networks based on vibration measurements, damage identification, and laser vibrometry techniques on national and international proceedings. Besides he was the referent for the Department of Aeronautical Engineering of University of Naples "Federico II" with the FENet project (Finite Element Network); he is a consulting engineer for the MESEMA research project (Magnetoelastic Energy Systems for Even More Electric Aircraft) and he is a member of the staff for the AERO–NEWS research project (Health monitoring of aircraft by Nonlinear Elastic Wave Spectroscopy).

Leonardo Lecce is born in Manduria, Taranto, Italy on March 13th, 1947 and he graduated in 1971 with honours in Aeronautical Engineering at University of Naples "Federico II". In 1973 was appointed Assistant Professor of Aircraft Structures at the Engineering Faculty of the University of Naples. In 1982 becomes Associate Professor of Aeroelasticity at the same University. In 1993 becomes Full Professor and from 1994 he is teaching the course on Aircraft Structures. From November 2000 he is Head of the Department of Aeronautical Engineering of the University of Naples "Federico II". After his graduation he applied for several courses, seminars and stages at important research centers in Italy and abroad. He has been supervisor of about 150 graduation thesis. From 1986 is member of the Doctorate Committee in Aerospace Engineering and supervisor of several Doctoral Thesis. From 1989 to 1994 he has been Scientific Consultant for the Vibration and Acoustic Group at CIRA (Italian Aerospace Research Center). From 1994 to 1997 he has been member of CIRA Scientific Committee. In many occasions has been member of the UE Expert Commission for the Evaluation of the research proposal in the field of the Vibration and Acoustics. He has been Scientific Responsible of several research contracts sponsored by UE in the field of the noise active control in aircraft and helicopters, and European Scientific Coordinator for the MADAVIC (Magnetostrictive Actuators for Damage Analysis and Vibration Control, 1998-2001) and now is Coordinator of the follow-on program MESEMA (2004-2006). He holds contacts and relationships with several and qualified Universities and Research Centers in USA and in Europe (ISVR, DLR, University of Glasgow, WVU, etc.). The results of his research activities (more than 120 papers) have been published on national and international journals and Proceedings. He is member of numerous scientific associations: AIDAA, AIAS, AIV, AIAA, SEM, etc.

CFD SIMULATIONS OF THE FLOW THROUGH A WATERJET INSTALLATION

N.W.H. Bulten, Wärtsilä Propulsion Netherlands BV / Technical University Eindhoven, The Netherlands
R. Verbeek, Wärtsilä Propulsion Netherlands BV, The Netherlands

SUMMARY

With the aid of modern computational fluid dynamics (CFD) tools it has become possible to calculate and visualise the flow through a complete waterjet installation. This can help to enhance the understanding of the occurring physical phenomena and this will lead to further improvements of the complete installation. Calculations are made with a steady-state multiple frame of reference (MFR) approach and with fully transient moving mesh method. Inlet and pump CFD models have been validated separately with available experimental data. Agreement between measurements and CFD results is good for both. The numerical models of inlet and pump are combined to form a complete waterjet installation. The thrust and torque predicted with this model agree very well with the values obtained with traditional software. A concentration scalar has been introduced to visualise the suction streamtube. With this scalar the averaged velocity of the ingested water, and consequently the wake fraction, is determined. Results of transient calculations are used to determine pressure fluctuations on the impeller blade and in the statorbowl.

NOMENCLATURE

IVR	Inlet velocity ratio = v_{ship}/v_{pump}	[-]
P	Power	[W]
Q	Volume flow	[$m^3\ s^{-1}$]
R	Resistance	[N]
T	Thrust	[N]
t	Thrust deduction factor	[-]
v	Velocity	[$m\ s^{-1}$]
w	Wake fraction	[-]
ε	Inlet loss coefficient	[-]
ϕ	Nozzle loss coefficient	[-]
η	Efficiency	[-]
μ	Jet velocity ratio = v_{in}/v_{noz}	[-]
ρ	Density	[$kg\ m^{-3}$]

1. INTRODUCTION

With the ongoing increase of installed power in a waterjet propulsion system, a high level of accuracy in performance prediction is required.

The overall propulsive efficiency is defined as:

$$\eta_{op} = \frac{R \cdot v_{ship}}{P_{shaft}} = \frac{(1-t) \cdot T \cdot v_{ship}}{P_{shaft}} \qquad (1)$$

where R is the resistance, v_{ship} the ship speed, P_{shaft} the power at the waterjet shaft, T the thrust and t the thrust deduction factor. The thrust is calculated according to:

$$T = \rho Q (v_{noz} - v_{in}) \qquad (2)$$

where ρ is the density, Q the volume flow through the installation, v_{noz} the jet velocity out of the nozzle and v_{in} the mass averaged velocity of the ingested water. The ratio between the two is denoted as jet velocity ratio μ:

$$\mu = \frac{v_{in}}{v_{noz}} = \frac{(1-w) \cdot v_{ship}}{v_{noz}} \qquad (3)$$

The factor, which represents the difference between the undisturbed velocity and the averaged ingested velocity, is denoted as wake fraction w.

The overall propulsive efficiency can be rewritten to the following expression, with aid of a standard expression for pump shaft power [1]:

$$\eta_{op} = \frac{(1-t)}{(1-w)} \cdot \eta_{pump} \cdot \frac{2\mu(1-\mu)}{(1+\phi) - \mu^2(1-\varepsilon)} \qquad (4)$$

where η_{pump} is the hydraulic pump efficiency and ε and ϕ the loss coefficients for the inlet and outlet. Performance prediction programs make use of some empirical factors for the thrust deduction and for the width of the boundary layer suction shape, which is required for the determination of the wake fraction. Waterjet manufacturers have used feed-back from sailing vessels to fine tune the empirical factors.

In general, the pump efficiency is based on operation with a uniform inflow. A waterjet operates most of the time in a strong non-uniform velocity field [2]. This non-uniform velocity distribution might have an influence on the pump efficiency and consequently on the overall propulsive efficiency.

In order to eliminate the influence of the different interaction effects a complete waterjet installation should be analysed. For a given ship speed and pump RPM, determination of the averaged inlet velocity and the volume flow is sufficient for the prediction of thrust. For the overall propulsive efficiency the shaft torque and the thrust deduction factor are required.

Determination of the averaged inlet velocity, or the wake fraction, is quite difficult in an experimental set-up. Within a numerical analysis of the flow, it is just a simple post-processing step. Volume flow and shaft torque can be determined both experimentally as well as numerically. The thrust deduction factor will be the only remaining empirical factor.

Recently, model tests of a complete waterjet installation have been carried out at Marin [3]. The waterjet was mounted on top of the large cavitation tunnel. Quite surprisingly, a small increase of pump efficiency was claimed due to the non-uniform inflow. This is contradictory to common perception, and it might be contributed to the measurement accuracy or to the determination of the wake fraction. In order to investigate the effect of non-uniform inflow on the pump performance in more detail, it has been decided to make a number of computational fluid dynamics (CFD) calculations of the complete waterjet installation.

First the two major components will be analysed separately; the inlet duct in section 2 and the mixed-flow pump in section 3. Available experimental data is used to validate the numerical models of the inlet duct and the pump. Once the accuracy of the predictions of the flow through the inlet and the pump is satisfactory, both models can be combined to form a complete waterjet installation. This will be discussed in section 4.

The CFD code provides two solution methods to take the centrifugal forces due to rotation into account: a steady state multiple frame of reference (MFR) approach and a fully transient method with moving mesh each time step.

The results of the steady state calculations will be used to visualise the suction streamtube and the determination of the wake fraction. These results are used for the calculation of thrust and efficiency. Performance prediction based on CFD analyses will be compared with data from the waterjet selection program of the authors' company. Accuracy of the predictions based on this selection program has been proven with actual sailing vessels, so the overall performance parameters should be accurate.

Transient calculations have been made for the design condition. Pressure fluctuations at the impeller tip and in the statorbowl are reviewed. Finally, conclusions will be formulated in section 5.

2. WATERJET INLET FLOW

2.1 NON-UNIFORM VELOCITY DISTRIBUTION AT IMPELLER PLANE

The flow through the waterjet inlet is calculated with a commercial CFD program, based on a Reynolds-Averaged Navier-Stokes (RANS) method. The inlet CFD analysis tool has been validated extensively [4]. The CFD results have been compared with model scale measurements. Agreement of the pressure distribution at the inlet ramp and the cavitation inception at the cut-water was excellent. The CFD waterjet inlet analysis tool is used in the standard design procedure nowadays. Application specific optimisation can increase the performance significantly. Some typical examples are published (see for example [5]).
In this subsection the calculated velocity field will be compared with experimental data for 2 conditions with different inlet velocity ratio (IVR). IVR is defined as:

$$IVR = \frac{V_{ship}}{V_{pump}} \qquad (5)$$

These plots give a clear impression of the amount of non-uniformity of the inflow. Figure 1 shows the measured and calculated velocity distribution for a medium IVR. This is a normal free sailing condition for a fast ferry. High speed motor yachts (+60 knots) operate at higher IVR. An example of a high IVR is shown in figure 2. All velocities have been normalised with the averaged axial velocity. The plots of the experimental results might give the impression of a quite large shaft diameter. However the actual shaft diameter is much smaller. The plots with the CFD results show the used shaft diameter.

Agreement between measurements and calculations is good for both conditions. At medium IVR a maximum velocity of 135% of the average velocity is found and a minimum of 65%. This gives an amplitude of 35% of the average inflow velocity. The amplitude increases to 50% for the high IVR. The velocity distribution in the upper part shows some resemblance with a ship propeller wake field. However, the waterjet non-uniformity is twice as strong due to the conservation of mass in the pump; all low speed regions have to be compensated with high speed regions.

2.2 ORIGIN OF NON-UNIFORMITY

From measurements in windtunnels as well as cavitation tunnels and CFD calculations it has been found that the non-uniform velocity distribution is very strongly related to the IVR and to a much smaller extent to the shape of the inlet ducting.

Analysis of the flow through the waterjet leads to the conclusion that four phenomena are contributing to the non-uniform velocity distribution. These four phenomena are:

1. ingestion of water out of the hulls boundary layer
2. deceleration of the flow at higher IVR
3. bend in inlet duct
4. obstruction of the flow due to the shaft

At high IVR the suction depth of the dividing streamtube decreases. As a consequence a larger part of the water comes out of the boundary layer. Often all ingested water comes completely out of the boundary layer. The velocity profile is strongly non-uniform in the boundary layer. This is the first contribution to the non-uniform velocity distribution further downstream.

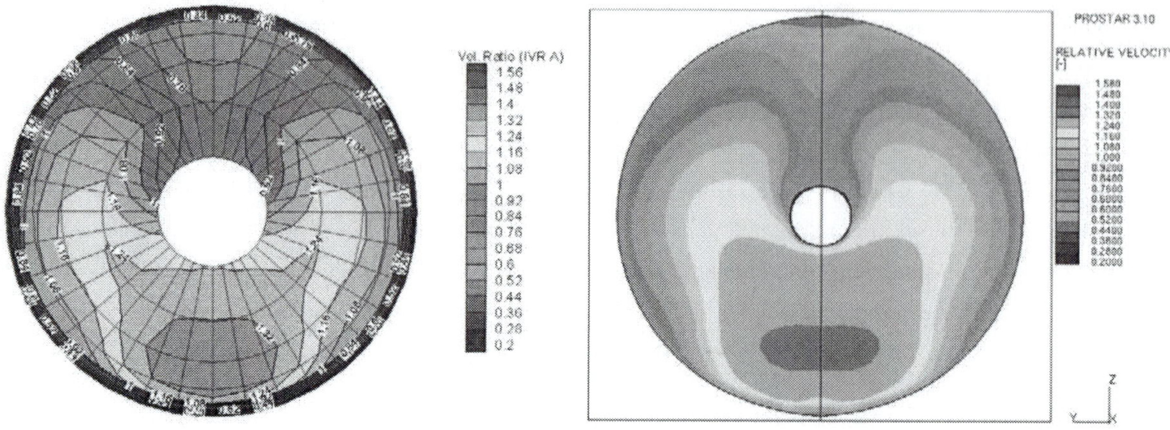

Figure 1: Velocity distribution for medium IVR (left is measurement, right is CFD calculation)

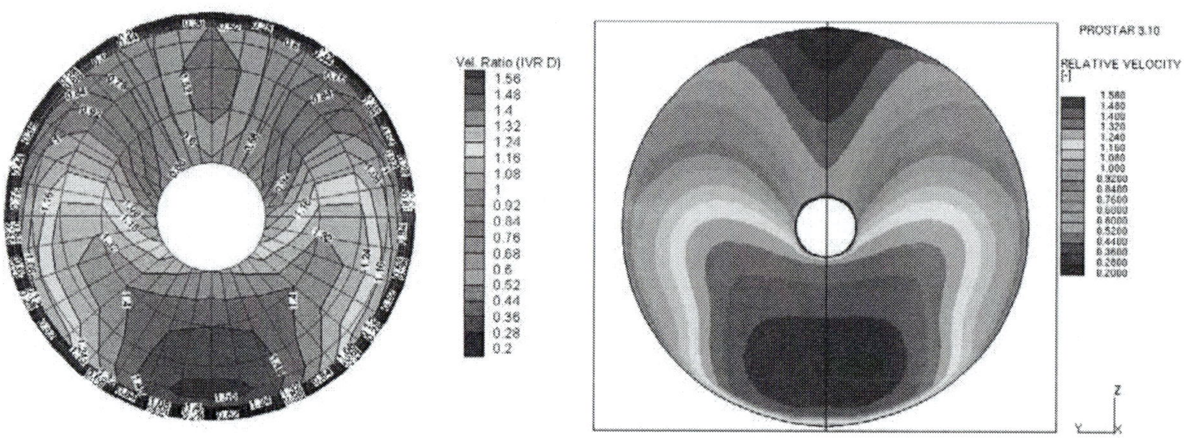

Figure 2: Velocity distribution for high IVR (left is measurement, right is CFD calculation)

At high IVR (>>1) the flow has to decelerate by definition. The part of the inlet duct where this occurs can be regarded as an artificial diffuser. The shape of this diffuser is not known a priori; the theory about diffusers is applicable however. Schlichting [6] shows the velocity distributions in diffusers with different diffuser angles. It is shown that the velocity profiles become more uniform for convergent channels and less uniform for divergent channels. If the diffuser angle becomes too large, then flow separation is observed. Both these phenomena also occur in the inlet of a waterjet.

Betz [7] gives an analysis of diffuser flow with non-uniform inlet conditions. It is shown that the low momentum fluid occupies much more space of the cross-sectional area of the outlet than the high momentum fluid. The low speed boundary layer flow will thus decelerate much more than the high speed fluid.

These two phenomena which contribute to the non-uniformity are both related to the IVR of the inlet. The other two phenomena occur in the last part of the inlet, where the flow can be regarded as normal pipe flow.

Effects of bends in pipes have been investigated in detail by Ito [8]. It is known from simple potential flow theory that a high speed region is present at the inner part of a bend. This effect is not so significant in a waterjet inlet duct however, because of the relative small angles of 20 to 30 degrees.

Obstruction of the flow due to the shaft leads to a small region with lower velocities in the upper part of the impeller plane.

With a stationary pipe around the shaft, two sets of vortices will be created in the shadow of the shaft. Shedding of these vortices will lead to higher vibrations in the installation. Rotation of the shaft stabilises this effect. This has been investigated by Seil [9] and confirmed by CFD calculations at the authors' company.

The four described phenomena are typical for flush mounted waterjet inlets. Therefore it can be concluded that the non-uniform velocity distribution is inevitable in commercial waterjet propulsion systems.

2.3 CONCLUDING REMARKS

CFD calculations of the waterjet inlet flow give very accurate predictions of the static pressure distribution in the inlet at model scale and the velocity distribution at the impeller plane. Moreover the cavitation inception pressure is predicted well. Consequently the method can be used to predict full scale losses in the inlet and to determine the shape of the streamtube of the ingested water.

3. MIXED-FLOW PUMP CFD ANALYSES

3.1 NUMERICAL MODEL FOR CFD ANALYSIS OF MIXED-FLOW PUMPS

In a similar way as the CFD validation project for the waterjet inlet ducts, a project has been defined for the validation of the numerical model of the mixed-flow waterjet pump. This validation process provided valuable information about the set-up of the numerical domain, the required mesh density and the settings of the numerical solution algorithms.

The numerical domain includes the complete impeller and the statorbowl in order to get good agreement with measurements over a large operational range. It is found that an analysis of one impeller blade only gives acceptable agreement in the design point.

In order to create a high quality mesh, it is decided to define an O-grid around the blades. This topology is continued along the hub surface. The channels between the blades are filled with a multi-block hexagonal mesh. The leading edge regions are meshed with a very fine mesh in order to capture the suction peak accurately. Special attention is paid to the tip clearance region. The gap is filled with some layers of extrusion cells, to allow flow over the blade tips. Comparisons with calculations without tip gap showed a notable difference; therefore inclusion of the tip gap cells has become the default approach.

For off-design predictions the length of the numerical domain in upstream direction turned out to be quite sensitive. If the inlet boundary conditions are placed too close to the pump, then the typical behaviour of a mixed-flow pump at medium flow can not be captured well.

The surface mesh of the impeller and the stator bowl is shown in figure 3. The final multi-block volume mesh contained about 1.1 M hexagonal cells.

For all calculations the well known k-ε turbulence model is used, though there are known deficiencies in this model. This will be discussed in more detail in the next subsection, where the results are compared with experimental data.

For spatial discretisation a second order method is applied. Implementation of the rotation of the impeller can be achieved in two ways: with a relative simple steady-state approach with Multiple Frames of Reference (MFR) or with a fully transient method with moving mesh. The first method freezes the rotor in a fixed position and adds extra source terms to the momentum equations, which represent the effect of rotation. With this approach the effect of rotor-stator interaction is neglected. Only one converged solution is found for a certain rotor position. The second method updates the rotor position at every time step during a calculation. Obviously this is a much more CPU time consuming calculation. The calculation can be stopped once the solution shows stable periodic behaviour. For a first estimation of the pump performance in uniform flow the MFR method is applied.

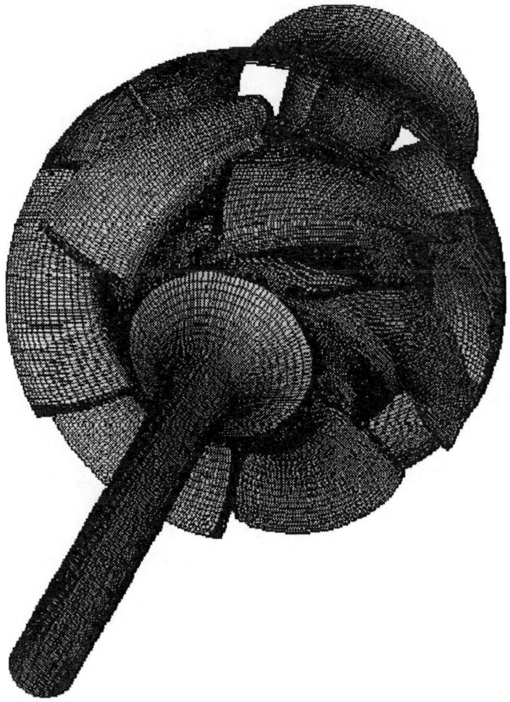

Figure 3: surface mesh of impeller and statorbowl

3.2 VALIDATION OF CFD MODEL OF MIXED-FLOW PUMP WITH UNIFORM INFLOW

A lot of experimental data was available from a model scale waterjet impeller. This data has been used for comparison with the CFD results. The numerical model is made according to the dimensions of the model scale pump. Figure 4 shows the comparison of the head and the efficiency of the pump. Both head and efficiency have been normalised with the values at the design point.

Agreement between measurements and calculations is good for the prediction of the pump head, the efficiency is however underestimated. Even at medium flow the typical drop in a mixed-flow pump head curve is predicted well.

The underestimation of the efficiency is due to a slight overprediction of the impeller torque. This error is almost completely contributed to the deficiencies in the used k-ε turbulence model. Separate studies of the flow through cascades showed an overprediction of the profile drag with about 30% compared to NACA measurements [10]. If this overpredicted profile drag is transformed into thrust and torque, then a torque overprediction of about 3-5% is found. Effect on thrust can be neglected because it is less than 0.5%.

The calculations have been repeated with a k-ω turbulence model, but these results were slightly worse. This k-ω turbulence model is not applied anymore.

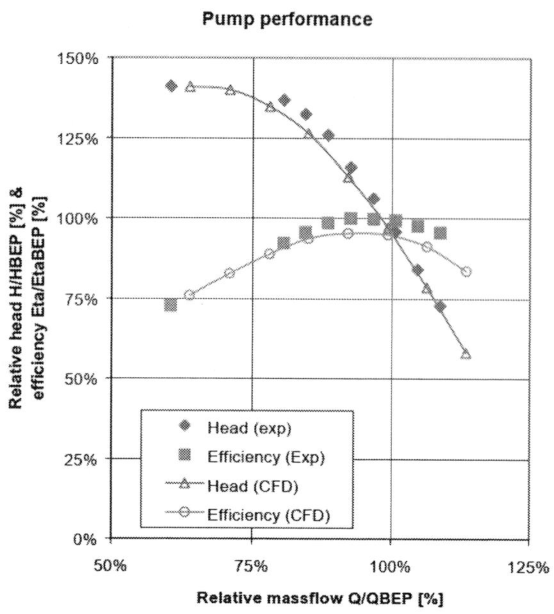

Figure 4: comparison of calculated and measured pump performance with uniform inlet velocity distribution. Head and efficiency are normalised with the design point values.

For one flow condition a fully transient moving mesh calculation has been made. Comparison of the static pressure of the steady state calculation with the moving mesh results gives an indication about the necessity of running in fully transient mode. Figure 5 shows the dimensionless static pressure in the statorbowl as function of the impeller blade location.

The steady state MFR result is plotted for one rotor-stator setting. The blade frequency can be recognised clearly in this plot. Time averaging of the transient results gives a dimensionless bowl pressure of 7.035, where the steady state approximation gives a dimensionless pressure of 6.986. Difference between both calculations is less than 1%. The quick steady state method gives a very good approximation of the time averaged pressure. So it is suitable for mixed-flow calculations with uniform inflow.

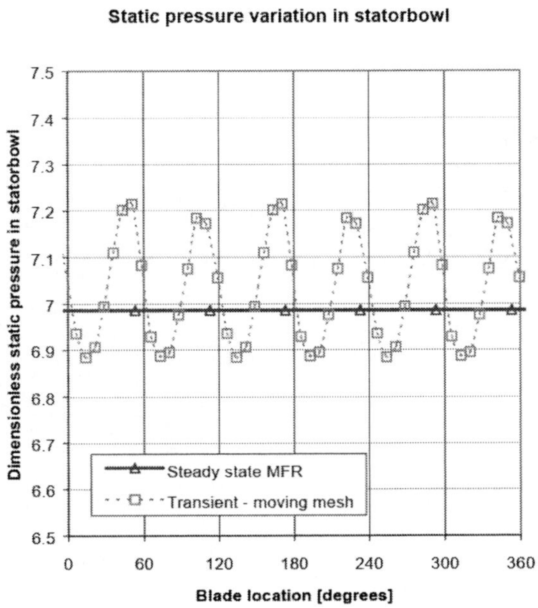

Figure 5: pressure fluctuations due to rotor-stator interaction in mixed-flow pump with uniform inflow velocity

4. COMBINATION OF INLET AND MIXED-FLOW PUMP

The two main components of a waterjet installation have been validated separately. The next step in the process is the combination of both meshes. At a straight plane just after the bend in the inlet duct an arbitrary coupling method is used to connect the inlet mesh to the impeller mesh. In this way both validated numerical models do not have to be modified. Figure 6 shows a plot of the complete mesh.

At the nozzle outlet surface a constant pressure boundary condition is applied. Consequently, the volume flow through the system becomes a part of the flow solution, whereas it is prescribed as boundary condition in the separate inlet and impeller calculations.

For solution of the flow again two methods have been applied: steady state MFR and transient moving mesh.

4.1 STEADY-STATE MFR APPROACH FOR COMPLETE OPERATIONAL RANGE

The first series of calculations are done with the steady state MFR method. Given the good results for the uniform inflow case, this seems to be justified. Each condition requires a few hours of computation time, so a number of operating points can be calculated overnight.

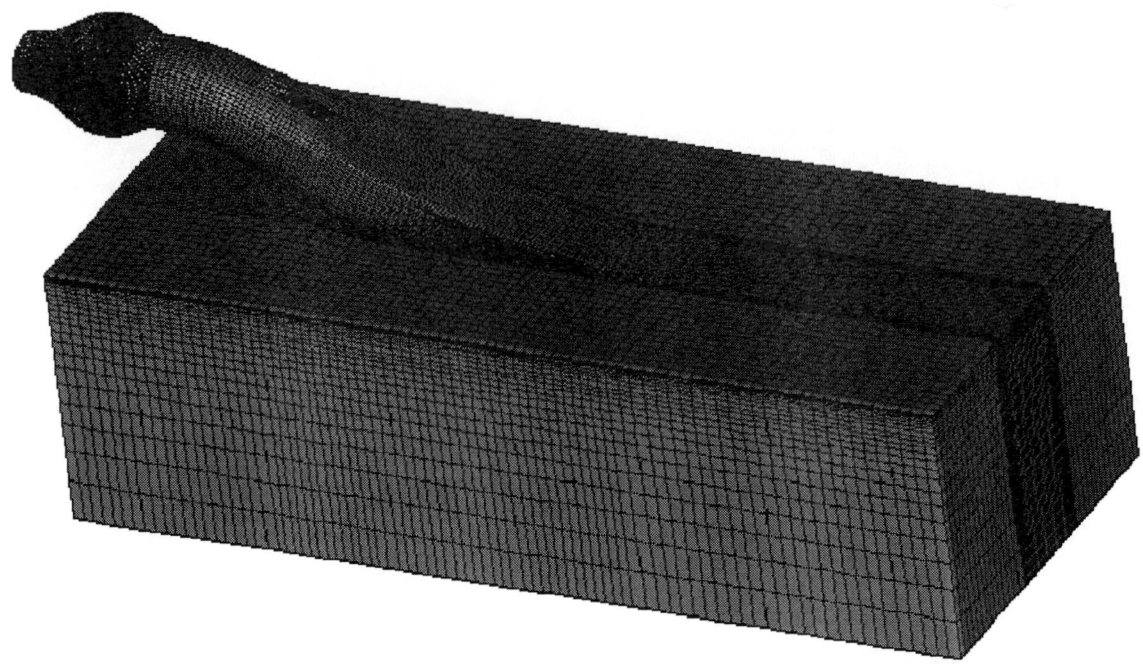

Figure 6: mesh of complete waterjet installation

Variation in operating condition can be obtained by variation in ship speed or pump RPM for certain geometry. In this paper results are presented of calculations with a constant pump RPM and varying ship speed. It is to be expected that the required power and the produced thrust will decrease with increased ship speed.

First a comparison of the predicted volume flow through the waterjet installation is made. The flow is determined by integration of the velocity at the nozzle exit plane. Table 1 shows the expected volume flow as given by the traditional software and the calculated flow with the CFD method. Differences in flow prediction are less than 1% for all conditions. Agreement is thus very good.

V_ship [knots]	Q_select [m^3/s]	Q_CFD [m^3/s]	Difference [%]
31	12.96	12.93	-0.23 %
35	13.11	13.08	-0.19 %
39	13.26	13.19	-0.54 %
43	13.41	13.31	-0.72 %
47	13.55	13.49	-0.45 %

Table 1: volume flow prediction based on selection software and CFD calculations

Determination of thrust can be done in two ways. The first method is based on integration of all momentum fluxes and the second method is based on a summation of the forces acting on the waterjet structure. Integration of the momentum fluxes requires a well defined control volume. This can be set equal to the complete numerical domain, or it can be restricted to the streamtube cells. The fluid cells, which contain ingested water, are marked with a separate scalar. This scalar acts as a weight function in the post-processing: 1.0 for ingested water, 0.0 for the other fluid. The streamtube can be visualised easily during this process with an isosurface plot. In such plot a surface is created, based on all cells with a constant value, in this case a concentration factor of 50% is used. Figure 7 shows the waterjet installation and a part of the streamtube. The shape of the streamtube has a semi-elliptical shape at the start of the numerical domain.

The thrust based on the momentum fluxes (eq. 2) requires the averaged inlet and outlet velocity and the flow through the system. Volume flow through the system (and thus indirectly the outlet velocity) is presented in table 1. The averaged inlet velocity is determined with aid of the scalar weight function. The resulting thrust is compared with the thrust from the selection software in figure 8.

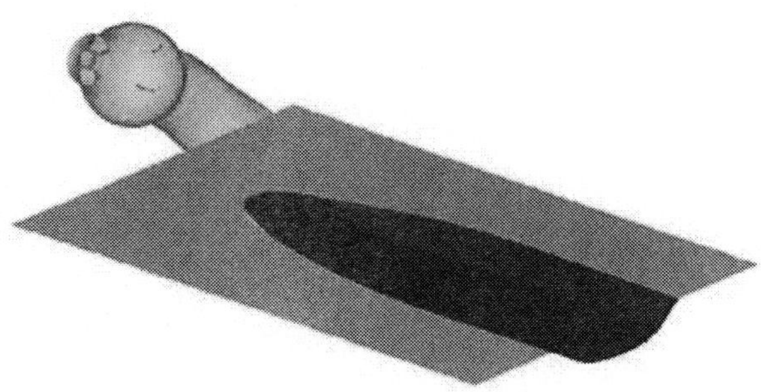

Figure 7: dividing streamtube of ingested water

Thrust calculation based on direct summation of all forces acting on the structure is the second method to determine the thrust. If all wall cells are taken into account, then the thrust of the waterjet system will include the hull drag of the bottom plating. This effect should be compensated for. It is also possible to select only the wall cells attached to the dividing streamline. In the end both methods should be independent of the dimensions of the numerical domain. Figure 8 shows the results of the thrust calculation for both methods of force summation.

All analysed conditions are calculated with the same RPM. Therefore the highest thrust, is found at the lowest ship speed.

The method, which summarises all wall cell forces gives the best agreement with the selection software. Differences are within 1% over the complete range.

If the two streamtube methods are compared, then an interesting phenomenon can be recognised. The thrust based on the momentum fluxes is lower than the thrust based on summation of the wall cell forces. This might be due to numerical errors, but it is also possible that there is a contribution to the thrust from the pressure acting on the dividing streamtube. Unfortunately this force can not be quantified accurately. A rough estimation, however, gives a positive contribution to the thrust of 4 to 5% of the total thrust.

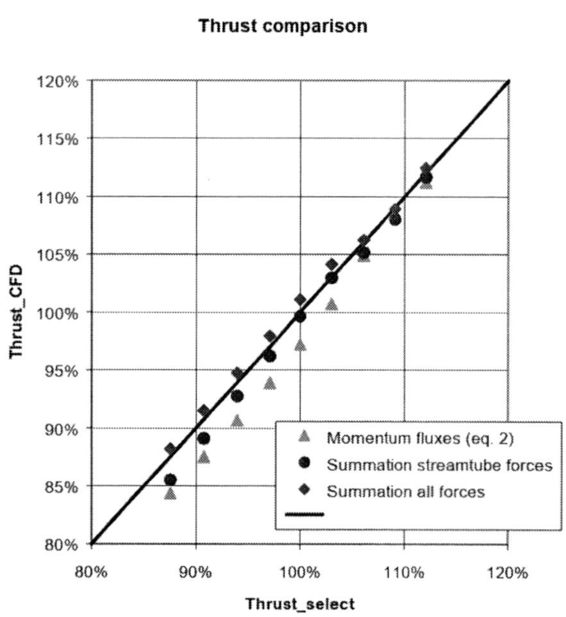

Figure 8: comparison of thrust from selection software with thrust derived from CFD results with 3 different methods

Shaft power for the waterjet is based on the required shaft torque. First the torque is directly calculated from the impeller wall cell forces in each cell. Second, it is derived from the tangential momentum upstream and downstream. Both methods are shown in figure 9, where the power is compared to the selection software.

The differences in power prediction show a more or less constant offset for the method of wall force summation for all conditions. It is according to the expectations, due to the numerical deficiencies in the turbulence modelling. In general the torque, and thus the power, is overpredicted with a few percent. A small correction factor for this known phenomenon can therefore be justified.

The method based on tangential momentum fluxes shows better agreement with the selection program for some conditions, but here the differences show more variations. This might be caused by the location of the integration planes.

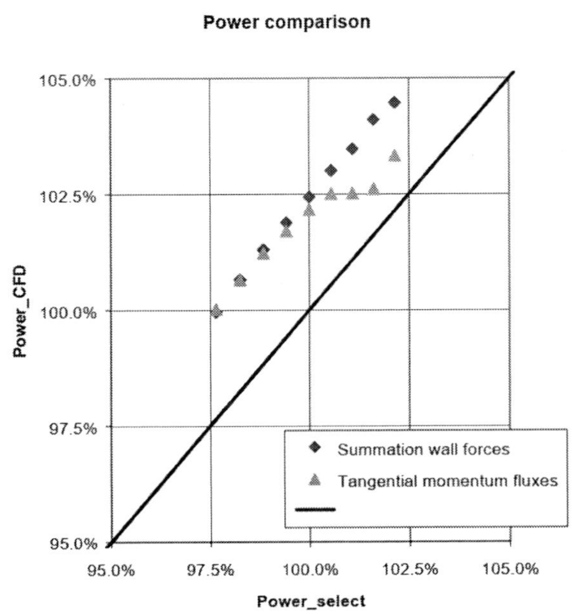

Figure 9: comparison of power from selection software with power derived from CFD results with 2 different methods.

4.2 FULLY TRANSIENT CALCULATION WITH MOVING MESH FOR DESIGN CONDITION

The design condition has been chosen for a fully transient calculation with moving mesh. After 2 impeller revolutions the solution shows quite stable harmonic behaviour. From this point on, the solution is assumed to be converged. Every 2 milli-seconds a results file is created for post-processing. With the currently available hardware it takes about two days to calculate 1 impeller revolution. The post file size is about 3 Gb for 1 revolution. So it should be obvious that this type of calculations still requires massive computational power and data storage.

Figure 10 shows the development of the static pressure at a certain point on an impeller blade during a revolution. There is a noticeable variation in pressure during one revolution. This is due to the non-uniform inlet velocity distribution.

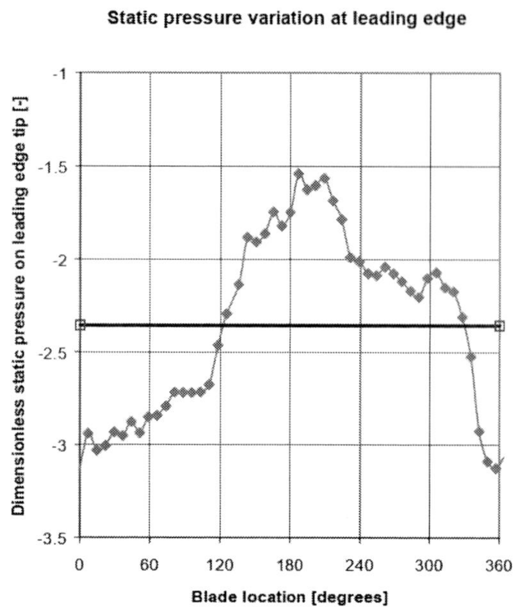

Figure 10: static pressure fluctuation at leading edge blade tip of mixed-flow pump with non-uniform inflow velocity

Figure 11 shows the static bowl pressure at two locations on the stator bowl surface. The two positions are both located at the same distance between two stator blades on the top and bottom of the stator bowl. Both pressure signals have similar mean and amplitude. This indicates that the effect of non-uniform inflow is dampened downstream of the impeller.

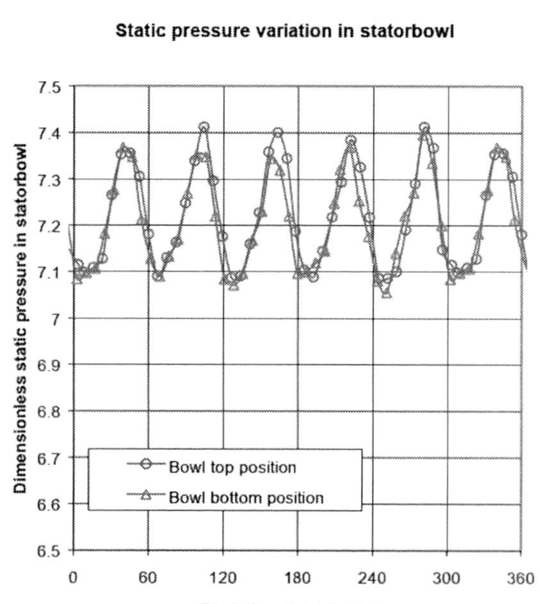

Figure 11: static pressure in statorbowl of mixed-flow pump with non-uniform inflow velocity

A comparison of the pressure levels of figure 5 and figure 11 shows a slightly higher pressure for the case of the complete pump. This is due to a small difference in volume flow through the pump.

5. CONCLUSIONS AND LOOK AHEAD

- CFD methods can be used to analyse a complete waterjet accurately within acceptable time scales.
- Accuracy of thrust and torque prediction is quite good with a steady state MFR solution method.
- Application of a concentration scalar for the ingested flow leads to a simple calculation of the averaged inlet velocity and determination of the shape of the streamtube of the ingested water.
- Transient analyses give more insight in pressure fluctuations. For example the loading variations on the seatring during an impeller revolution can be quantified more accurately.
- In the future a cavitation model will be implemented in the numerical analysis. This will give an impression of the development of the cavities. Variations in cavity shape are to be expected due to non-uniform inflow.

6. REFERENCES

1. VORST, H.J.A. VAN DER, VERBEEK, R., HENDRIKS, A.J.A.M., 'Design considerations of waterjet propulsion systems', 13th Fast Ferry International Conference, Singapore, 1997

2. VERBEEK, R. & BULTEN, N., 'Recent developments in waterjet design', RINA Waterjet Propulsion II conference, Amsterdam, 1998

3. KOOIKER, K., VAN TERWISGA, T., VERBEEK R., VAN TERWISGA, P., 'Performance and cavitation analysis of a waterjet system on a cavitation tunnel', proceedings FAST 2003 conference, session A1 pp 57-62, Ischia, Italy, 2003

4. BULTEN, N. & VERBEEK, R., 'Interpretation of model scale test results with aid of CFD calculations', RINA Waterjet Propulsion III conference, Gothenburg, 2001

5. BULTEN, N. & VERBEEK, R., 'Design of optimal inlet duct geometry based on vessel operational profile', proceedings FAST 2003 conference, session A2 pp 35-41, Ischia, Italy, 2003

6. SCHLICHTING, H., 'Boundary layer theory', McGraw Hill, New York, 1968

7. BETZ, A., 'Introduction to the theory of flow machines', Pergamon Press, Oxford, 1966

8. ITO, H., 'Pressure losses in smooth pipe bends', ASME Journal Basic Engineering, Vol. 82, p. 131., 1960

9. SEIL, G. J., 'The effect of the shaft, shaft rotation and scale on the flow in waterjet inlets', RINA Waterjet Propulsion III conference, Gothenburg, 2001

10. EMERY, J.C., HERRIG, L.J., ERWIN, J.R., FELIX, A.R., 'Systematic two-dimensional cascade tests of NACA 65-series compressor blades at low speeds', NACA report 1368, 1958

11. TERWISGA, T. VAN, 'Waterjet - hull interaction', PhD thesis, Delft, 1996

7. AUTHORS BIOGRAPHY

Norbert Bulten works at the Propulsor Technology - Hydrodynamics (PTH) department of Wärtsilä Propulsion Netherlands BV. He is responsible for CFD and FEM projects for waterjets and propellers. He is preparing a PhD thesis at the Eindhoven Technical University (TU/e) about the effects of non-uniform inflow to a waterjet pump.

Rob Verbeek is responsible for product development at the Thruster and Jet segment of Wärtsilä Propulsion Netherlands BV, formerly known as Lips Jets. He has worked in the waterjet area for more than 10 years already.

STRUCTURAL RESPONSE OF A FAST PATROL VESSEL WHEN OPERATING IN SEVERAL SEA AREAS: COMPARISON BY MEANS OF A FATIGUE ASSESSMENT.

C. Cartalemi, S. Ferraris, Fincantieri C.N.I. Naval Vessel Business Unit, Italy
A. Moggia Cappelletti, L. Sebastiani, CETENA, Italian Ship Research Centre, Italy
D. Boote, DINAV, University of Genoa, Italy

SUMMARY

Medium size light vessels are operated worldwide to perform patrol and rescue activities. The characteristics of such vessels, together with the various environmental conditions they might have to cope with all around the world, lead to significantly different sea-keeping behaviour and structural response when operating in open oceans or in closed seas.

A first comparison had been carried out a few years ago [1], stating that, from a static point of view, the structural behaviour of such vessels can be better in open seas, where waves are generally higher and longer, rather than in closed seas, characterised by smaller but shorter waves. Further to that paper, aim of present study is to deepen the subject and try to make the comparison more exhaustive. In fact reliability of the structural behaviour of the vessel has been assessed by means of a comprehensive calculation procedure, which allows determining the stress fields acting on the ship structure and the "Structural Response Operator" (SRO) and to perform an exhaustive fatigue assessment in way of significant structural details.

1 INTRODUCTION

Medium size light vessels have to accomplish, worldwide, several operations in very random environmental conditions.

Hull structure of such vessels is designed to achieve weight and cost targets, which lead to meet minimum scantling requirements and to adopt simple structural details. This optimization, motivated by commercial demands, has been achieved through the use of higher strength materials (mainly steel) and more sophisticated design tools, which allow a better knowledge of stress distributions and the definition of all necessary modifications to structural details to keep stress concentrations as low as possible, thus ensuring a longer fatigue life.

Aim of this paper is to deeply investigate the structural behaviour of the same vessel set out in already published benchmarks [1,2], performing the fatigue assessment of a few critical details, with the ship operating in three typical areas respectively characterised by long, medium and short wave lengths. Fatigue analysis has been carried out by means of a full spectral approach based on the "unit pressure facet method" [2,3] in way of significant structural details. The fatigue life has been evaluated by using the S-N curve fatigue approach under the assumption of linear cumulative damage (Palmgren-Miner's rule), stating an operating profile in the different geographical areas.

2 ENVIRONMENTAL CONDITIONS

The analysis has been performed in three different geographical areas having different sea characteristics. Not only are such zones deemed to represent most of geographical areas of possible operation, but also those of major interest from a commercial point of view.
According to BMT's Global Wave Statistics [4] the following areas have been considered:

- Area 9 **Northern Atlantic ocean** as long wave sea (similar to areas 16, 90, 83, 87)
- Area 26 **Western Mediterranean sea** as medium length wave sea (similar to areas 27, 47)
- Area 37 **Red sea** as short length wave sea (similar to areas 38, 5)

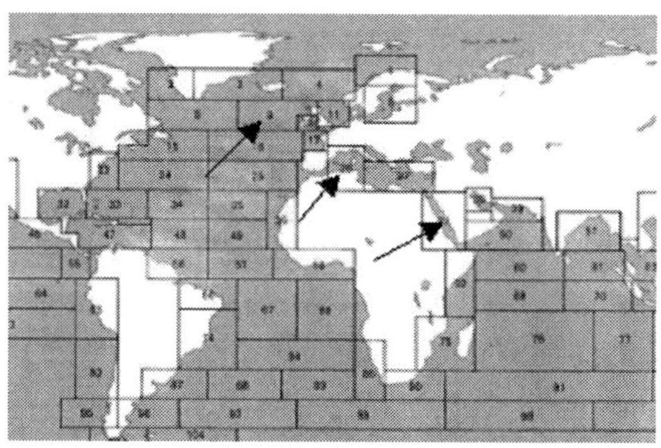

Figure 1: Geographical areas

3 FATIGUE ASSESSMENT

3.1 REFERENCE VESSEL

A medium size fast patrol vessel has been taken as reference ship for the execution of the fatigue analysis. Table 1 shows the main ship data.

Length between perpendicular – L(m)	47,2
Moulded Breadth – B(m)	8,1
Depth – D(m)	5,4
Draught – T(m)	2,2
Full Load Displacement – Δ(t)	392

Table 1: Main ship data

3.2 CRITICAL DETAILS

The fatigue assessment has been performed for three typical details, having different hot spot points. In the literature [5], the hot-spot is defined as the critical location at the weld toe (or weld end) where a fatigue crack can be expected to initiate. Hot spots can be classified as two types "a" and "b", as shown in Figure 6.

In type "a" the weld is located on a plate surface, while in type "b" the weld is located on a plate edge.

The selected details are the following:

- Primary stiffeners intersection;
- Primary and secondary stiffeners intersection;
- Bulkhead vertical girder and primary stiffener intersection.

Following Figures 2, 3 and 4 show the structural details as part of the global finite element model of the reference vessel used in the analysis (Figure 5).

The first detail regards the cross joint between a transversal beam and a longitudinal girder of the main deck. The chosen hot spot points are located at the weld toe and at the weld end (Figure 2) of the welded joint between the flanges of the beams. The longitudinal girder (web and flange) is continuous through the intersection.

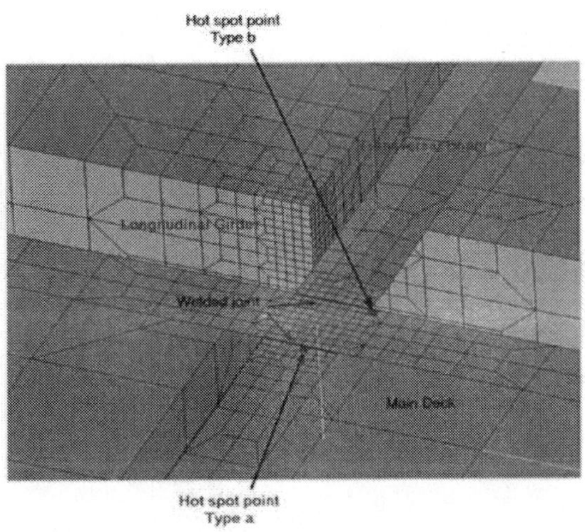

Figure 2: Detail 1 - Primary stiffeners intersection

The second detail (Figure 3) concerns a L-shaped longitudinal stiffener crossing the web of a transversal beam. The longitudinal (both web and flange) is continuous through the transversal beam. The hot spot point is at the flange end of the stiffener.

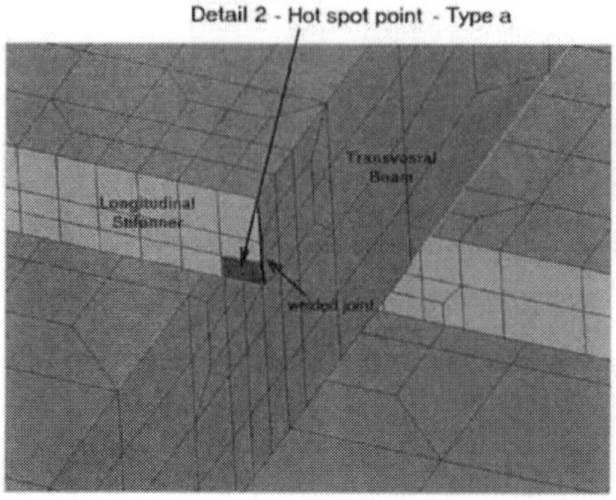

Figure 3: Detail 2 - Primary and secondary stiffeners intersection

The third detail is located under the aft bulkhead of the superstructures where a vertical primary stiffener ends on the cross between a longitudinal girder and a transversal beam of the main deck. The chosen hot spot point is on the welded joint between the web of the primary stiffener and the deck plate (Figure 4).

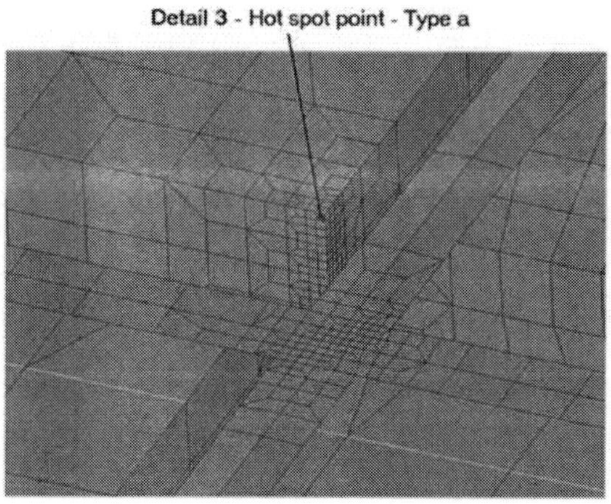

Figure 4: Detail 3 - Bulkhead vertical girder and primary stiffeners intersection

3.3 APPLIED PROCEDURE

Fatigue analysis has been carried out by means of a full spectral approach based on "unit pressure facet method" [2,3], whose main steps can be summarized as follows:

- Hydrodynamic analysis;
- Stress transfer function evaluation;
- Hot-spot stress range calculation at the crack initiation site;
- Individuation of the S-N curve relevant to the details;
- Long term hot-spot stress range distribution;
- Fatigue damage summation according to Palmgren-Miner's rule.

3.4 HYDRODYNAMIC ANALYSIS

Sea-keeping calculation has been carried out by using CRS-PRECAL, a 3D hydrodynamic code developed by MARIN within the Co-operative Research Ships [6]. Following reference conditions have been considered:

- Ship speeds: 9, 16, 20 kn
- Heading angles: from 0° to 360°, step 15°
- Wave frequencies: 23 values between 0.2 and 1.5 rad/s

Taking any combination of such parameters as input, the following characteristics have been evaluated:

- Vertical bending moment RAOs
- Vertical shear force RAOs
- Horizontal bending moment RAOs
- Horizontal shear force RAOs
- Torsional Moment RAOs
- Longitudinal force RAOs
- Motion RAOs
- Center of gravity accelerations

In the same conditions, hydrodynamic pressures acting on the wet hull have been calculated. This output is fundamental as a pressure translator can allow to turn such loads into an input for the FEM model.

3.5 STRESS TRANSFER FUNCTION CALCULATION

The structural analysis used for the evaluation of the stress transfer function has been performed by using a detailed 3D finite element model representing the whole hull girder and the superstructure (Figure 5). The model represents the actual geometric shape of the hull with reasonable accuracy. In the following this model will be identified as "coarse" model.

Figure 5: Finite element Model

Stress transfer function has been calculated for the selected details taking into account the combinations of ship speed, heading angle and wave frequency according to the procedure described in [2]. Calculation has been based on the "unit pressure facet method" [3], a linear superposition assumption of the load effects on the FE structural model. The method has required the application of a unit pressure on each facet of the wet hull, creating single load cases. The single stress responses have been then combined with the pressure transfer function (evaluated by means of the hydrodynamic analysis), giving the stress transfer function for each wave loading condition, as shown in following expression:

$$\sigma_{jm}(V,\beta,\omega_e)=\sum_{k=1}^{N_{fac}}\sigma_{jmk}\cdot p_k\cdot\cos(\omega_e t+\varepsilon_k) \qquad (1)$$

where:

σ_{jm} is the m-component of the stress transfer function in the reference j-element

σ_{jmk} is the m-component of the stress in the reference j-element due to the unit pressure applied to the k-facet of the wet hull

p_k is the amplitude of the pressure transfer function related to the k-facet of the wet hull for the specified speed V, heading angle β and wave frequency ω_w

ω_e is the encounter frequency

ε_k is the phase angle of the k-pressure transfer function.

N_{fac} is the number of the wet hull facets

t is the time

The stress transfer function evaluated on this FE model can be assumed as the *nominal stress*, taking into account the gross geometry of the details. The nominal stress range, as first approximation, may be assumed as the double of the transfer function value. Next step in the fatigue assessment procedure is the structural hot-spot stress evaluation.

3.6 HOT-SPOT STRESS EVALUATION

The geometrical stress (or hot spot stress) includes all stress raising effects of a structural detail excluding the stress concentrations due to the weld profile itself. The hot-spot stress has been evaluated starting from the nominal stress provided by the "coarse" global model and computing the stress intensity factor K_s according to the IIW Fatigue recommendations [5], by means of the refined finite element mesh of the details. As the stresses in the finite elements are dependent on the mesh density and the element properties, it is necessary to follow some guidelines in the choice of element types and sizes as well in stress evaluation at the extrapolation points. The IIW proposes two different types of mesh requirements for shell element modeling; the former is called "relatively coarse mesh" and the latter is named "relatively fine mesh". In the FE analysis of large structures (e.g. like ship hulls) it is not practical to use the fine element mesh required for accurate resolution of the stress field in way of the weld toe. Therefore, a "relatively coarse" mesh has been used [7], requiring a fixed mesh size of the elements of **txt** for hot spot of type "a" (where **t** is the thickness of the element) and of 10mm x 10mm for hot spot of type "b" in way of the extrapolation points. Such points are located in the middle of the elements (type "a") or in the mid-side of the elements (type "b") adjacent to the weld toe, as shown in Figure 6.

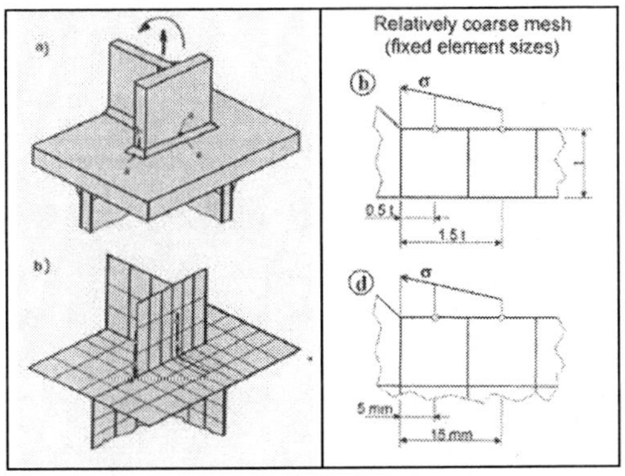

Figure 6: Hot spot type and extrapolation points

The linear extrapolation has been performed by using stresses $\sigma_{0,5t}$ and $\sigma_{1,5t}$ located respectively at $0.5t$ and $1.5t$ from the weld toe for hot spot of type "a". As regards type "b" hot spot, the linear extrapolation has been performed by using stresses at 5 mm and at 15 mm from the weld toe. The locations of extrapolation points are measured from the intersection of shells (shell element modeling without welds).

Type "a": $$\sigma_{hs} = 1.50\sigma_{0.5t} - 0.50\sigma_{1.5t} \qquad (2)$$

Type "b": $$\sigma_{hs} = 1.50\sigma_{5mm} - 0.50\sigma_{15mm} \qquad (3)$$

The stress value to be used in the computation is the maximum principal stress component normal to the weld toe, whichever is dominant.

The stress concentration factor has been evaluated comparing the hot spot stress provided by the refined mesh of the detail and the nominal stress calculated in the related "coarse" mesh in a reference loading condition. Assumed reference loading condition was a hydrodynamic sagging wave (usually known as "equivalent wave" [8,9,10]), having phase angle and length such to maximise the vertical bending moment response operator at 20 kn. The applied loads consisted of hydrodynamic pressures and inertial accelerations in an equilibrium condition. The stress intensity factor K_s has been then calculated as the ratio between the hot spot stress and the related nominal stress.

$$K_s = \frac{\Delta\sigma_{hs}}{\Delta\sigma_{nom}} \qquad (4)$$

Following figures (7, 8, 9, 10) show the extrapolation points in the chosen details. The hot spot points in detail 2 and detail 3 are assumed of type "a" while in detail 1 two different hot spot of type "a" and "b" have been studied.

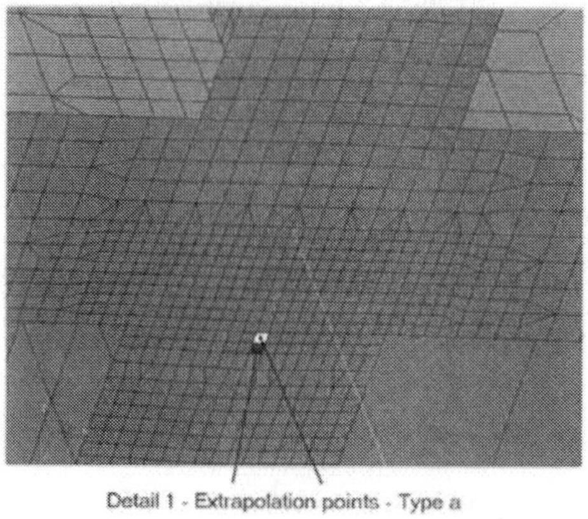

Figure 7: Detail 1 – Extrapolation points – Type "a"

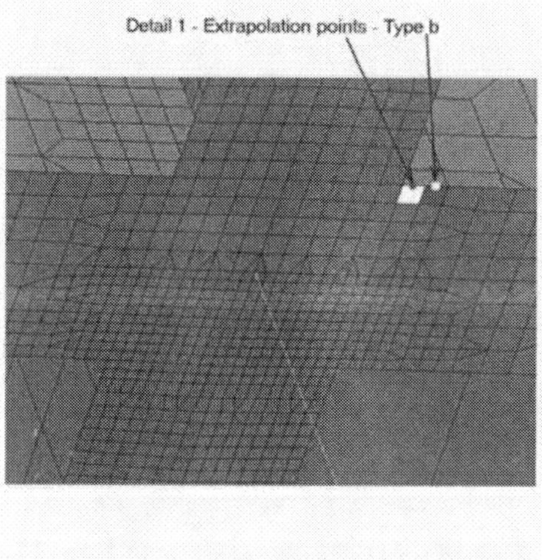

Figure 8: Detail 1 – Extrapolation points – Type "b"

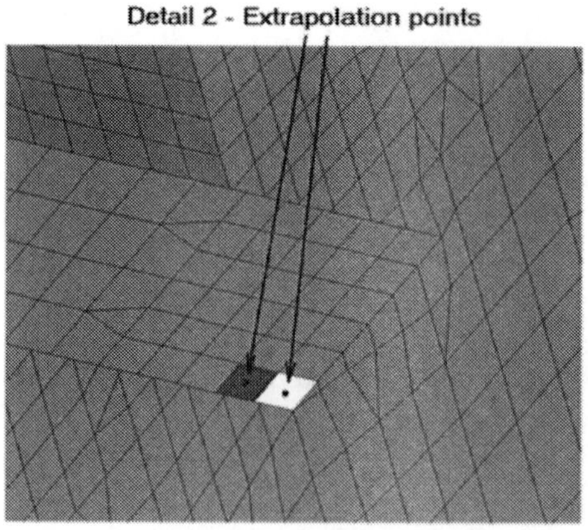

Figure 9: Detail 2 – Extrapolation points – Type "a"

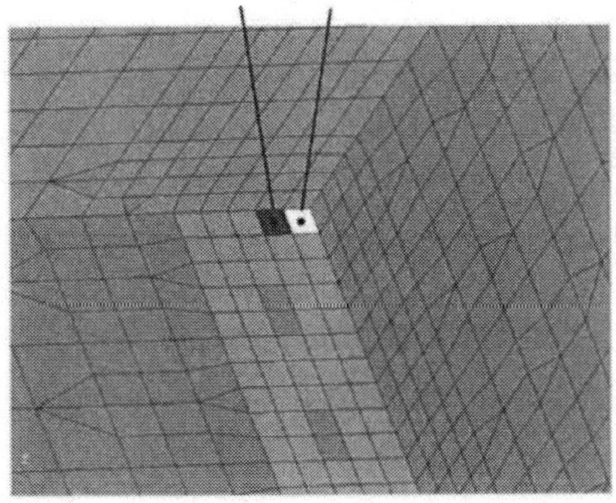

Figure 10: Detail 3 – Extrapolation points – Type a

The stress concentration factors for the chosen details are reported in the following tables (2, 3, 4, 5):

Hot-spot stress	**5.87**	MPa
Nominal stress	4.87	MPa
Ks	1.204	

Table 2: Detail 1 – type "a" – Stress concentration factor

Hot-spot stress	**16.95**	MPa
Nominal stress	16.7	MPa
Ks	1.015	

Table 3: Detail 1 – type "b" – Stress concentration factor

Hot-spot stress	**17.25**	MPa
Nominal stress	13.3	MPa
Ks	1.297	

Table 4: Detail 2 – type "a" – Stress concentration factor

Hot-spot stress	**33.8**	MPa
Nominal stress	25.4	MPa
Ks	1.331	

Table 5: Detail 3 – type "a" – Stress concentration factor

The stress concentration factor has been then applied to the "nominal" stress range transfer function previously evaluated by means of the "coarse" FE model, in every wave loading condition, obtaining the related hot spot stress range transfer function. Next step of the calculation procedure has been the definition of a long-term distribution for the hot-spot stress range.

3.7 LONG TERM HOT-SPOT STRESS RANGE DISTRIBUTION

The long-term stress range distribution has been evaluated through a full spectral analysis in the previously defined geographical areas (BMT 9, 26 and 37) assuming a Rayleigh probability density function [2,3,11,12] within each short term period for the different loading conditions. A design life of 20 years has been assumed and the following ship life fractions related to each operating speed have been used in each area:

V = 9 kn - 0.25; V = 16 kn – 0.50; V = 20 kn – 0.25

The Jonswap two-parameter formula has been used to evaluate the wave spectrum for the sheltered seas as area 26 and 37, while the Bretschneider or two parameter Pierson-Moskowitz spectrum has been adopted for the North Atlantic (BMT area 9).

Long crested and short crested seas have been separately considered. The directional sea spectrum in short crested sea has been defined using a cosine-squared spreading function. Each heading angle between 0 and 360 deg has been assumed equally probable.

As a first approximation, the hot spot stress range has been assumed as twice the hot spot stress values.

The resulting stress range cycle distributions in long crested and short crested sea are plotted in the following figures (11, 12, 13, 14) for each detail:

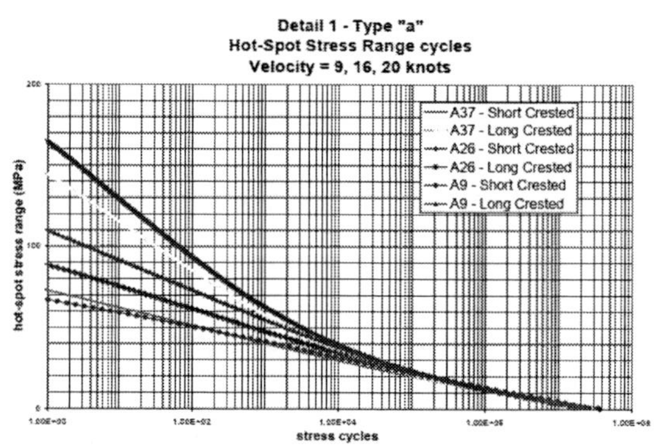

Figure 11: Detail 1 – Type "a" – Long-term stress cycles distributions

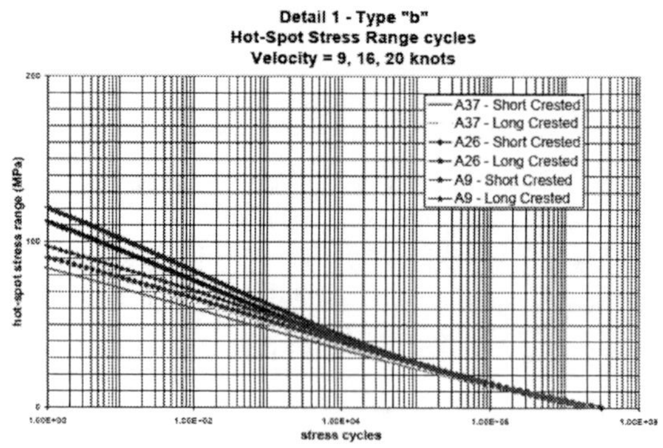

Figure 12: Detail 1 – Type "b" – Long-term stress cycles distributions

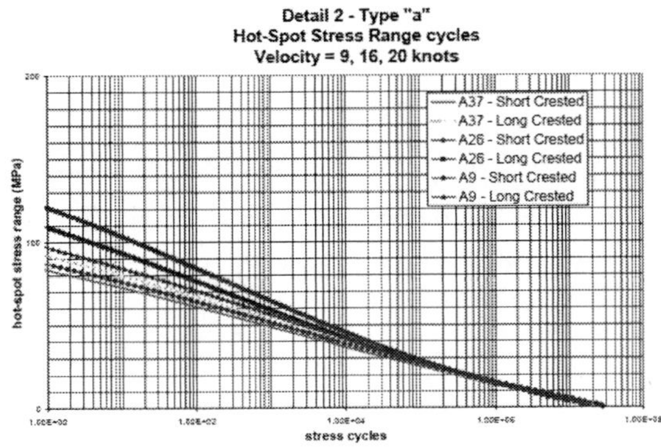

Figure 13: Detail 2 – Type "a" – Long-term stress cycles distributions

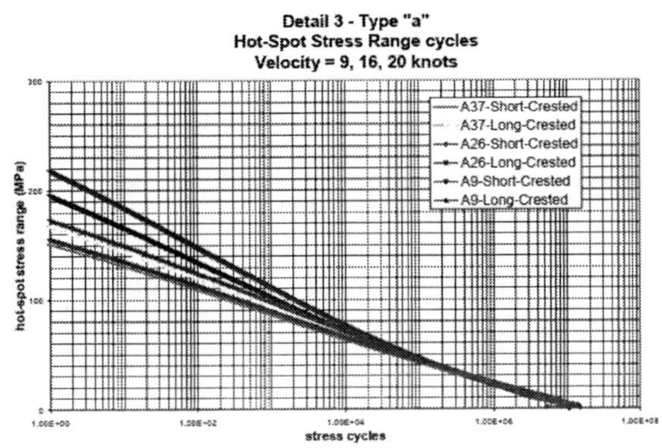

Figure 14: Detail 3 – Type "a" – Long-term stress cycles distributions

3.8 S-N CURVES

S-N curves for fatigue resistance vs. hot spot stress reported by IIW [5] have been considered. The resistance values refer to the as-welded condition and the effects of welding residual stress are included. The S-N curves are identified by the allowable stress range for a fatigue life of $2{\cdot}10^6$ cycles, that is named FAT number. Relatively to variable amplitude loading, two forms of S-N curve can be used for cumulative damage calculations, one more conservative (one slope) and the other with two different slopes (bilinear). The one slope S-N curve can be defined as:

$$N_i(\Delta\sigma_i) = \frac{C}{(\Delta\sigma_i)^m} \tag{5}$$

where:

N_i is the number of cycles to failure at constant stress range $\Delta\sigma_i$

C and m are the one slope S-N curve parameters.

In the case of constant amplitude loading, a fatigue limit $\Delta\sigma_{R,L}$ appears in the S-N curve. If all applied stress ranges remain below the fatigue limit, no fatigue damage is assumed to occur and an infinite fatigue life is expected. The constant amplitude fatigue limit is also valid for variable amplitude loading, but only if all stress ranges are below the constant amplitude fatigue limit.

S-N curves recommended for the weld types of the joints selected, T-joint or cruciform joint (type "a" detail 1, 2 and 3) and fillet weld with short attachment (type "b" detail 1) correspond to a fatigue class FAT 100, having a slope m of 3.00, and a constant fatigue limit $\Delta\sigma_{R,L}$ of 73.7 MPa at $5\cdot10^6$ cycles (Figure 15).

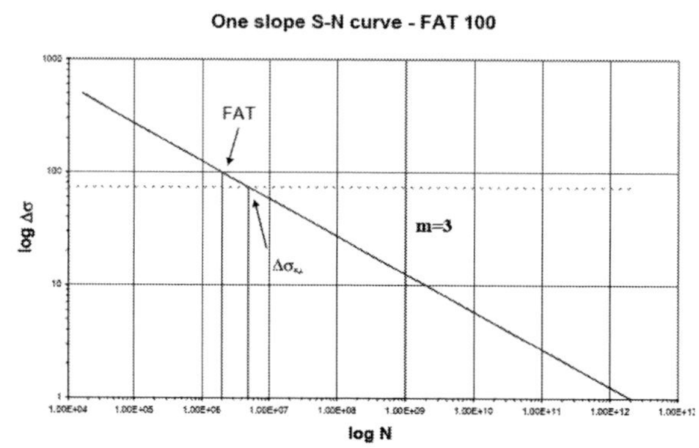

Figure 15: S-N curve – FAT 100

3.9 FATIGUE DAMAGE EVALUATION

The fatigue life has been evaluated by using the S-N fatigue approach under the assumption of linear cumulative damage (Palmgren-Miner's rule). The accumulated fatigue damage D can be expressed by the summation of each single stress range contribution as:

$$D = \sum_{i=1}^{k} \frac{n_i(\Delta\sigma_i)}{N_i(\Delta\sigma_i)} \tag{6}$$

where k is the number of stress constant amplitude stress range $\Delta\sigma_i$, n_i is the number of cycles of the i-stress range and N_i is the number of cycles to failure at constant stress range $\Delta\sigma_i$. The fatigue criterion requires an accumulated damage D<1. Assuming a Rayleigh distribution within each short term period and a one-slope S-N curve, the accumulated fatigue damage results [12]:

$$D = \frac{\upsilon_0 T_d}{C}\Gamma\left(1+\frac{m}{2}\right)\sum_{n=1}^{Nload} p_n \sum_{v=1}^{NVel} \sum_{h=1}^{Nhead} \sum_{i=1,j=1}^{nHs \atop nTz} p_v p_h r_{ijhvn}\left(2\sqrt{2m_{0ijhvn}}\right)^m \tag{7}$$

where:

nHs is the number of significant wave heights H_s

nTz is the total number of zero crossing periods T_z

NHead is the total number of heading angles

NVel is the total number of ship speeds

Nload is the total number of load conditions

p_n is the fraction of design life in load condition n, $\sum p_n \leq 1$, here assumed as 1.

p_v is the ship life fraction related to each operating speed

p_h is the probability of occurrence of a given heading

r_{rijhvn} is the relative number of stress cycles in short-term condition i,j,h,v,n

m_{0ijhvn} is the zero spectral moment of the hot spot stress response

v_0 is the long-term average response zero-crossing-frequency

T_d is the design life of the vessel in seconds (20 years)

C, m are the one slope S-N curve parameters

$\Gamma(1+m/2)$ is the Gamma function

The following fatigue damage summations (Tables 6, 7, 8, 9) have been obtained for each examined detail:

	Detail 1 - Type "a" – Fatigue Damage		
	Area 9	**Area 26**	**Area 37**
Short crested	2.15E-02	2.14E-02	1.48E-02
Long crested	2.78E-02	2.92E-02	2.10E-02

Table 6: Detail 1 – Type 'a' - Fatigue Damage

	Detail 1 - Type "b" Fatigue Damage		
	Area 9	**Area 26**	**Area 37**
Short crested	4.27E-02	4.08E-02	2.71E-02
Long crested	4.76E-02	4.54E-02	3.02E-02

Table 7: Detail 1 – Type 'b' - Fatigue Damage

	Detail 2 - Fatigue Damage		
	Area 9	**Area 26**	**Area 37**
Short crested	4.24E-02	4.58E-02	3.23E-02
Long crested	4.85E-02	5.19E-02	3.64E-02

Table 8: Detail 2 – Fatigue Damage

	Detail 3 - Fatigue Damage		
	Area 9	**Area 26**	**Area 37**
Short crested	2.68E-01	2.99E-01	2.13E-01
Long crested	3.07E-01	3.40E-01	2.41E-01

Table 9: Detail 3 – Fatigue Damage

The fatigue lives evaluated in the North Atlantic area and in the Mediterranean sea appear very similar and in most cases area 9 appears more favorable than area 26. This behaviour is probably due to the shorter waves of area 26.

The presented procedure allows not only to evaluate the fatigue life of a detail by means of wave loading direct calculation, but also to study the contribution of each sea state taking into account each heading angle and ship speed. It is therefore possible to plot the portion of fatigue damage relevant to each cell (Hs, Tz) in the wave scatter diagram expressed as fraction of the total accumulated damage. The following figures (Figures 16 ÷ 21) show this damage ratio related to Detail 2, in the three geographical areas and in long crested sea. The first three figures (Figures 16 ÷ 18) show the damage ratio in a scatter diagram form while the other figures (Figures 19 ÷ 21) report these values in a plot form.

Significant wave height (m)	Detail 2 - A9 - Long Crested Zero crossing period (s)									
	4.50	5.50	6.50	7.50	8.50	9.50	10.50	11.50	12.50	13.50
0.50	1.59E-06	6.32E-06	6.69E-06	2.40E-06	4.09E-07	5.84E-08	0.00E+00	0.00E+00	0.00E+00	0.00E+00
1.50	0.00E+00	1.40E-04	4.34E-04	3.64E-04	1.33E-04	2.68E-05	3.73E-06	5.70E-07	0.00E+00	0.00E+00
2.50	0.00E+00	2.16E-04	1.17E-03	1.73E-03	1.02E-03	3.29E-04	6.47E-05	1.06E-05	1.67E-06	0.00E+00
3.50	0.00E+00	1.97E-04	1.38E-03	2.85E-03	2.35E-03	1.00E-03	2.61E-04	5.07E-05	9.15E-06	0.00E+00
4.50	0.00E+00	0.00E+00	9.75E-04	2.82E-03	2.98E-03	1.58E-03	5.28E-04	1.23E-04	1.94E-05	0.00E+00
5.50	0.00E+00	0.00E+00	4.45E-04	2.21E-03	2.72E-03	1.71E-03	6.89E-04	1.69E-04	3.55E-05	1.15E-05
6.50	0.00E+00	0.00E+00	7.35E-04	1.22E-03	2.02E-03	1.54E-03	6.83E-04	1.86E-04	5.86E-05	1.91E-05
7.50	0.00E+00	0.00E+00	0.00E+00	6.22E-04	1.38E-03	1.18E-03	5.83E-04	2.14E-04	4.50E-05	0.00E+00
8.50	0.00E+00	0.00E+00	0.00E+00	9.06E-04	1.01E-03	8.61E-04	5.09E-04	2.07E-04	6.55E-05	0.00E+00
9.50	0.00E+00	0.00E+00	0.00E+00	0.00E+00	7.02E-04	8.01E-04	4.74E-04	1.45E-04	9.15E-05	0.00E+00
10.50	0.00E+00	0.00E+00	0.00E+00	0.00E+00	0.00E+00	5.41E-04	3.20E-04	1.96E-04	0.00E+00	0.00E+00
11.50	0.00E+00	0.00E+00	0.00E+00	0.00E+00	0.00E+00	0.00E+00	4.20E-04	0.00E+00	0.00E+00	0.00E+00

Figure 16: Detail 2 – Scatter diagram of the damage ratio within sea states – area 9

Significant wave height (m)	Detail 2 - A26 - Long Crested Zero crossing period (s)						
	4.00	4.50	5.50	6.50	7.50	8.50	9.50
0.50	6.28E-05	1.47E-04	6.40E-05	1.04E-05	1.09E-06	8.49E-08	0.00E+00
1.50	4.24E-04	2.72E-03	2.34E-03	6.16E-04	1.01E-04	1.15E-05	1.30E-06
2.50	3.92E-04	4.21E-03	5.18E-03	1.85E-03	3.88E-04	6.36E-05	6.00E-06
3.50	3.59E-04	3.25E-03	5.17E-03	2.34E-03	5.85E-04	1.16E-04	1.65E-05
4.50	0.00E+00	2.30E-03	3.66E-03	1.95E-03	5.65E-04	1.24E-04	3.50E-05
5.50	0.00E+00	1.40E-03	2.51E-03	1.58E-03	4.13E-04	1.13E-04	0.00E+00
6.50	0.00E+00	0.00E+00	1.38E-03	1.30E-03	3.41E-04	1.86E-04	0.00E+00
7.50	0.00E+00	0.00E+00	2.12E-03	1.00E-03	5.24E-04	0.00E+00	0.00E+00

Figure 17: Detail 2 – Scatter diagram of the damage ratio within sea states – area 26

Significant wave height (m)	Detail 2 - A37 - Long Crested Zero crossing period (s)						
	4.00	4.50	5.50	6.50	7.50	8.50	9.50
0.50	1.07E-04	1.63E-04	5.30E-05	6.77E-06	6.16E-07	8.43E-08	0.00E+00
1.50	9.82E-04	3.89E-03	2.39E-03	5.01E-04	7.07E-05	6.83E-06	1.29E-06
2.50	9.09E-04	5.61E-03	4.99E-03	1.44E-03	2.50E-04	3.16E-05	5.96E-06
3.50	3.56E-04	3.22E-03	3.85E-03	1.31E-03	3.17E-04	5.78E-05	0.00E+00
4.50	0.00E+00	1.52E-03	1.82E-03	8.59E-04	2.25E-04	6.14E-05	0.00E+00
5.50	0.00E+00	0.00E+00	8.29E-04	3.92E-04	2.05E-04	0.00E+00	0.00E+00

Figure 18: Detail 2 – Scatter diagram of the damage ratio within sea states – area 37

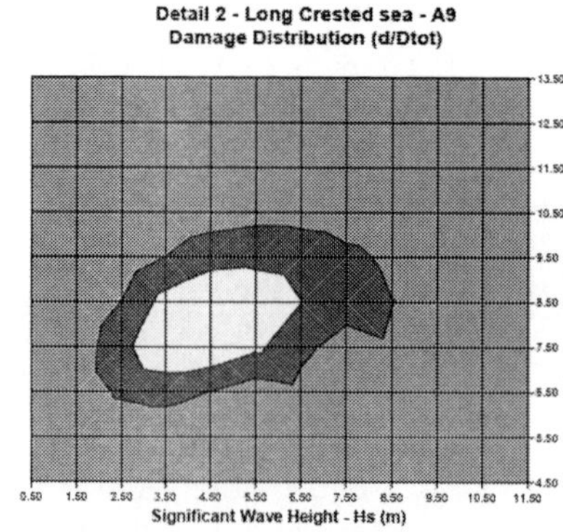

Figure 19: Detail 2 – Damage ratio – Long crested sea – area 9

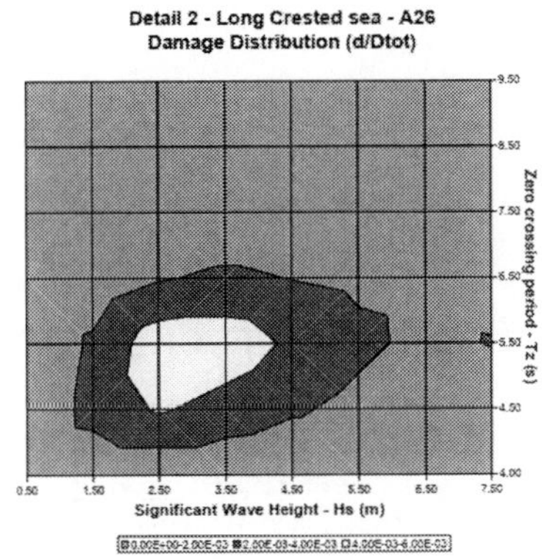

Figure 20: Detail 2 – Damage ratio – Long crested sea – area 26

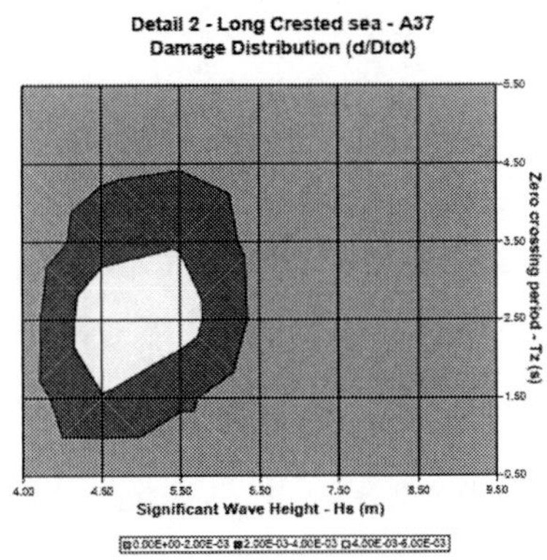

Figure 21: Detail 2 – Damage ratio – Long crested sea – area 37

4 CONCLUSIONS

The structural fatigue assessment of a medium size fast patrol vessel, operating in different geographical areas, has been performed by using a full stochastic analysis.

This approach is based on the wave loading direct calculations for a wide set of heading angles and wave frequencies, covering in this way all possible wave conditions that the ship can experience during its life. The hot spot stress transfer function has been evaluated through the stress concentration factor, by means of a refined mesh of the analysed details, according to the most recent literature indications provided by IIW and Classification Societies [5,12]. The long-term distribution of stress cycles has been performed assuming a simple but realistic operational profile, consisting of three different speeds and relevant probability of occurrence. This exhaustive procedure has allowed comparing the structural behaviour of the ship in different areas, confirming what partially highlighted by the previous works [1,2]. In fact, the Mediterranean sea, characterized by shorter waves, affects the structural detail life more than the North Atlantic area, even though the two examined environments do not lead to substantial differences. Sheltered seas (like BMT area 37), characterized by shorter wave lengths, appear to be less significant in terms of fatigue damage than the other areas considered.

Furthermore, the computed fatigue lives show that the analysed structural details are well designed and do not seem to expect fatigue cracks during the assumed ship life. Thus, the results confirm that the referenced ship can be suitably operated both in open and closed seas, without any significant modification in structural scantling. In addition, the presented procedure leads to a better understanding of the parameters affecting the fatigue life of a structural detail, as the influence of the geographical area, of the sea spectrum formulation, of the operational profile and of each heading angle, speed and sea state, which can be analysed separately.

5 REFERENCES

1. FOLSØ R., FERRARIS S., GRILLO R., ALBERTONI F., FOLSØ L., 'Sea-keeping and structural behaviour of a fast patrol vessel in closed and open sea', *Proc. of FAST 2003*, Ischia, Italy, 2003.

2. CARTALEMI C., MOGGIA C. A., FERRARIS S., SEBASTIANI L., BOOTE D., 'Structural Response Operator for the Fatigue Assessment of a Fast Patrol Vessel in the Mediterranean sea', *FAST 2005,* St. Petersburg, Russia, 2005.

3. MOGGIA C. A., IACCARINO R., SEBASTIANI L., 'A spectral approach to fatigue assessment based on unit pressure facet method', *ICMRT'05,* Ischia, Italy, 2005.

4. BRITISH MARINE TECHNOLOGY (Primary contributors Hogne N., Da Cunha L.F. and Oliver H.N.) 'Global Wave Statistics', *Unwin Brothers Limited*, London, England, 1986.

5. HOBBACHER A., 'Recommendations for fatigue design of welded joints and components', *IIW*, 2002.

6. MARIN, 'PRECAL User's guide and Theory Manual', *CRS (Cooperative Research of Ships)*, 2003.

7. NIEMI E., FRICKE W., MADDOX S.J., 'Structural hot spot stress approach to fatigue analysis of welded components', *IIW, Delegation of Finland.*

8. MOGGIA C. A., SEBASTIANI L., PINO E., PETTERSSON J., 'A Rational Approach to the Structural Design of Large Fast Ships', *Proc. of HSMV 2002*, Naples, Italy, 2002.

9. GROSSI L., CAMPORESE E., DOGLIANI M., FOLSØ R., 'Ship structural design based on first principles dynamic load evaluation: a case study', *CETENA's International Seminar on Dynamic Loads*, Genoa, Italy, 2002.

10. MOGGIA C. A., PETTERSSON J., 'First Principle Approach for the Ship Longitudinal Strength Assessment', *CETENA's International Seminar on Dynamic Loads*, Genoa, Italy, 2002.

11. HUGHES O., 'Ship Structural Design', *SNAME*, 1988.

12. DNV, 'Fatigue assessment of ship structure', *Classification Notes N.30.*, 2001.

13. SSC, ' Fatigue resistant detail design guide for ship structure', *SSC 405*, 1999.

SSC, 'Fatigue technology assessment and strategies for fatigue avoidance in marine structure', *SSC 367*, 1999.

STUDY OF TWO-PHASE BUBBLY WATER RAMJET PERFORMANCE

Alon Gany and Marat Mor, Technion - Israel Institute of Technology, Haifa, Israel

SUMMARY

This study presents a theoretical performance evaluation of two-phase bubbly water ramjet propulsion. Conservation equations of the two-phase flow are solved, and expressions for flow properties such as Mach number and gas volume fraction (void fraction) are discussed. Performance maps are used to show the dependence of ramjet thrust and propulsion efficiency on vessel speed and air mass flow rate, for both ideal and practical cases. Propulsion efficiencies in the range of 70%-80% are demonstrated.

NOMENCLATURE

A	Cross-section area (m^2)
C_D	Interphase drag coefficient
C_{p_a}	Specific heat of gas ($J/kg\,K$)
d	Diameter (m)
$\dot{E}$	air compression power
f	Wall friction coefficient
L	Length (m)
M	Mach number
m'	Mass flow rate ($kg/\sec$)
T	Thrust (N)
T_a	Air temperature (K)
P	Pressure (Pa, atm)
R	Specific gas constant ($J/kg\,K$)
r	Radius of bubble (m)
U	Velocity ($m/\sec$)
x	Location along engine (m)
a	Void fraction (air volume fraction)
g	Specific heat ratio
h	Flow capture coefficient
l	Virtual mass coefficient
m	Air to water mass flow rate ratio
r	Density (kg/m^3)

t Wall stress (N/m^2)

Subscripts

0 Stagnation

¥ , inf Free stream

a Air (gas) phase

$diff$ Diffuser

e Nozzle exit

i Diffuser entrance

m Mixing chamber entrance

mix Mixing chamber

n Nozzle entrance

$nozzle$ Nozzle

t Nozzle throat

vap Water vapor

w Water (liquid) phase

1. INTRODUCTION

The objective of this work is to provide a performance prediction of two-phase bubbly water ramjet propulsion, by introducing theoretical formulation and numerical solutions of the air-bubbles/water mixture flow within a two-phase ramjet-type propulsion unit.

The concept of a two-phase marine ramjet has been described in a number of articles [1-8]. The general idea is presented in Fig. 1. The propulsion unit is basically a hollow duct submerged in the water, containing no moving parts. Free stream water (condition ∞) enters the diffuser inlet (cross-section i). The pressure increases while the water flow velocity decreases along the diffuser. At the mixing chamber (m), air is added to the water, forming a two-phase bubble-water mixture. The bubbly mixture is a compressible fluid because of the gas bubbles (in contrast to a water-only flow). It expands in the nozzle because of the decrease in pressure, forming a high speed jet expelled from the nozzle exit, hence producing thrust.

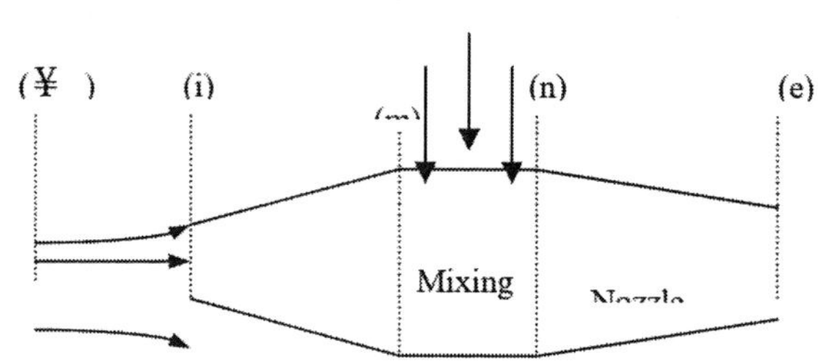

Figure 1: Description of the two-phase bubbly water ramjet engine.

In a previous work by Gany [8] the ideal thermodynamic cycle of the two-phase water ramjet was presented (Fig. 2). The cycle consists of air compression (isentropic in the ideal case) from the ambient pressure (P_a) to P_{comp} , which is equal to the pressure at the mixing chamber P_{mix} , and isothermic expansion of the air bubbles from P_{mix} back to the ambient pressure.

Although the analysis of the ideal cycle can yield an estimation of the propulsion concept, it does not account in detail for different two-phase flow phenomena involved in the process. The present paper consists of a comprehensive analysis of the two phase flow within the propulsion unit, yielding performance maps based on different properties and operating parameters.

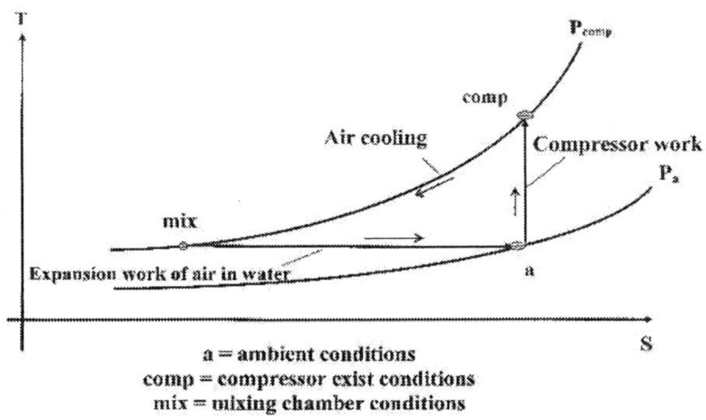

Figure 2: The ideal thermodynamic power cycle of the two-phase marine ramjet presented on a temperature –entropy (T-S) plot (after Gany [8]).

The quasi one-dimensional two-phase flow equations are summarized below. Spherical bubbles without coalescence or breakup and with no mass transfer between phases are assumed:

- Water mass conservation

$$(1- a)r_w U_w A = m'_w \qquad (1)$$

- Air mass conservation

$$a r_a U_a A = m'_a = mm'_w \qquad (2)$$

- Constant mass of a single bubble

$$r_a r^3 = const \qquad (3)$$

- Total momentum conservation

$$a r_a U_a \frac{d}{dx} U_a + (1- a) r_w U_w \frac{d}{dx} U_w = \\ = - \frac{4t}{d} - \frac{d}{dx} P \qquad (4)$$

- Bubble momentum conservation[3,9-16]

$$
r_a U_a \frac{d}{dx} U_a + \frac{l}{2} r_w \frac{d}{dx}\left(U_a^2 - U_w^2\right) = - \frac{d}{dx} P - \frac{3}{8r} C_D r_w \left(U_a - U_w\right)\left|U_a - U_w\right| \qquad (5)
$$

- Equation of state of air

$$
P = r_a R T_a \qquad (6)
$$

Here the indices a and w denote air and water, respectively. The terms and variables are: a - volume fraction of air (or void fraction), U - velocity, m' - mass flow rate, m - air/water mass flow rate ratio, A - cross-section area, r - density (the water phase is considered incompressible, $r_w = const$), r - radius of bubble, $4t/d$ - drag term due to wall friction, C_D - interphase drag coefficient [2-5,14-17], l - virtual mass coefficient (usually assumed $1/2$) [18-20]. The water temperature inside the propulsion unit remains practically constant. Since the mass of the air phase is negligible (compared to the water), the entire process is assumed to be isothermal, i.e., the temperature T_a remains constant.

The thrust is defined by the following expression:

$$
T = m'_w\left(U_{w_e} - U_{¥}\right) + m'_a\left(U_{a_e} - U\right) + (P_e - P_{¥})A_e \qquad (7)
$$

where the subscript ¥ refers to the free flow and *e* denotes the nozzle exit conditions (see Fig. 1).

2. PERFORMANCE MAPS OF AN IDEAL ENGINE

It is possible to obtain the upper limit for both thrust and efficiency (the ideal engine) without solving the entire flow inside the engine. In such case one assumes that the radius of injected bubbles is infinitesimally small, imply-ing the same velocity of the water and bubble phases (homogeneous flow [9,21-23]), $U_a = U_w = U$. In addition, for an ideal cycle, there is no loss in stagnation pressure*. Then Eq. (7) becomes

$$
T = m'_w(1 + m)(U_e - U_{¥}) + (P_e - P)A_e \qquad (8)
$$

Using Eq. (1) for the water mass conservation, one can write Eq. (8) in the following form

* Based on Ref. [23] it can be shown that there is always loss in stagnation pressure during the mixing process, unless the velocity inside the mixing chamber is zero.

$$\frac{T}{A_e} = (1+m)(1-a)r_w U_e (U_e - U_\infty) + P_e - P_\infty \qquad (9)$$

Since the stagnation pressure is kept constant

$$P_0 = P_\infty + \frac{1}{2} r_w U^2 = const \qquad (10)$$

Using the two mass conservation equations (Eqs. (1-2)) we derive the air volume fraction (void fraction):

$$a = \frac{1}{1 + \frac{r_a}{m r_w}} \qquad (11)$$

From this equation, together with the equation of state of air (Eq. (6)), one can calculate the void fraction at the stagnation point.

Introducing Mach number[23] we obtain an important relation, connecting between Mach number and void fraction only:

$$\left(\frac{1}{M^2} - 1 + \frac{1}{a}\right) d \ln\left(\ldots - 1 + \frac{1}{a}\right) d \ln(M) = 0 \qquad (12)$$

where Mach number in two-phase flow is defined as:

$$M = \frac{U}{a} = \sqrt{\frac{1+m}{mRT_{aw}}} a U \qquad (13)$$

Here a is the speed of sound in a two-phase flow (which is rather low) [23].

For a converging nozzle only, an unchoked nozzle is always adapted to the ambient pressure, therefore at its exit $P_e = P_\infty$ .On the other hand, for a choked nozzle the flow at the exit is sonic, $M_e = 1$, and the static pressure is higher than the ambient atmospheric pressure.

Solving Eq. (12) with the stagnation void fraction as initial condition, one can find the void fraction and the pressure as a function of Mach number. Using M_e and a_e, the exit velocity U_e and then the ratio T / A_e (from Eq. (9)) can be calculated.

Figure 3 presents the performance map of an ideal bubbly water ramjet engine. The nozzle is first choked at a vessel speed of $U_\infty \approx 14 m/\sec$, which increases with reduction in the mass flow rate values. The dotted curves in Fig. 3 are T / A_e contours for adapted Laval nozzle, where the connection between the throat (choking plane) and the nozzle exit cross-section can be obtained using Eq. (1) at those points:

$$A_e = \frac{1 - a_t U_t}{1 - a_e U_e} A_t \qquad (14)$$

The addition of a diverging part to the nozzle slightly improves the performance. Design of Laval nozzles in such systems may be rather complex, since one needs to correctly formulate the supersonic two-phase bubbly flow, accounting for possible shock waves and their behavior, location and thickness.

The thick lines of $\alpha_\ell = 0.74$ in Fig. 3 and later on also in Fig. 4, present a level of air volume fraction that corresponds to the maximum void fraction of packed individual monodisperse bubbles in a continuum of a water phase.

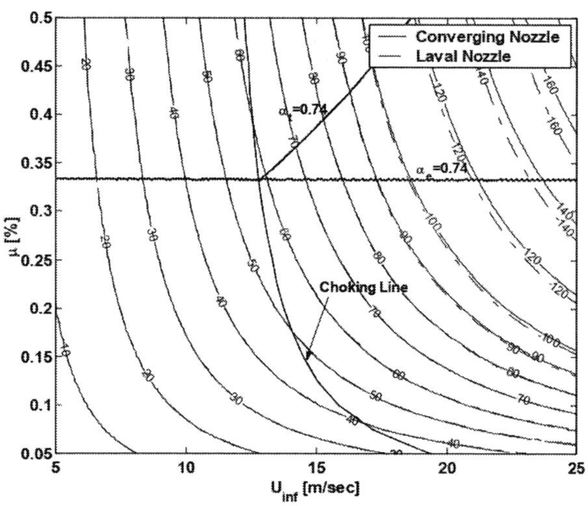

Figure 3: T / A_e (in kN / m^2) contours for the ideal water ramjet engine with converging and Laval nozzles as a function of the air to water mass flow rate ratio μ and the vessel speed U_{inf}.

Another important parameter characterizing engine performance is its propulsion efficiency, $TU_\infty / \dot{E}$ (as presented in Fig. 4), where $\dot{E}$ is the power required to compress the injected air from the ambient pressure to the mixing chamber pressure. Note that even for the ideal engine there are energy losses resulting from the inherent decrease in delivered energy (isothermal expansion) compared to compression energy. As mentioned before, optimal mixing occurs at negligibly small flow velocity (typical to stagnation conditions).

In such cases the isentropic air compression power is:

$$\dot{E} = \dot{m}_a C_{p_a} T_a \left[\left(\frac{P_0}{P_\infty} \right)^{\frac{g-1}{g}} - 1 \right] \qquad (15)$$

From Fig. 4 it is obvious that the efficiency is mostly influenced by the amount of injected air. The increase in the cruise velocity (for the same air to water mass flow rate ratio μ) slightly reduces the propulsion efficiency. On the other hand, it increases the thrust as can be seen in Fig. 3

.

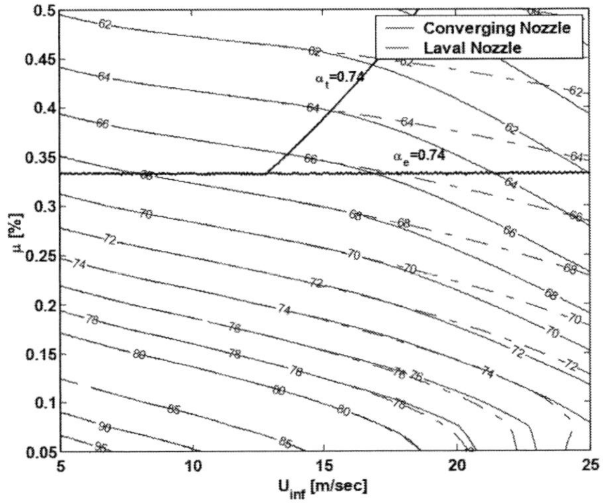

Figure 4: Efficiency (in $\%$) contours for the ideal water ramjet engine with converging and Laval nozzles as a function of the air to water mass flow rate ratio μ and the vessel speed U_{inf}.

3. PERFORMANCE OF A PRACTICAL WATER RAMJET

The following procedure and performance calculation are related to a water ramjet of a general shape (and notation) as presented in Fig. 1.

The diffuser is solved analytically. The mass flow rate of water entering the diffuser is

$$m_w' = r_w U_{¥} A \tag{16}$$

where $A_{¥}$ is the capture area of the water entering the ramjet inlet. Due to spillage

$$A_{¥} = h A_i \tag{17}$$

where h is a flow capture coefficient and A_i is the inlet geometrical cross-section.

Solving a momentum conservation equation for the water only (including a wall friction term), for a diffuser circular cross section shape, one obtains

$$U_m = U_{¥} \frac{A_{¥}}{A_m} \tag{18}$$

and

$$P_m = P_i + \frac{1}{2} r_w U^2 \left[1 - \left(\frac{A_{\infty}}{A_m} \right)^2 - h^2 f \frac{L_{diff}}{d_m} \left(1 + \frac{A_i}{A_m} \right) \left(1 + \sqrt{\frac{A}{A_m}} \right) \right] \quad (19)$$

The solution in the mixing chamber and nozzle is obtained solving Eqs. (1-6), which define a mixed set of algebraic and differential nonlinear equations for the variables a, r_a, U_a, U_w, r, P. Using the finite differences method to expand the differential equations, together with the rest of equations, a method based on a semi-analytical approach to solve such system is derived. The flow in the mixing chamber is solved similarly to the one in the nozzle with a modification due to the injection of bubbles. The air/water mass flow rate ratio m increases along the mixing chamber, since the bubbles are injected with a certain distribution along the chamber, as defined a-priori by the designer. In a case of a linear distribution throughout the length of the mixing chamber one can write

$$m_{mix}(x) = m \frac{x}{L_{mix}} \quad (20)$$

where $m_{mix}(x)$ is the mass flow rate ratio inside the mixing chamber and is changed from zero to the final value m. This variable is used in the mentioned system of equations (Eqs. (1-6)) instead of the constant mass flow rate m inside the nozzle.

Since the variable h (flow capture coefficient) is unknown, an iterative procedure is needed to obtain a solution for the entire engine at some working point specified by given mass flow rate and cruise speed. Guessing initial value for h, one solves the flow in the engine as explained above. In the case of a converging nozzle, either one of the two conditions is satisfied, namely $P_e = P_{\infty}$ or $M_e = 1$, dictating a certain capture coefficient h (through the iterative process).

Figure 4 presents thrust behavior in a sample water ramjet engine, whose geometry is given in Table 1, as a function of cruise speed and air mass flow rate. In the engines working region the thrust obviously goes up when either of those parameters increases.

Table 1: Sample water ramjet engine geometry data.

A_i [cm^2]	A_m, A_n [cm^2]	A_e [cm^2]	L_{diff} [cm]	L_{mix} [cm]	L_{nozzle} [cm]
77	351	147	30	15	40

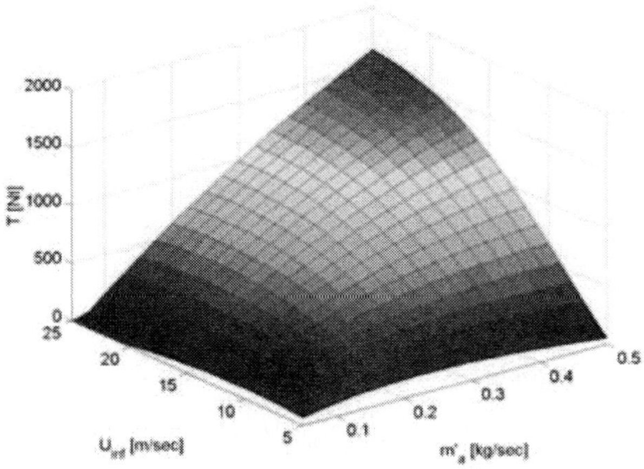

Figure 5: Water ramjet engine thrust as a function of cruise speed and air mass flow rate with geometry given in Table 1 and initial bubbles radius $r_{in} = 2.5\,mm$.

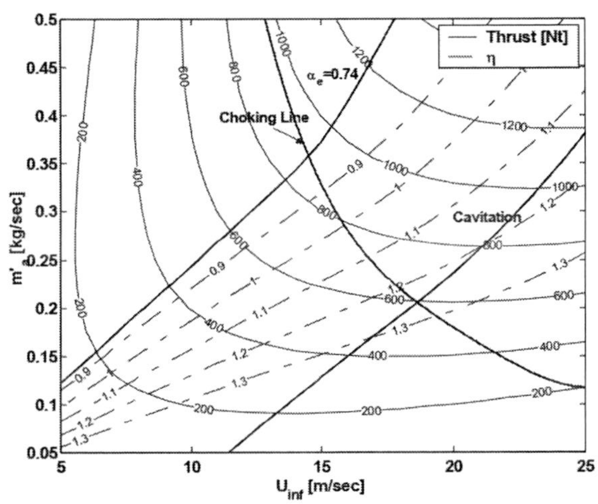

Figure 6: Water ramjet engine performance map with geometry given in Table 1 and initial bubbles radius $r_{in} = 2.5\,mm$.

Figure 6 presents the performance map of the engine. The solid lines are the thrust contours, while the dashed lines represent values of the capture coefficient.

The "cavitation line" in Fig. 6 reflects situations where the pressure at the diffuser inlet P_i decreases to the value of the water vapor pressure (which is very low):

$$P_i = P_0 - \frac{1}{2} r_w U_i^2 = P_{vap} \qquad (21)$$

It is connected to the specific values of the diffuser inlet and nozzle exit cross-section areas, and is always associated with flow capture area coefficients greater than unity. Another curve presented in the figure is

$a_e = 0.74$. During the flow acceleration the void fraction increases (see Ref. [23]) and reaches its maximal value at the exit of the nozzle. The maximal volume fraction for the packing of spherical bubbles (with the same radii) is 0.74 [24]. Above this value the flow becomes with mixed bubbles in water and drops of water in a continuous air phase. The equations above treat only bubbly flows and may yield inaccurate results for $\alpha > 0.74$. The results presented in Figs. 5 and 6 are obtained for an initial radius of $2.5\,mm$ for the injected bubbles.

Figure 7 presents a map of thrust for the bubbly water ramjet specified in Table 1. The effect of initial bubbles radius r_{in} is also included in Fig. 7, presenting the lines for both $r_{in} = 0.5mm$ and 10 mm. One can see that smaller bubbles require somewhat less air to produce the same thrust. In the very extreme cases* this effect is significant. It is obviously better to have small bubbles, however, the smaller the radius is the less efficient is the mixing process, since the bubbles are less able to penetrate the water phase in the mixing chamber.

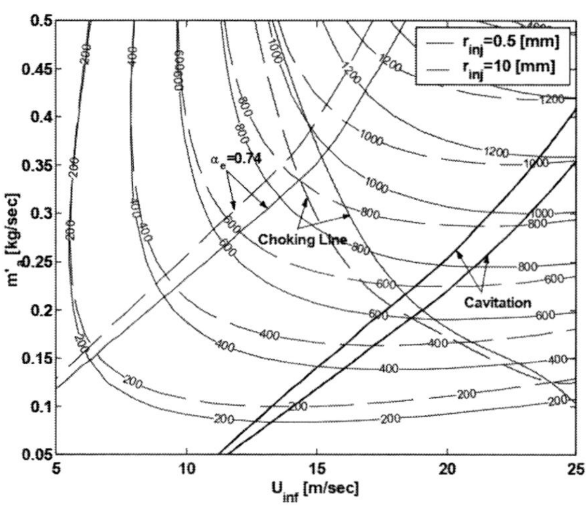

Figure 7: Water ramjet engine thrust lines (in N) for the geometry given in Table 1 and two initial bubble radii, $r_{in} = 0.5$ and $10\,mm$.

4. CONCLUSION

The performance of ideal and practical water ramjet engines is presented. The ideal engine's Thrust/Exit Area contours and the contours of the propulsion efficiency define the limits of the possible working region for the practical case. The propulsion efficiency is relatively high, and for the most real cases in the working region it varies between 70%-80%. The performance maps for the bubbly water ramjet may be limited by possible cavitation at the inlet region and by air bubbles packing volume fraction lines. The limit of $a_e = 0.74$, however, is not physical, but is a result of bubbly flow modeling restrictions. The

---

*The two limit cases of homogeneous flow (with zero bubble radius) on the one hand and a flow without interphase drag (infinite bubble radius) on the other hand, provide very different results, especially in low speeds, where the difference in thrust could reach 100% between the two cases

performance of the water ramjet is affected by the radius of injected bubbles and is superior in the homogeneous flow case. In addition, the differences in performance results between the converging and Laval nozzles are small, leading to a simplification in design.

5. REFERENCES

1. Mottard, E.J., Shoemaker, C.J., "Preliminary Investigation of an Underwater Ramjet Powered by Compressed Air", NASA Report TND-1991, Dec. 1961.

2. Muench, R.K., Ford, A.E., "A Water-Augmented Air Jet for the Propulsion of High-Speed Marine Vehicles", NASA N D-991, Langley Research Center, 1961

3. Witte, J.H., "Predicted Performance of Large Water Ramjets", AIAA Paper 69-406, AIAA 2nd Advanced Marine Vehicles and Propulsion Meeting, Seattle, Washington, 1969.

4. Amos, R.G., Maples, G., Dyer, D.F., "Thrust of an Air-Augmented Waterjet", *J. Hydronautics*, Vol. 7, No. 2, pp. 64-71, 1973.

5. Albagli, D., Gany, A., "Analysis of a High Speed Bubbly Waterjet", AIAA Paper 92-3380, 28th AIAA/SAE/ASME/ASEE Joint Propulsion Conf., Nashville, TN, 1992.

6. Varshay, H., Gany, A., "Underwater Two-Phase Ramjet Engine", U.S. Patent No. 5,598,700, Feb. 4, 1997.

7. Varshay, H., Gany, A., "Underwater Two-Phase Ramjet Engine", U.S. Patent No. 5,692,371, Dec. 2, 1997.

8. Gany, A., "Analysis of a New Thermodynamic Power Cycle for a Two-Phase Waterjet", International Conference Waterjet Propulsion 4, RINA, London, UK, May 26-27, 2004.

9. Van Wijngaarden, L., "One-Dimensional Flow of Liquids Containing Small Gas Bubbles", Annual Review of Fluid Mechanics, vol. 4, pp.369-394, 1972.

10. Biesheuvel, A., Van Wijngaarden, L., "Two-Phase Flow Equations for a Dilute Dispersion of Gas Bubbles in Liquid", *J. Fluid Mechanics*, Vol. 148, pp.301-318, 1984.

11. Presperetti, A., Van Wijngaarden, L., "On the Characteristics of the Equations of Motion for a Bubbly Flow and the Related Problem of Critical Flow", *J. Eng. Math.*, Vol. 10, pp.153-162, 1976.

12. Nishikawa, H., Matsumoto, Y., Ohashi, H., "The Numerical Calculation of the Bubbly Two-Phase Flow Around an Airfoil", Comput. Fluids, Vol. 19, p. 453, 1991.

13. Yonechi, H., Suzuki, M., Ishii, R., Morioka, S., "Bubbly Flows Through a Convergent-Divergent Nozzle", Mem. Fac. Eng. Kyoto Univ., Vol. 45, p. 3, 1992.

14. Ishii, R., Umeda, Y., Murata, S., Shishido, N., "Bubbly Flows Through a Converging-Diverging Nozzle", *Phys. Fluids A*, Vol. 5, No. 7, pp.1630-1643, 1993.

15. Wang, Y., Chen, E., "Effects of Phase Relative Motion on Critical Bubbly Flows Through a Converging-Diverging Nozzle", *Phys. Fluids*, Vol. 14, No. 9, pp.3215-3223, 2002.

16. Albagli, D. and Gany, A., "High Speed Bubbly Nozzle Flow with Heat, Mass, and Momentum Interactions", *International Journal of Heat and Mass Transfer*, Vol. 46, pp. 1993-2003, 2003.

17. Soo, S.L., Multiphase Fluid Dynamics, Science Press, Beijing, 1990.

18. Van Wijngaarden, L., "Hydrodynamic Interaction Between Gas Bubbles in Liquid", *J. Fluid Mech.*, Vol. 77, pp.27-44, 1976.

19. Cook, T.L., Harlow, F.H., "Virtual Mass in Multiphase Flow", *Int. J. Multiphase Flow*, Vol. 10, pp. 691-696, 1984.

20. Biesheuvel, A., Spoelstra, S., "The Added Mass Coefficient of a Dispersion of Spherical Gas Bubbles in Liquid", *Int. J. Multiphase Flow*, Vol. 15, pp.911-924, 1989.

21. Tangren, R.F., Dodge, C.H., Seifert, H.S., "Compressibility Effects in Two-Phase Flow", *J. Applied Physics*, Vol.20, pp.637-645, 1949.

22. Muir, J.H., Eichhorn, R., "Compressible Flow of an Air-Water Mixture Through a Vertical, Two-Dimensional, Converging-Diverging Nozzle", Proc. 1963 Heat Transfer and Fluid Mechanics Institute, Stanford, pp.183-204, 1963.

23. Mor, M. and Cany, A., "Analysis of Two-Phase Homogeneous Bubbly Flows Including Friction and Mass Addition", *Journal of Fluids Engineering*, Vol. 126, pp.102-109, January 2004.

24. Gardner, M. "Packing Spheres." Ch. 7 in Martin Gardner's New Mathematical Diversions from Scientific American. New York: Simon and Schuster, pp. 82-90, 1966.

6. AUTHORS BIOGRAPHY

Alon Gany is a Professor, holding the Lena and Ben Fohrman Chair in Aeronautical Engineering, and Head of Sylvia and David I.A. Fine Rocket Propulsion Center in the Faculty of Aerospace Engineering, Technion – Israel Institute of Technology, Haifa. His research and expertise include rocket, ramjet and marine jet propulsion and related areas. His professional publications include patents on two-phase marine ramjets. He is one of the founders of Marine Zoomjets Ltd., a startup company holding IP and aimed at the research and development of two-phase marine jet propulsion.

E-mail: gany@tx.techion.ac.il

Marat Mor is currently a Postdoctoral Research Fellow in the Department of Aeronautics and Astronautics at the University of Washington, Seattle, WA, USA. He received his Ph.D. degree from the Technion – Israel Institute of Technology, Faculty of Aerospace Engineering, Haifa, Israel for his studies and research in the area of bubbly flows. His research and expertise include Multidisciplinary Design Optimization (MDO), which is implemented in flight vehicles as well as in marine propulsion systems.

E-mail: mor@aa.washington.edu

INTRUDER: A DEVICE TO IMPROVE HULL PERFORMANCES

A Mancini, **A Moriconi**, Insean - Italian Ship Model Basin, Italy

SUMMARY

The intruder is a device that has been used for a long time now on big fast monohulls like "Acquastrada". It consists of a small vertical thin plate positioned under the transom of a planing or semi-planing hull, jutting out just a few millimetres. It is able to improve hull performances assuring a smaller drag thus a higher speed for the same installed power. For this reason, for its working and installation simplicity, as well as for its low cost, the intruder is becoming increasing popular even with small working and sports crafts.

In order to improve the knowledge about the hydrodynamic effects caused by this device, some tests have been performed at INSEAN using three planing and semi-planing models.

As the results of tests show, the intruder changes significantly the dynamic trim of the hull and it raises the stern. These effects can reduce the wave pattern and thus the drag. The improvements are larger in the pre-planing phase ($1<F_{n\nabla}<2.5$) where drag reductions - of more than 20% - have been measured.

NOMENCLATURE

A_P	Planing bottom area (m^2)
AR	Aspect ratio
B_{PT}	Transom breadth
B_{PX}	Maximum breadth over chines (m)
C_F	Frictional resistance coefficient
C_R	Residuary resistance coefficient
C_T	Total resistance coefficient
$F_{n\nabla}$	Volumetric Froude Number
g	Acceleration of gravity (9.803 m/s^2)
L_{PR}	Projected chine Length (m)
L_{WL}	Length of Waterline (m)
P_E	Effective power (kw)
R_n	Reynolds Number
R_T	Total Resistance (N)
S	Area of wetted surface (m^2)
S_i	Protrusion of intruder (m)
V	Speed of the model (m/s)
β	Deadrise angle of planing bottom (degree)
β_T	Deadrise angle at transom (degree)
λ	Scale ratio
∇	Displacement Volume (m^3)
ν	Kinematic viscosity (N s m^{-2})

Δ Displacement (N)

ϑ Static trim angle (degree)

τ Dynamic trim angle (degree)

1. INTRODUCTION

A small vertical thin plate positioned under the transom jutting out the bottom a few millimeters, is what is very often called intruder in naval world today. The hydrodynamic effect it causes enables to use it instead of the more ordinary flaps. Like these they can stabilize the hull and correct the trim, but with the advantage that the performances are often better.

Up till now in literature there is still scarce information about a quantitative evaluation of its advantages. In order to add something more to what today known about the hydrodynamic effects caused by this device, some comparative tests with and without intruder have been performed at the Italian Ship Model Basin using three planing and semi-planing models.

2. ORIGIN AND HISTORY OF INTRUDER

The intruder does not have a unique definition yet: sometimes is also named interceptor or something else. It comes from a device used in aeronautics called "Gurney Flap", from a racing car pilot name, Dan Gurney, that, at the end of the sixties, installed it on his cars to increase ailerons lift. Beginning from the half of the seventies it was studied and developed in aeronautics research [1], and the first applications were on the wings of DC10. It was a vertical plate positioned on the trailing edge of the wing, jutting out a small quantity, generally included from 1 and 2% of the wing chord.

Experimental studies [2] [3] [4] have proved that the Gurney Flap positioned on traditional profiles can determine, for a specific angle of attach, an increase of the pressure on the face and, above all, an important decrease on the top surface, that is a high increase of the lift, that for some profiles can exceed 50% [5]. Besides, if the protrusion of the plate is less than 2% of the chord, the increase of resistance is negligible, because the device is completely dipped in the boundary layer. For the same lift coefficient produced, the increase of Gurnery Flap protrusion delays separation, causing a further increase of the profile efficiency.

These effects are caused by an alteration of Kutta-Joukowsky condition (flow uniformity downstream a profile): the Gurney Flap produces a little upstream separation region, and two downstream controrotating vortices, and then the chord profile is longer near the trailing edge.

This vorticity, added to the profile peculiar one, brings about a higher velocity gradient between the face and the top surface of the profile, and produces an increase of pressure gradient and then of lift [6].

Gurney Flap can be used even as stabilizer: a couple of theme, appropriately positioned, has been put on many helicopter types. [7].

This device has been used in naval field since many years on "deep V" high speed monohull, such as the "Acquastrada" ferryboat, as trim corrector, stabilizer and backing device for steering the ship, and it has been named "interceptor". Today, even if with some changes, it is often used on planing and semi-planing high speed little vessels, in order to improve their performances, and is often named "intruder".

3. TESTED MODELS.

At the Italian Ship Model Basin (INSEAN), intruder has been tested on three models: two standard planing ones and one catamaran, that we call for simplicity model A, B and C, with scale λ 5.5, 10 and 8 respectively.

The L_{PR}/B_{PX} ratios (Projected chine Length / Maximum breadth over chines) of the first two tested models, are typical of a 10-15 meters long yacht and a 20 meters big one respectively. We have to underline that these ratios are relative to the planing surface determined by the chine, how is usual for planing hulls. Model C, very different from the others, is a particular case, but it has shown similar results.

	A	B	C*
L_{PR}/B_{PX}	2.88	4.72	12.09
$A_P/\nabla^{2/3}$	6.15	3.54	-----
β	16°	14.8°	14.5°
β_T	12.7°	7.9°	14.5°
$L_{WL}/\nabla^{1/3}$	4.28	5.65	5.87

* : Parameters related to only one hull

(L_{WL} related to no intruder tests)

Table 1: A, B, C. models dimensionless parameters.

In table 1 typical dimensionless parameters are reported for the three models,

During the tests resistance and heaves have been measured for different speeds in accordance with Insean standard procedures [8]. The obtained results, in order to make the comparison easier, have been appropriately made dimensionless. By speed the Volumetric Froude Number has been calculated:

$$F_{n\nabla} = \frac{V}{\sqrt{g\nabla^{1/3}}}$$

where V is speed, g gravity acceleration and ∇ is the displacement volume. The resistance has been simply devided by displacement, obtaining the resistance coefficient R_T/Δ.

4. FIXED INTRUDER.

4.1 EXPERIMENTAL DATA

Model A has been tested from 2.15 to 3.45 Volumtric Froude Number, without intruder at ϑ=0°, whereas with it both at $\vartheta=0°$ and at $\vartheta=1.15°$ (trim by the stern), using the ratio $S_i/L_{PR}=1.03\times10^{-3}$, where S_i is the intruder protrusion.

In figure 1 R_T/Δ and dynamic trim angle τ curves versus the Volumetric Froude Number are represented.

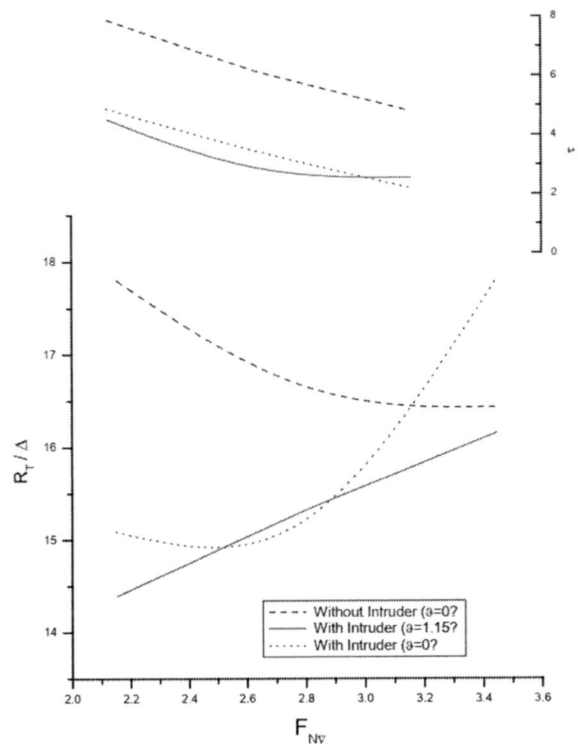

Figure 1: Mod A – Experimental data, with and without intruder.

Comparing the two broken lines ($\vartheta=0°$), intruder advantage is very clear at lower Volumetric Froude numbers, but it decreases when speed increases, as far as $F_{n\nabla}=3.1$, where there is a reversal of trend. In the same figure the continues line represents the coefficient resistance curve obtained at $\vartheta=1.15°$: the starting positive trim (by the stern) causes a higher improvement at very low speeds, about 19% at $F_{n\nabla}=2.2$, advantage that decreases if speed increases remaining obvious for all tested field anyway.

Tests on model B have been carried out from $F_{n\nabla}=1.73$ to $F_{n\nabla}=2.67$. Without device it has been tested at $\vartheta=-0.5°$ (trim corresponding to the smallest resistance at $F_{n\nabla}=1.9$), while with intruder ($S_i/L_{PR}=0.76\times10^{-3}$) at $\vartheta=0°$.

Curves trend in figure 2 is slightly different from the preceding case: the improvement starts from 9.5% at $F_{n\nabla}=1.7$, increases and grows steady, about 17%, for the remaining tested $F_{n\nabla}$ (from 2.0 to 2.7).

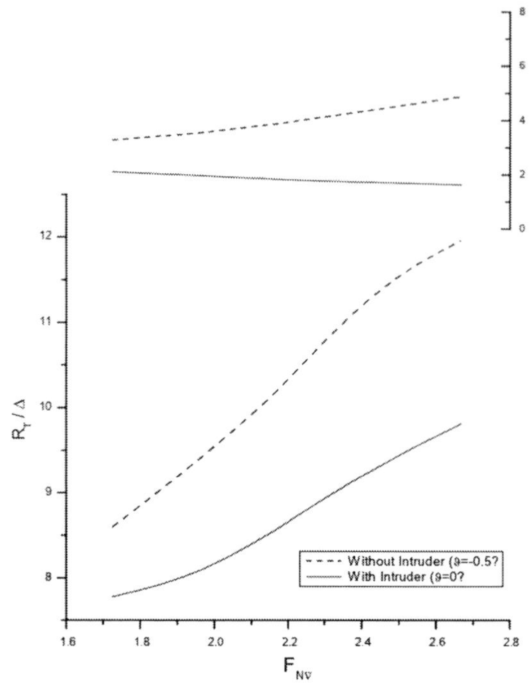

Figure 2: Mod B - Experimental data, with and without intruder.

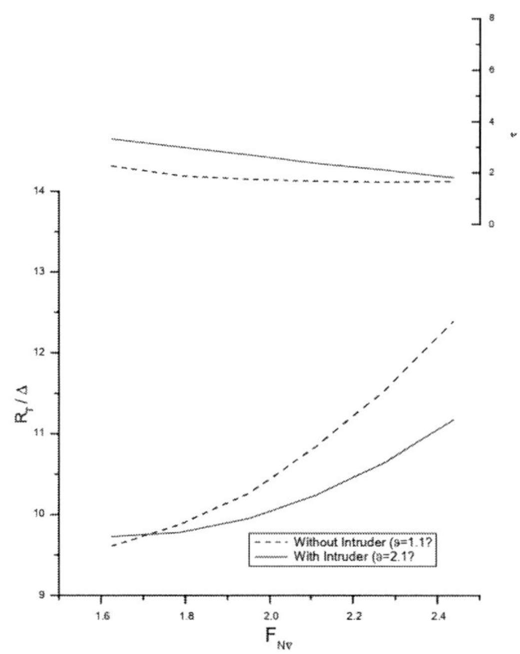

Figure 3: Mod C - Experimental data, with and without intruder.

As regards the catamaran (model C) the two different tested trims, with and without intruder, are respectively $\vartheta=1.14°$ and $\vartheta=2.12°$, with $S_t/L_{PR}=1.14\times10^{-3}$, from $F_{n\nabla}=1.63$ to $F_{n\nabla}=2.44$. Curves shown in figure 3 highlight that, at very low Volumetric Froude Numbers, intruder causes a slight worsening, trend that soon after changes when speed increases, getting on for 10% at $F_{n\nabla}=2.4$.

In described figures, dynamic trim angle curves furthermore are shown, always versus Volumetric Froude Number: in all tests with intruder τ is less.

4.2 DATA ANALYSIS .

Analysing the obtained experimental results, it is necessary to notice that the Volumetric Froude Number fields are very different, because they correspond to the real functioning speeds of full scale ships. For example model A - short and wide - has been tested up to $F_{n\nabla}=3.4$, (the starting planing $F_{n\nabla}$ is about 2.5), experiments with model B - longer and more narrow – have been done up to $F_{n\nabla}=2.6$, while the catamaran has been tested up to $F_{n\nabla}=2.44$. These differences do not prevent us from doing any comparisons and from drawing common conclusions, such as the clear advantage caused from intruder use, even if it is in percent very different in the three cases if Volumetric Froude Number varies.

There are no studies in depth that justify the positive effect of intruder from a quantitative point of view. But we may suppose the qualitative principle of operation: on the analogy of Gurney Flap [9] in aerodynamic field, we may think that a small separation region forms upstream intruder. It is a kind of dynamic wedge that modify significantly the pressure field just below the bottom of the hull, producing a higher hydrodynamic support.

This hypothesis is supported for example by figure 4, that shows model C at $F_{n\nabla}=2.44$ without and with intruder: a small plate jutting about one thousandth of the hull length raises the stern about 5% of length itself.

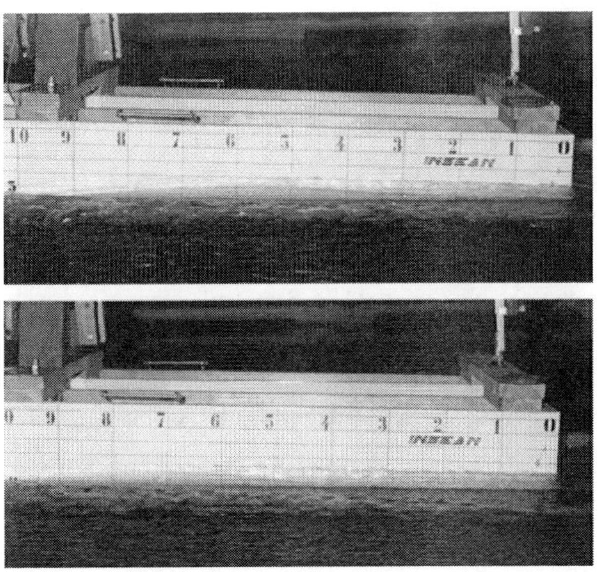

Figure 4: Model C: Profile of stern wave at $F_{n\nabla}=2.44$ without (upper photograph) and with intruder (lower one).

The overall effect is comparable to that produced by big flaps, but the added resistance is very lower. Intruder in fact – even if it is perpendicular to the flow – works completely deep in the boundary layer. Tests have shown that both wet length and surface are usually higher (even more than 20%); but the variation of length changes the friction resistance coefficient C_F very little in percent, whereas the increase of surface conditions greatly the total resistance coefficient C_T, that decreases considerably, where C_F calculated in accordance with ITTC '57 [10] is:

$$C_F = \frac{0.075}{(Log_{10}R_n - 2)^2}.$$

Mentioning

$$C_T = C_F + C_R = \frac{R_T}{\frac{1}{2}\rho S V^2}$$

it is evident that the advantage caused by intruder is due to a decrease of the residuary resistance coefficient (C_R), that, besides, is very important for a high speed hull, producing even more than 50% of total resistance.

This kind of analysis is confirmed by experimental observations; in figure 5, for instance, is clear that using intruder the waves are lower, that generally means lower C_R.

Figure 5: Model B at $F_{n\nabla}=2.20$ with intruder (upper photograph $S_i/L_{PR}=0$, $\vartheta=-0.5°$) and without (lower one $S_i/L_{PR}=0.76\times10^{-3}$, $\vartheta=0°$).

By the obtained graphs is clear that positive effect of intruder is more evident in pre-planing range ($1<F_{n\nabla}<2.5$). Looking at curves, three models have differences in behaviour, mainly because they have quite unlike geometry and so they plane at different Volumetric Froude Numbers. Model A for example at $F_{n\nabla}=2.2$ is in pre-planing range, therefore intruder produces a 20% gain nearly, soon after full planing starts, and the advantage decreases quickly. Model B instead, longer and more narrow, has a larger pre-planing range, indeed the advantage caused by intruder extends over $F_{n\nabla}=2.2$. If model B tests had been performed also at higher Volumetri Froude Numbers the two curves had got near and after intersected, but tests speeds had been choosen in accordance to the real speed of corresponding full scale ships.

5. MOBILE INTRUDER: COMPARISON WITH FLAPS.

5.1 EXPERIMENTAL DATA

As already said, intruder use can greatly improve a hull performances, but it is very often difficult to foresee its efficacy, and above all it can considerably depend on its protrusion, especially for hulls that works in a large speed range. It is sometimes suitable to use a variable protrusion of the device, as a function of speed, on the analogy of flaps, whose pressure plane is movable.

In order to evaluate the efficacy of intruder protrusion variation and to compare it with a couple of flaps, some tests have been performed on model A. It has been chosen because of its very large speed range. Tests have been done at $F_{n\nabla}$=*2.15*, *2.80* and *3.45*, at $\vartheta=0°$, varying both intruder protrusion and flaps angle. As regards tests with intruder they have been performed also at $\vartheta=1.15°$. Used flaps were commercial ones with $AR=2$ (width=27% of B_{PT}, chord=2.3% of L_{PR}).

Experimental data are gathered in graphs of figure 6 where, for every tested $F_{n\nabla}$, R_T/Δ variations in percent are shown, as regards starting condition without intruder nor flaps at $\vartheta=0°$. In the same figure dynamic trim angles in degrees are represented versus flaps angle or intruder protrusion S_i/L_{PR} (upper curves).

It is evident that intruder is more efficacious than flaps. It is useful to underline that protrusions of intruder are always lower than 20 mm for full scale ships, value over that a high decrease of performances is generally found.

At $F_{n\nabla}=2.15$ flaps use improves hull performances, up to about 5% at their maximum testable angle of incidence that is 8 degrees. At the same Volumetric Froude Number improvement caused by intruder is about 17%. With model trim by stern ($\vartheta=1.15°$) intruder efficacy produces the maximum resistance reduction – about 20% - but not at the maximum intruder protrusion. As a matter of fact, if S_i/L_{PR} increases a little more R_T/Δ increases very much. It's interesting to notice that if the resistance decreases the dynamic trim angle generally decreases too: from 8°, of the starting common condition without flaps nor intruder, to 7° with flaps and to about 4° with intruder at minimum R_T/Δ.

At $F_{n\nabla}=2.80$ performances improve up to about 3% by flaps, whereas resistance decreases about 8.5% at $\vartheta=0°$ and about 7% at $\vartheta=1.15°$ by intruder, and the corresponding τ is 4°. Also in this condition is evident that if intruder protrusion keeps on increasing, resistance increases too.

F_{nV}	TEST WITH FLAPS (θ= 0°)	TESTS WITH INTRUDER (θ= 0°)	TESTS WITH INTRUDER (θ=1.15°)

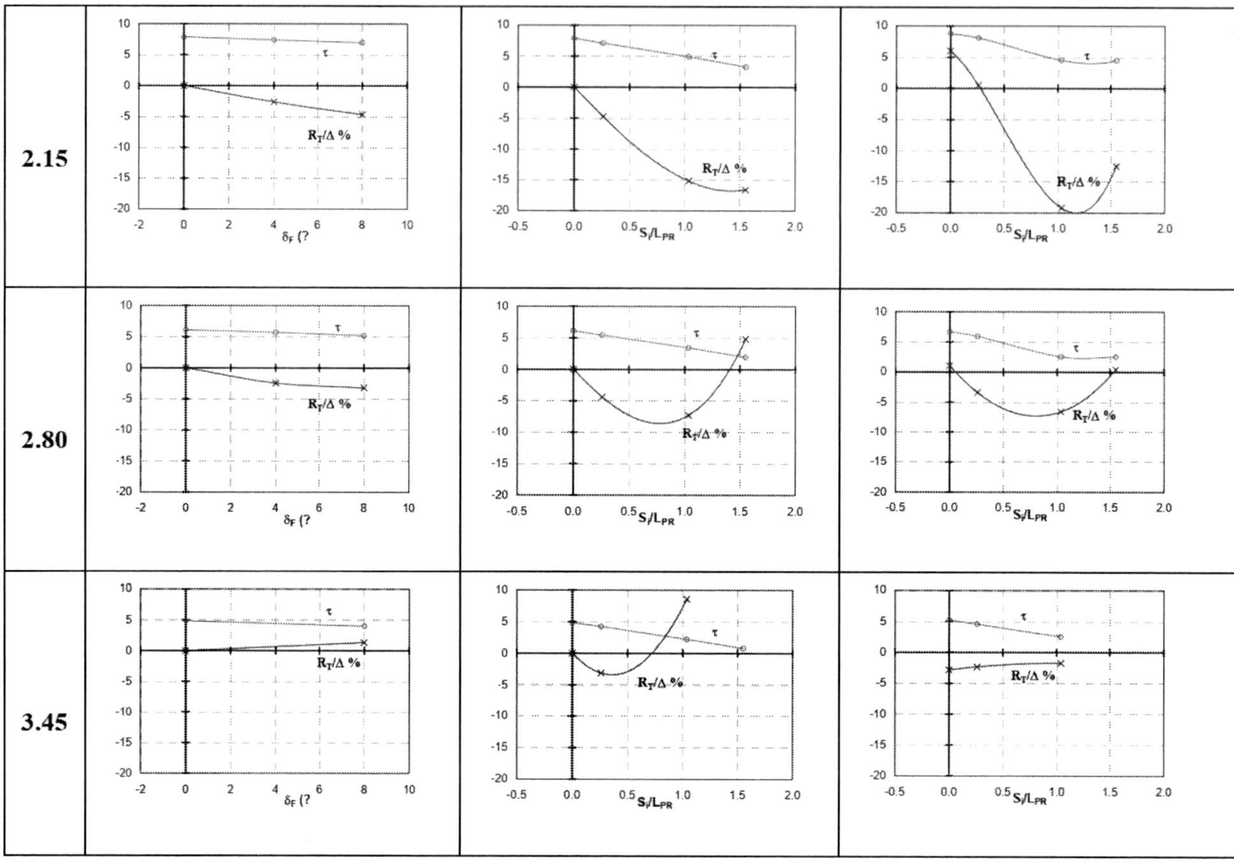

Figure 6: Model. A: Comparative tests flaps/intruder.

At $F_{n\nabla}=3.45$ a resistance worsening is obtained by flaps, whereas by intruder use, at $\vartheta=0°$, resistance decreases about 3%, and dynamic trim angle τ is 4° again. At $\vartheta=1.15°$ R_T/Δ reduction - about 2% - is anyway present because of trim by stern condition, and it persists even if there is a little intruder protrusion. Also in this case at minimum R_T/Δ dynamic trim angle is 4°. If protrusion increases τ still decreases but R_T/Δ increases.

5.2 DATA ANALYSIS

In graphs of figure 6 it is interesting to notice that τ about 4° - corresponding to the minimum R_T/Δ - is the angle of minimum resistance obtained by Savitsky for prismatic planing hulls [11] and in DTMB Systematic Series 62 [12] with similar geometrical characteristics. Besides it is appropriate to remark that the dynamic trim angles obtained by the model during the basin tests, because of not well-known reasons caused by scale effect, are generally higher than the corresponding full scale ship values [13].

Some graphs show that minimum resistance condition is verified at different S_i/L_{PR} values if $F_{n\nabla}$ changes.

On ships with large speed range, remote controlled mobile intruders are advantageous; they can be used like flaps to stabilize the boat and to improve performances. Plate protrusion – higher in pre-planning phase – must be reduced when the hump speed get near.

Photographs in figure 7 show the behaviour of the model at $F_{n\nabla}=3.45$ with $\vartheta=0°$. In the first one, without intruder, keel comes out from water at section 7 whereas in the second one, with $S_i/L_{PR}=0.26\times10^{-3}$ it comes out further on (at section 8), and the hull is more horizontal in comparison with the precedent condition. In this last case we have about the minimum resistance and the dynamic trim angle is 4°. The following photographs show tests with $S_i/L_{PR}=1.03\times10^{-3}$ and $S_i/L_{PR}=1.55\times10^{-3}$: because of the excessive protrusion, the point where keel comes out from water moves toward the bow, dynamic trim angle decreases and waves are higher and more visible. The result is a quick increase of resistance for a decrease of the dynamic trim angle in comparison with its optimum value, phenomenon already pointed out by studies on planing prismatic hulls [11] and by systematic series [12].

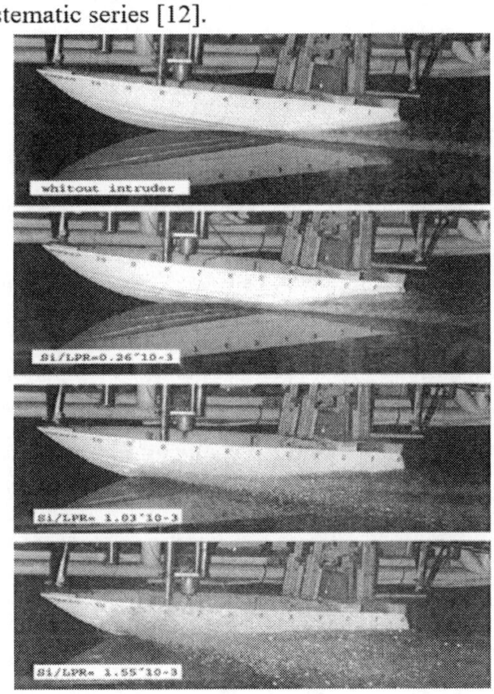

Figure 7: Differences of behaviour of mod A at $F_{n\nabla}=3.45$, $\vartheta=0°$ with and without intruder.

6. CONCLUSIONS

By tests performed at Insean on three different models is clear that intruder can induce significant changes of dynamic trim angle and can lift the stern. These changes can reduce waves and then hull resistance. On the other hand an excessive protrusion, or it's use if not necessary, can strongly increase resistance. Besides:

- Fixed intruder is useful above all on big planing and semi-planing hulls, generally the ones with high L_{PR}/B_{PX} (≥4) or the ones which navigate at Volumetric Froude Numbers lower than 2.5 (hulls that never navigate completely planing). In these cases intruder is useful to correct a too high trim by stern, which sometimes is characteristic for this kind of boats that tend "to sit on wave" they produce. It's installation on these hulls is very cheap because it consists of a plain plate positioned under the transom jutting out just a few millimetres. Experimental results have shown a resistance reduction even more than 20%. Protrusion must be carefully chosen by practical tests.
- Mobile intruder is more recommended for ships with a larger speed range, ($F_{n\nabla}>2\text{-}2.5$). In these cases in fact as speed gets near the hump one, plate protrusion has to be shortened, because intruder reduces dynamic trim angle too much. On these boats, that typically have low *L/B* ratios, is often enough to install commercial devices, even if they have a partial length in comparison with the transom one.
- Intruder is useful to correct the behaviour of a boat that navigates trimmed by stern too much and to improve it's performances. Moreover experimental tests have shown that – in these cases – the efficacy of this device increases and it is valid on a larger speed range. Therefore, if intruder is used, can sometimes be useful to trim by stern statically the boat.
- By experimental data analysis intruder is generally more efficacious than flaps, both as regards improvement of performances and as regards the larger functioning speed range. In fact to obtain by flaps an effect as intruder one – with respect to dynamic stern support and to improve the trim – big flaps with high angles of incidence are indispensable, but they cause a high resistance increase and reduce - till they undo them – the advantages produced by themselves. By intruder, instead, a few millimetres protrusion is enough to obtain the same dynamic effects, against a negligible added resistance because it completely works in the boundary layer.
- Matched devices intruder/flap have been already tested and they are spreading, with intruder placed downstream flaps on their trailing edge or upstream. In this last case practically flaps are not coplanar with the bottom but they are a few millimetres lower. [14]. If the two devices are joined, to balance their characteristics very well is necessary.

The rapid spreading of intruder, caused by it's cheapness and it's simple installation and use, is giving rise to some changes of it's morphology, and then of it's working, to adapt it to particular hull forms or appendages and/or to existing flaps. Therefore further studies and tests are useful and desirable to advise planners and market about the working and the potential of this device, as well as about it's correct installation.

7. REFERENCES

1. LIEBECK, R.H., 'Design of Subsonic Airfoils for High Lift', *Journal of Aircraft, Vol.15, No. 9*, September 1978.

2. STORMS, B.L., JANG, C.S., 'Lift Enhancement of an Airfoil Using a Gurney Flap and Vortex Generators', *Journal of Aircraft, Vol. 31, No. 3*, June 1994.

3. JEFFREY, D., ZHANG, X. and HURST, D.W., 'Aerodynamics of Gurney Flaps on a Single-Element High-Lift Wing', *Journal of Aircraft, Vol. 37, No.2*, 2000.

4. ZERIHAN, J., ZHANG, X., 'Aerodynamics of Gurney Flaps on a Wing in Ground Effect', *AIAA Journal, Vol. 39, No.5*, May 2001.

5. KATZ, J., LARGMAN, R., 'Effect of 90 degree flap on the aerodynamics of a two element airfoil', Journal of Fluids Engineering, 111:93-4, 1989.

6. PROUTY, R.W., 'Aerodynamics'", *Aviation today -Rotor & wing,* October 2003.

7. ROESCH, P., VUILLET, A., 'New designs for improved aerodynamic stability on recent Aerospatiale helicopters', Proceeding of the 37th Annual Forum, American Helicopter Soc., 1981.

8. IX ITTC, 'Ninth International Towing Tank Conference – Proceedings', Paris 1960.

9. SEWALL, W.G., MCPHEE, R.J. and FERRIS, J.C., 'Wind tunnel test results of airfoil modifications for the EA-6B', *AIAA Paper 87-2359*, August 1987.

10. VIII ITTC, 'Eight International Towing Tank Conference', Madrid 1957.

11. SAVITSKY, D., 'Hydrodynamic Design of Planing Hulls', *Marine Tecnology*, 1964.

12. CLEMENT, E.P., BLOUNT, D.L., 'Resistance Tests of a Systematic Series of Planing Hull Forms', *Transaction SNAME, Vol. 71, pp. 491-579*, 1963.

13. PENSA, C., 'La Sperimentazione in Vasca Navale delle Imbarcazioni da Diporto', *SEA-MED 2004, Messina*, July 2004.

14. TSAI, J.F., HWANG, J.L., CHOU, S.K., 'Study on the Compound Effect of Interceptor with Stern Flap for two Fast Monohulls with Transom Stern'. *FAST 2003, Ischia (Italy),* October 2003.

8. AUTHORS BIOGRAFY

Andrea Mancini and **Alessandro Moriconi** hold the current position of experimental technicians at Insean, the Italian Ship Model Basin. Their previous experience includes resistance, self-propulsion and sea-keeping tests on a lot of types of hulls, such as ferry boats, catamarans, planing and semi-planing ships, motorboats, IACC yachts etc.

Experimental study of the influence of turbulence on the performance of a blade of hydrofoil

[1]HAMMADI L ; [1]BELHADRI M ; [2]IMINE O ; [2]ADJLOUT H .

[1] Laboratory of Rheology Transport and Treatment of the Complex Fluids (U.S.T.O)
Faculty of architect and civil Engineering Department of hydraulics U.S.T.O,
B.p 1505 El-Maneur, Oran, Algeria
E-mail: Hammadi7280@yahoo.fr

[2] Department of marine engineering U.S.T.O, B.p 1505 El-Maneur, Oran , Algeria.

Abstract

An experimental study for the investigation of the flows in the wake of profile of hydrofoil was carried out. The study in question consists in testing out of blower a profile of the Eppler type 193 G.U.K modifies, the distributions of pressure on the surface of the profiles were measured for different levels from turbulence. The distribution of the C_P showed that the increase in the level of turbulence causes a limiting separation of layers on the suction face of the profile.

The distribution speeds in the wake showed on the one hand that the flow is two-dimensional and on the other hand highlighted the signature of the wake.

Nomenclature

C: Twist, P: Pressure room on the profile, Pvap: Vapor pressure, X: Displacement of the probe on (OX) , Y: Displacement of the probe on (OY) , Z: Displacement of the probe on (OZ) ,

CP $== \dfrac{P_i - P_\infty}{1/2\rho_{air}U^2_\infty}$: Pressure coefficient , P_i: Pressure on the under-surface and the suction face, Ux,

Uy, Uz: Speeds, U_∞: of the free flow has a temperature T=24.2°C, T_u: Level of turbulence , Cz:

Coefficient of bearing pressure, Re: Reynolds number. , Ux: Speed according to (OX) , Uy: Speed according to (OY), Uz: Speed according to (OZ) , U ∞ : speed of the free flow.

1. Introduction

The generation of wings subcavitantes of hydrofoils starting from airfoils, such as the NACA 16, remains valid until rates of travel of order 50 to 55 nodes realizing of the particular precautions .

For speeds higher than 55 nodes, it becomes impossible to avoid the demonstration of a cavitation thickness on the preceding profiles from where the idea to truncate their part postpones in order to stabilize a cavity on this level to decrease the trail.

The studies relating to this broken down mode, called by time improperly trancscavitant are based in theory on the use of parabolic profiles, study in particular by V.E.JOHNSON [1] which have the advantage of reconciling a good inertia of section with a distribution of suitable pressure along their cord.

These profiles are to be regarded as the prolongation of the profiles subcavitants beyond the cavitation thickness. Beyond 70 nodes, it is imperative to use profiles such as the cavity is established leading edge of acute form and extends on the unit from the suction face like behind from the trailing edge. Theoretical work of M.P. TULIN et al. [2] makes it possible to define bearing profiles whose interest is limited by low intimacy of section and an unquestionable brittleness of the leading edge. Currently, it seems very improbable to design airfoils ventilated or supercavitating in the whole of the field of the flight of a hydrofoil. The researchers tried to cure in state by providing them with camber flaps as it is of use to equip the modern planes with them.

Therefore it was imagined profiles functioning, if possible in mode subcavitant at the low speeds with a lift increasing on takeoff, the release of ventilation or cavitation having to occur in a more or less regular way with the increase in speed , [3] .

This study consists in testing out of blower a plane profile, relative thickness of 18%, its cord is 156 mm, the maximum thickness being in the medium of the cord Fig.6.

The experimental checks consisted with:

a) The measurement of the coefficient pressure on the levels of the suction face and the under-surface of the profile louse three levels of turbulence

b) The estimate rates of the flow in the wake of the profile for three levels of turbulence

2. Presentation of the blower of the institute of marine engineering [4]

It is made up of four principal parts :

- **a collector** : convergent tunnel in which the flow is accelerated starting from a rather low speed of diameter to the downstream is 300 mm and the diameter with the upstream is of 500 Misters .
- **an intermediate diffuser** : passage of the square section to the circular section length of 1000 Misters
- **a vein of test** : it is the narrowest part of the tunnel or the tests are carried out and or speed is maximum. Its geometrical form is square dimensions (300 × 300 mm).
- **a conical diffuser** : it is the part or the flow is slowed down its length is 660 mm.

The draught which circulates in the blower is generated by a ventilator with five blades supplied with an electric motor standard **VZ80 h4** of power of **0,75 kW** and number of revolutions 1420 tr/min , figure (10).

3. Presentation of the profile

The profile to be realized, has a relative thickness of 18% (sec Fig.6) being in the medium of the cord. This one of 165 Misters is the cord . The coordinates of the profile are presented in tab.1

X/c	Y/C (intrados)	Y/C (Extrados)
1	0	0
0.93	0	0.033
0.86	0	0.06
0.8	0	0.08
0.73	0	0.1
0.66	0	0.12
0.6	0	0.14
0.53	0	0.153
0.46	0	0.166
0.44	0	0.166
0.33	0	0.17
0.26	0	0.166
0.2	0	0.148
0.13	0	0.12
0.06	0	0.086
0	0	0

Tab1. Co-ordinates of the profile

3.1 Distribution of the pressure under-surface suction face

On the suction faces and under-surface are placed openings to measure the pressure . These openings are connected to different openings placed on the with dimensions one of wing in order to be able to measure the pressure of under-surface and suction face is measured respectively on the right and on the left (see Tab.2).

X/C(intrados)	X/C(extrados)
0,195	0,811
0,290	0,702
0,414	0,622
0,536	0,510
0,597	0,366
0,658	0,268
0,756	0,195

Tab.2. Co-ordinates of the pressure taps

4.1 Tube of Pitot

The velocity measurement of one flow using the tube of Pitot is an application of the law of Bernoulli. In a steady flow of an incompressible true fluid, along a thread of current, the quantity:

$$\frac{1}{2}U^2 + P + \rho gz = Cte \qquad (1)$$

The tube of Pitot comprises two pressure C_P: the flow (impact pressure P a) the other perpendicular with the flow (static pressure P) . For a horizontal thread of current, or for air whose weight can be neglected, one has for the thread of current which arrives at the point of stop [5]:

$$\frac{1}{2}.U^2 + P_S = P_a \quad (2)$$

The rate of the flow is then given by:

$$U_\infty = \sqrt{\frac{2.(P_S - P_a)}{\rho}} \quad (3)$$

Speed thus depends on the difference between the impact pressure and the static pressure. This difference is called dynamic pressure.

The calculation rate of flow starting from the relation (3) deduced from the law from **Bernoulli** supposes that the fluid is incompressible. Thus speeds obtained are:

With a grid of turbulence placed at X1=340mm has a temperature T=24,2°C and Pa = 1011 mbr, U_∞ = 14,84m/s

With a grid of turbulence placed at X2=525mm has a temperature T=24,2°C and Pa = 1011 mbr, U_∞= 13,84m/s

With two grids of turbulence placed at X_1=340mm and X_2=525 mm has a temperature T=24,2°C and Pa = 1011mbr, U_∞ = 11,58 m/s

4.2 The probe with five holes [6]

The probe with five holes presented on figure (10) is of a diameter external of **10mm** and a length of **425mm** it is pierced with five holes, which set out again as follows:

- A hole, on the axis of the probe.
- Four holes left again around the axis of the probe (high, low, left, right-hand side) and forming an angle of **45°** with the axis of the probe. This probe makes it possible to measure the rates of the flow. The use of the probe requires a calibration to allow the conversion of the measured pressures.

4.2.1 Calibration of the probe

4.2.1.1 Principle of the calibration

The calibration is carried out with a speed of the constant blower and value of (θ) and (α) included/understood between -20° with +20° with a step of 5°. The device of the calibration is schematized on the figure (10). For each point of measurement (θ, α) , one deduces the pressures (P_1 , P_2 , P_3, P_4 , P_5) starting from these pressure one determines the function following:

$$yh_1 = \frac{(P_3 - P_2)}{(P_1 - P_5)} \quad \text{for} \quad \alpha \le 0° \text{ et } - 20° \le \theta \le 20°$$

(4)

$$yh_2 = \frac{(P_3 - P_4)}{(P_1 - P_2)} \quad \text{for } (\alpha \ge 0°) \text{ and } (20° \le \theta \le 20°) \tag{5}$$

$$yh_3 = \frac{(P_1 - P_2)}{(P_1 - P_5)} \quad \text{for } (\alpha \le 0°) \text{ and } (\theta \ge 0°) \tag{6}$$

$$yh_4 = \frac{(P_1 - P_5)}{(P_1 - P_2)} \quad \text{for } (\alpha > 0°) \text{ and } (\theta \ge 0°) \tag{7}$$

$$yh_5 = \frac{(P_5 - P_1)}{0.5 \times \rho \times U^2} \quad \text{for } (-20° \le \alpha \le 20°) \text{ and } (20° \le \theta \le 20°) \tag{8}$$

Once the calibration carried out, one determines the functions indicated, which will be smoothed by the polynomials of six degrees and which are as follows:

$$yh_i = \sum_{n=1}^{7} a_n \theta^{n-1}, \quad i=1,2$$

(9)

$$yh_i = \sum_{n=1}^{7} b_n \alpha^{n-1}, \quad i=3,4 \tag{10}$$

$$yh_i = \sum_{n=1}^{7} C_n \theta^{n-1}, \quad i=5 \tag{11}$$

The functions yh_1, yh_2, yh_3, yh_4 and yh_5 are selected in order to have curves representing of the monotonous functions (increasing or decreasing), come for example the fig.2 the figures (3 ÷ 5) represent examples charts as of the these functions. These charts make it possible to calculate the angle of the flow and the three components speed, [6],[7].

4.2.1.2 Program calibration

From the charts of calibration one little to carry out a program for the calculation of the angles of the flow. This program uses as data the pressures (P_1, P_2, P_3, P_4, $_5$) calculation of the functions yh_1, yh_2, yh_3, yh_4 and yh_5 by carrying out the reading of the coefficients polynomials of these functions. In continuation, it tests the value of ($P_1 - P_2$). If the difference between P_1 and P_2 is negative, it takes the reports/ratios yh_1 and yh_3 in the contrary case, it takes the reports/ratios yh_2 and yh_4 One using a linear interpolation and following an intersection of the two polynomials, can about it obtain the two angles of the flow (α and θ) to see the flow chart [6]

5 Generation turbulence

To generate of turbulence, it is to cause a disturbance while making unstable a flow. This, can be to generate either using a particularly turbulent fluid current, or like made **PRANDTL. L** (1904) [4], by creating a zone tourbillion- surface in front of the body while placing a ring of wire or while making rough the surface of the body. The grids were used as means of generation of turbulence by **ROACH** (1986) [5]. In our case, the grids used are of type S.M.S [7]. Experiments As it is a question of studying the distribution of velocity the wake for three levels of turbulence:

- Turbulence animated by a grid (S.M.S) at a distance 340mm of the trailing edge of the wing, Tu=6,3%.
- Turbulence animated by a grid (S.M.S) at a distance 525mm of the trailing edge of the wing, Tu=4,6%
- Turbulence animated by a grid (S.M.S) posed at a distance 340mm and the other at a distance 525mm of the trailing edge of the wing.

6.1.1 Distributions of the CP for differ levels from turbulence

The distributions of the CP for differ levels from turbulence are shown in the figures (7, 8, 9).

- The values of (C_P) for under-surface and suction faces all are negative. It should be noted that the position of flow on the suction face which is with a positive gradient of pressure are constant.
- The value of the absolute ($C_{P\,min}$) decreases with the increase in the level of turbulence of the flow.
- The value of the C_P for the first point measured on the under-surface decreases with an increase in the level of turbulence

It was noted consequently a relative agreement between the study resulting from the experimental method and that defined by the theory. [7], [8],[9]

6.2 Distributions speeds In the wake

The distributions of the components speed in the wake for three plan x=2 cm, x=4 cm, and X = 6 cm for three levels of turbulence are presented on the figures (11 to 17).

a) Axial speed

The presence of a hollow in the distribution of axial velocity is signature on the wake whose value minimal speed decreases with the increase in the level of turbulence. The speed of the hollow increases with the increase in the position (x) probe with borders of escape of the profile, [13]

b) Uy speed

For each plan one will note that Uy speed following the axis is almost null there, which assure us that the flow will be only two-dimensional.

C) Speed Uz

One notices that the distributions speeds according to (Z) are divided into three zones:

1) the case Uz > 0 or the flow is accelerated (negative gradient of pressure).
2) the case Uz=0 or the flow is uniform.
3) the case Uz < 0 where the flow is decelerated (positive gradient of pressure).

It is also noticed that the fluctuations Ux speeds are much more significant than those Uz speeds in the wake.

The coefficient of drag is calculated by the following formula (tableau.1), [10],[11]:

$$Cx = \frac{2}{C}\int_{Sillage} \frac{U}{U_\infty}\left(1-\frac{U}{U_\infty}\right)dz$$

X/C	0°	10°	-10°
1.061	0.07864	0.09137	0.07789
1.12	0.082	0.12	0.0753
1.18	0.082	0.13769	0.116
1.24	0.08113	0.1695	0.10383
1.305	0.061	0.16409	0.1404
1.366	0.011	0.16261	0.0172

Tab3. Coefficient of drag of profiles

6. Errors of measurement

The determination of the complete assessment of the errors made during the experimentation be pressed on the measuring equipment presented as follows:

6.1 Blower

The blower can present itself a fortuitous error in particular due to:

a) Positioning

The plan on which the blower is posed can present a defect of longitudinal slope or transverse.

b) The geometrical shape of the vein of test

It must be square but there can be a defect of deformation undergone during displacement for example.

6.2 The electric motor

It generates vibrations which disturb the flow inside the blower and contributes to the increase in the level of natural turbulence.

6.2 Pressure gauge

6.2.1 Misreading

The measurement of the differential pressure is necessarily sullied with a misreading estimated at 0,64mm.

6.2.2 Slope of the pressure gauge

the error of slope is estimated at ± 1° (45° ± 1°).

6.2.3. Effect of instability of the level of alcohol:

Measures to the zones of strong turbulence, example: wake, the level of alcohol in the pressure gauge is unstable. This difficulty of reader requires the use of the following criterion:

6.2.4 Density of alcohol

The alcohol used in the pressure gauge is methanol, it is manufactured by Kanto chemical Coinc (Tokyo Japan), its density is 794 Kg/m^3 with a tolerance of 20/20C°.

6.3 The probe

The probe is used to measure the pressures in an unspecified point, is of volume equal to $523{,}33mm^3$. This involves necessarily an error on the position of the effective point of measurement. In fact, the probe carries out an integration in the volume of measurement. The disturbances of which the effects are not known until now.

a) Displacement of the probe

The error of displacement of the probe is estimated at 0.5mm in the three directions.

b) Fastening of the probe

During the use of the probe, there can be a defect of fixing when the axis is not directed exactly according to the average direction of the flow. To correct this error, one makes sure so that the pressures P_3 and P_4 is equal.

However, in the case of the lace this technique of equalization of pressure is impossible to realize because of the presence of the checkmate. An error estimated at 0,3° for the angle α remains in spite of the precautions taken for the fixing of the probe.

6.4 Barometer and Thermometer

The errors made during the use from these two install of measurement are:

For the barometer Δ h=2,5mbarPour the thermometer ΔT=0,1 Ç°

7. Conclusion

- It requires only one calibration of the probe with five holes on a field of + / - 20°.
- No office plurality of errors in the calculation of the various physical sizes (angles of incidence).
- Use of a criterion of rejection of the points out of the field independent of calculation **(Interpolation)**
- Possible extension of measurements to the flows compressible at high temperatures.
- Lower error than 1° on the angles and 1% on speed, whatever the aperture.

The study The influence of level of turbulence of the coefficient pressure on the levels of the under-surface and the suction face of the profile showed that :

- the value of (Cp) for the under-surface and the suction face are all negative.
- the position separation on the suction face is constant.
- the value of $|C_{P_{min}}|$ decrease with the increase in the level of the turbulence of the free flow.

The study of the distributions speeds in the wake of the profile showed that the flow is tow dimensional , and the component speed according to (x) has the shape of a bell, which means that there is a formation of a wake around profile, a depth which decreases when the plan of pressure C_P moves away at the edge of escape from profile. The component speed along the axis (z) is disturbed.

References

[1] V.E Johnson, '' Low-drag base Vented hydrofoils designed for near the free surface'', hydrodnatis technical, Report 001-15.

[2] M.P Tulin et Al, '' linearized theory for flow about lifting foils at zero cavitation nombre'' DTMB Report c 638, February 1995.

[3] M. Pouillot, '' Ailes Ventilées Mixtes'', Association Technique Maritime et Aéronautique, No. 72 Session de 1977, PP. 433-455.

[4] Hammadi , L ; Imine , O ; Ajllout , H .'' Etude expérimentale de l'influence de l'incidence sur les caractéristique d'une pale d'hydroptère'' . 6^{éme} Congrès de mécanique Tanger , Maroc 15-18Avril 2003

[5] R, Commolet. '' Mécanique expérimental des fluides'' TOM II, 3eme Edition Masson.

[6] Roach. '' The generation of nearly isotropic turbulence by means of grids''

International Journal of heat and fluid flow, 1986.

[7] Jean Lefebvr . '' Mesure des débits et des vitesses des fluides'' . Edition Masson 1979 .

[8] L. Jacqun et Al, ''Utilisation de sondes a 4 fils chauds pour l'étude d'écoulements turbulents '' , 26 colloque de l'aérodynamique appliquée Toulouse, 23-25 octobre 1989, ONERA.

[9] Jena- Pierre Franc, et AL '' La cavitation mécanismes physiques et aspects industriels'', Presses Universitaires de Grenoble 1995 .

[10] B. Satyanarayana, '' Unsteady Wake Measurements of Airfoils and Cascade'', AIAA Journal Vol. 15, N°. 5 May 1977. pp. 613-618

[11] Jean Marc Peallat and Al , ''Experimental validation of tow- and tree- dimensional numerical analysis of partially cavitating hydrofoils'', Journal of Ship Research, VOL. 40, No. 3, Sept. 1996, pp. 211-223.

[12] Jacques Verron and Al, '' Base-Vented Hydrofoils of Finite Span Under a Free Surface : An Experimental Investigation Journal of ship Research, VOL. 28, No. 2, June 1984, pp. 90-106

[13] Lawrence J. Doctors, Effects of a Finite Froude Number On a Supercavitating hydrofoil Journal of ship Research, VOL. 30, No. 1, March 1986, pp. 39-50

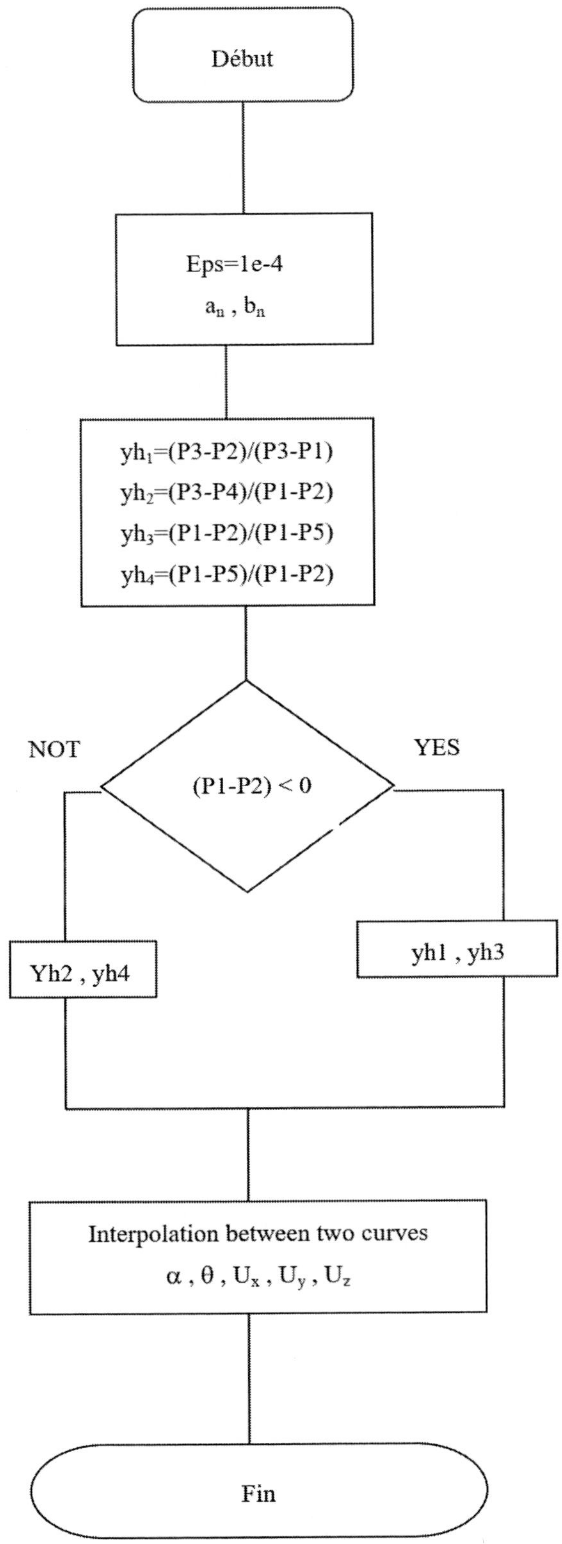

Fig.1. Flow chart principal of the calibration

Fig.2 Variation of the ratio yh_1 in function angle θ and α

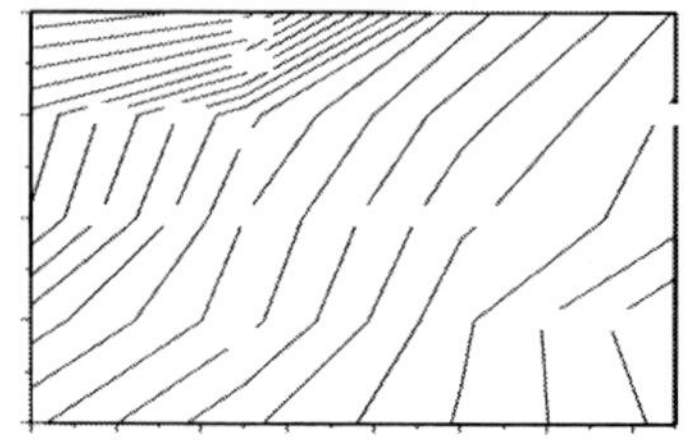

Fig.3. The chart of the ratio yh_2

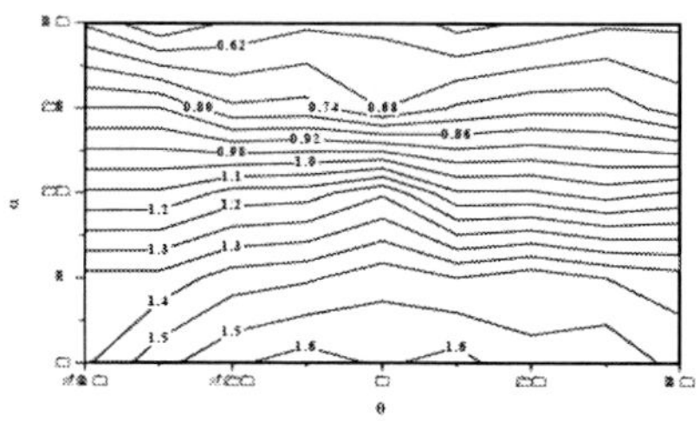

Fig.4. The chart of the ratio yh_5

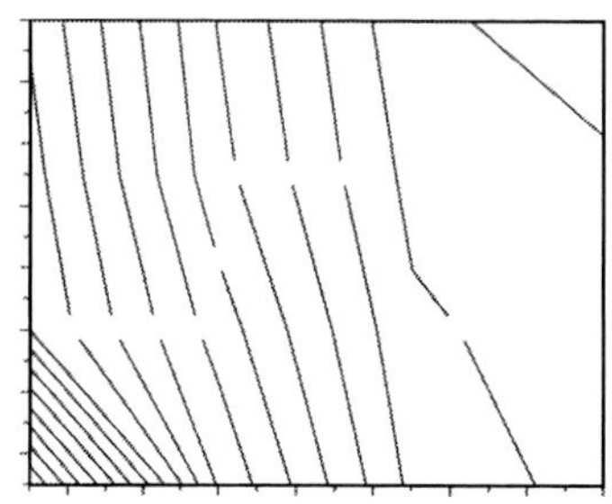

Fig.5. The chart of the ratio yh 3

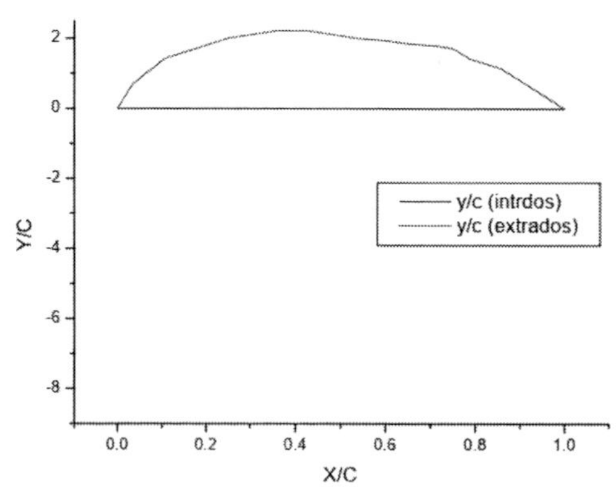

Fig.6.Geometrical form of the profile

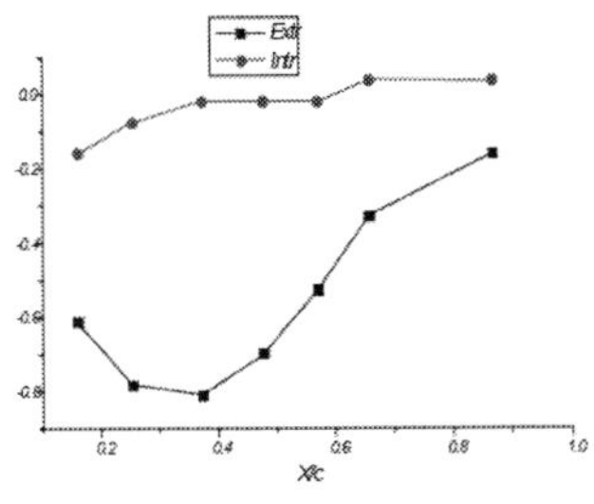

Fig.7 Variation du Cp experimental for Tu= 6.3%

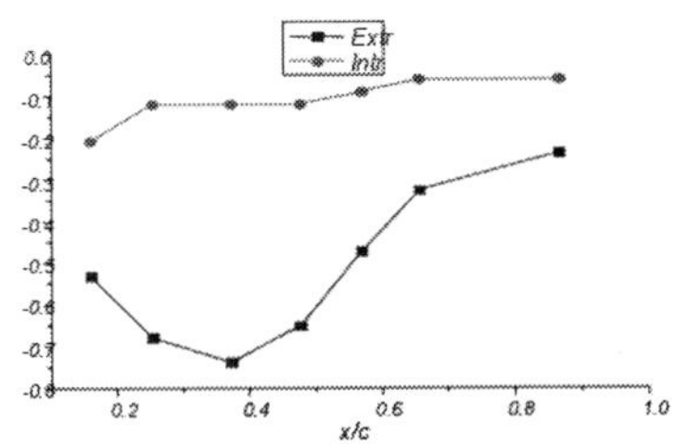

Fig.8 Variation du Cp experimental for Tu= 4,6 %

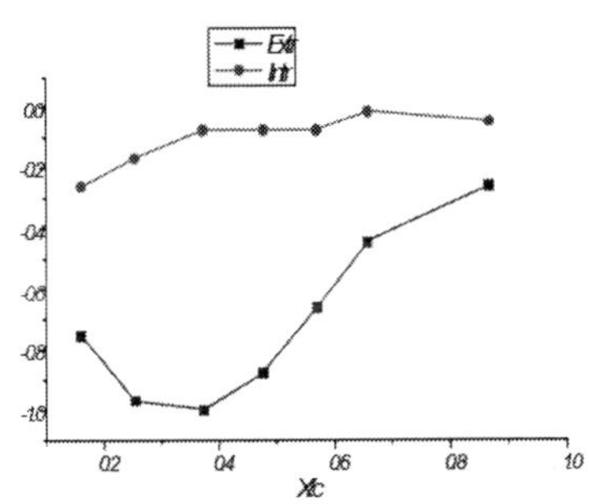

Fig.7 Variation du Cp experimental With two grids of turbulence placed

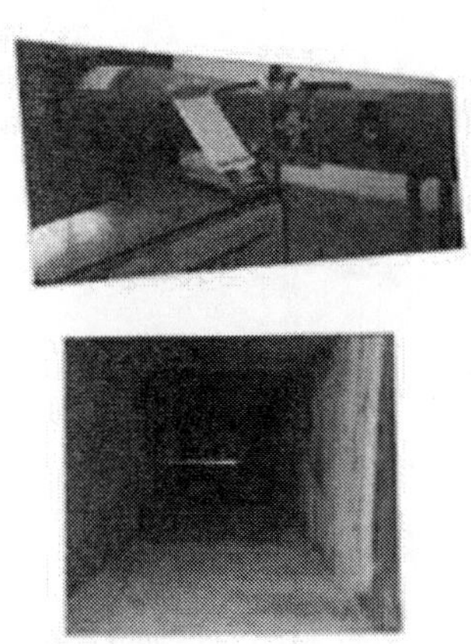

Fig.10. Device of calibration

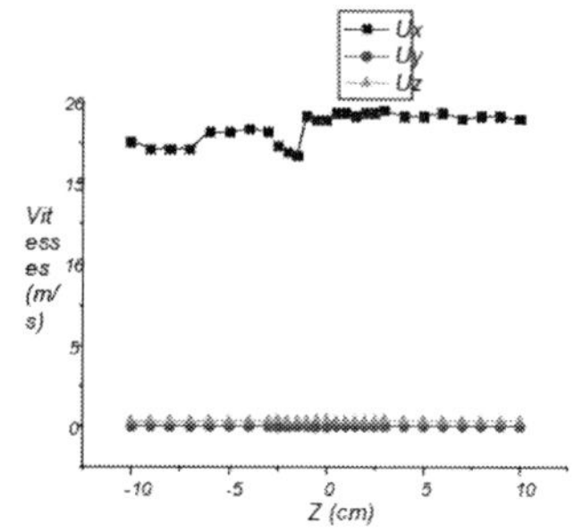

Fig.11 The distributions of the components speed in the weak for x= 2 cm and y=0 cm by a grid (S.M.S) at a distance x_1= 340 mm

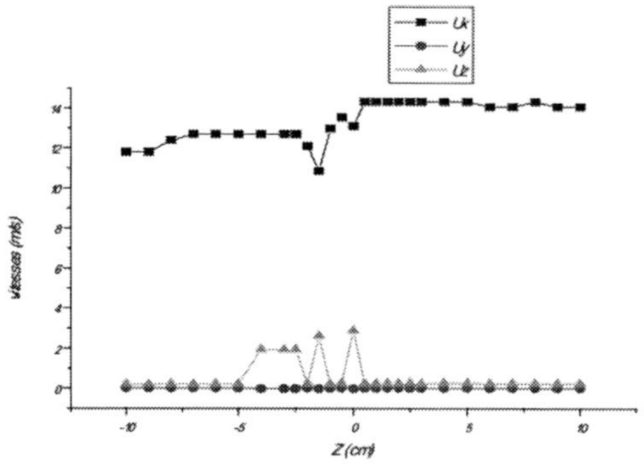

Fig.12 The distributions of the components speed in the weak for x= 6 cm et y=0 cm by a grid a x_1= 340 mm

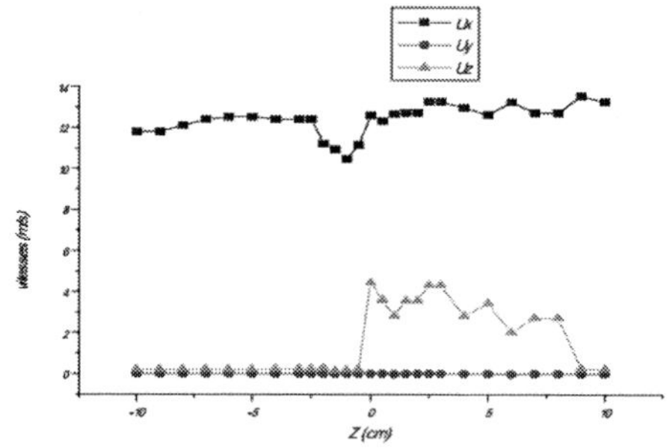

fig.13 The **Distribution of speed for in x= 2 cm and y=0 cm with grid in a x_2= 525 mm**

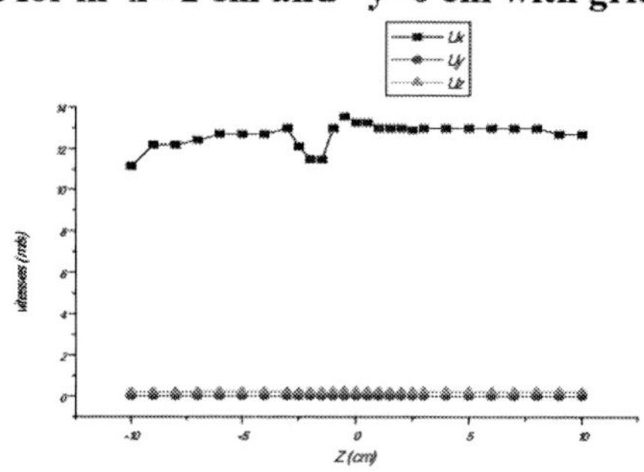

Fig.14 The distributions of the components speed In the weak for x= 6 cm an y=0 cm by a grid (S.M.S) at a distance x_2= 525 mm

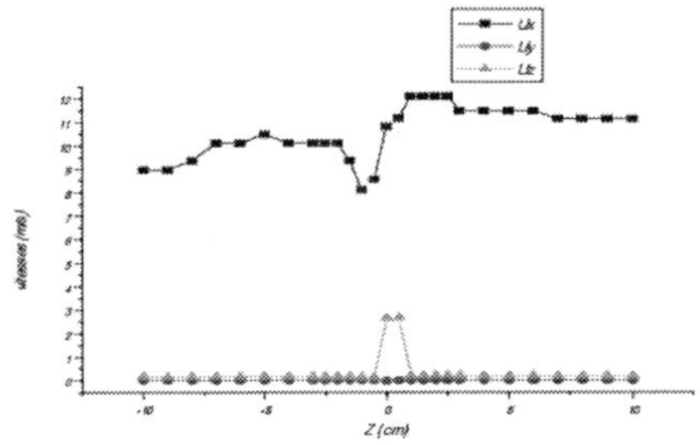

Fig.15 the Distribution of speed for x= 2 cm an y=0 cm pour turbulence with tow grid a x_1= 340mm and x_2= 525 mm

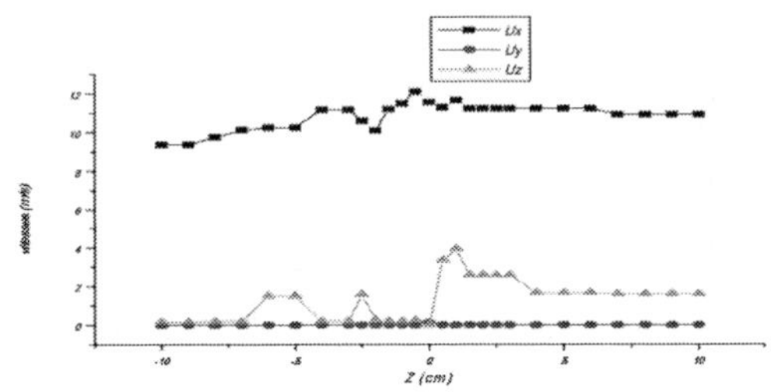

Fig16 The **Distribution of speed in a x= 4 cm an y=0 cm with tow grids at in x_1= 340mm and x_2= 525 mm**

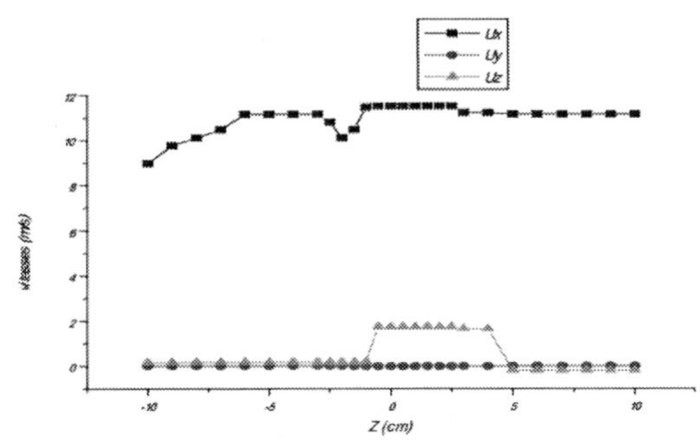

Fig.17 The Distribution of speed for x= 6 cm and y=0 cm for tow grid's in x_1= 340mm and x_2= 525 mm

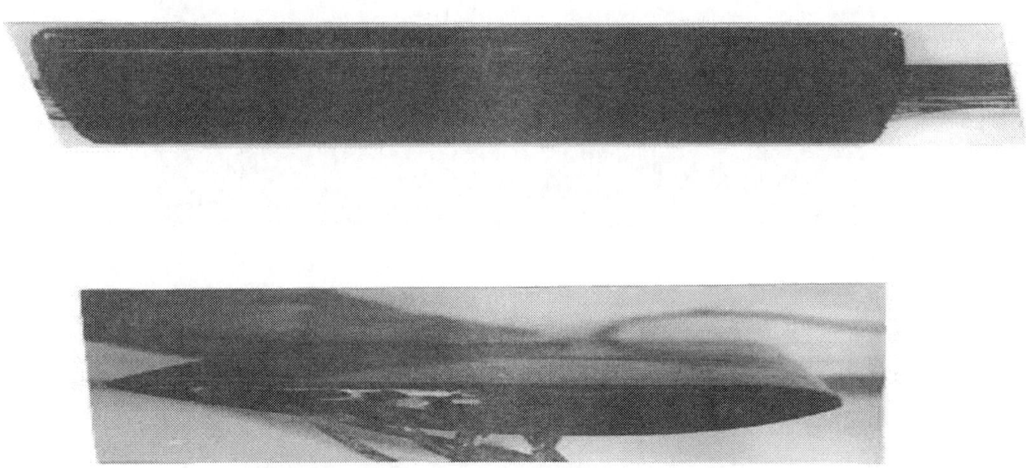

photographs.1 . Profile of hydrofoil carried out

Design Optimization in High-Performance Boats

Antonio Del Prete, Molly Hong

1.0 Altair® OptiStruct® and Optimization

Altair OptiStruct is a highly advanced finite element based software for both structural analysis and design optimization. OptiStruct is used to design, evaluate and improve performance of mechanical structures. Unlike classical finite element solvers, OptiStruct software proposes new and better solutions throughout the entire design process. Powerful optimization routines are deeply integrated with popular analysis types including linear statics, buckling and frequency analysis - making design optimization easy, robust and accurate.

OptiStruct is tighly integrated into the HyperWorks environment. Thus, users can perform complete problem set-up, submission and post-processing quickly and easily.

OptiStruct benefits the design process by:

- Cut development time and costs by providing clear design direction early in the design cycle
- Develop competing design themes for alternative methods of manufacture
- Create unique, high performance designs that satisfy required performance criteria
- Lead the industry with the ability to solve the largest and most complex optimization problems
- Enable innovation - increasing product competitiveness
- Generate reliable and highly accurate results - industry standard technology

1.1 Topology Optimization

This award-winning design synthesis technology uses the topology optimization approach to generate innovative concept design proposals.

In the initial phase of the development process, the user enters package space information, design targets and manufacturing process parameters. OptiStruct generates a design proposal which is optimized with respect to the given design targets.

The results of a topology optimization can be imported into a CAD system via IGES surfaces, generated by OSSmooth, a utility included in the OptiStruct package.

1.2 Topography Optimization

This novel technology allows the design of stamped beads in thinwalled structures. The problem setup requires only the definition of the design region, the maximum bead depth and the bead draw angle.

OptiStruct's topography optimization technology will generate an innovative design proposal for the ideal bead pattern of reinforcement.

Also, the results of a topography optimization can be imported into a CAD system using geometry generated with OSSmooth.

1.3 Shape Optimization

Shape optimization is a methodology to optimize the shape of existing components. For example, local stress can be reduced in critical areas by modifying the contour of the structure. OptiStruct uses HyperMesh's morphing technology to prepare finite element meshes for optimization. As a result, dramatic shape changes are possible without mesh distortion. OptiStruct can easily propose design modifications without underlying CAD data and with minimum user interaction.

1.4 Size Optimization

Size optimization defines ideal component parameters such as material values, cross section dimensions and thicknesses. Within the OptiStruct environment, optimization parameters can be defined with only a few mouse clicks.

1.5 Finite Element Analysis

Altair OptiStruct's analysis module uses the latest element formulations available and a fast, robust sparse matrix solver for linear static, frequency, buckling or simple contact problems. With its large spectrum of solutions, material models and element types, OptiStruct covers the majority of all analysis activities performed by today's manufacturers.

OptiStruct's high accuracy has been proven by numerous industry standard comparisons such as "Standard NAFEMS Benchmarks".

2.0 Optimization Applications in High Performance Recreational Boats

Altair has been working with many different companies in the marine vessel industry like, Sea Ray, Swath, Group Finot, to name a few. Altair has also successfully done a series of America's Cup race boat projects.

Pual Cayard, skipper and CEO of AmericaOne, Says, "Technology plays an increasingly important role in the America's Cup design and optimization process, particularly the structural design and performance of the boats. Altair Engineering's optimization services were readily and seamlessly integrated to our design environment and provided our team with some unique insights for improving the structural efficiency of the hull, resulting in a stiffer, lighter and faster boat."

"OptiStruct," notes Bruce Nelson, chief designer of the AmericaOne, "saved us countless hours of investigating nonoptimal solutions and focused our engineers on the best distribution of material. It also allowed us to develop an extremely optimized, lightweight and rigid structural framework"

Following is an example of optimization application on a racing sailboat—the FILA.

Sailboat racing requires safe, fast, easy-to-handle boats. Just ask Jean-Marie Finot, renowned for building such high-performance vessels. One such sailboat, the FILA, typifies advancement made in boat design: this 60-ft, open yacht futures the use of new materials and components refined through the use of computer simulation.

OptiStruct enabled engineers to redesign the keel of the FILA sailboat. The optimized keel, using extra high-strength and ultra-high-strength steel, was 100kg lighter than the original.

2.1 Mission

In a sailboat the keel is perhaps the most stressed part of the boat when subjected to a wide range of differing loading conditions, e.g. fatigue loading, dynamic loading, impulsive loading. You can therefore imagine that on a racing sailboat like FILA (Fig.1), which is expected to perform in the most extreme conditions with wind speeds of 50 knots or more (in 1998 FILA rolled over in the Atlantic Ocean and dismasted during a storm with winds up to 80 knots) the stress on the keel is extremely severe.

In addition FILA's keel is a canting keel i.e. it is used to increase the torque reaction against the wind's heeling effect (Fig. 2).

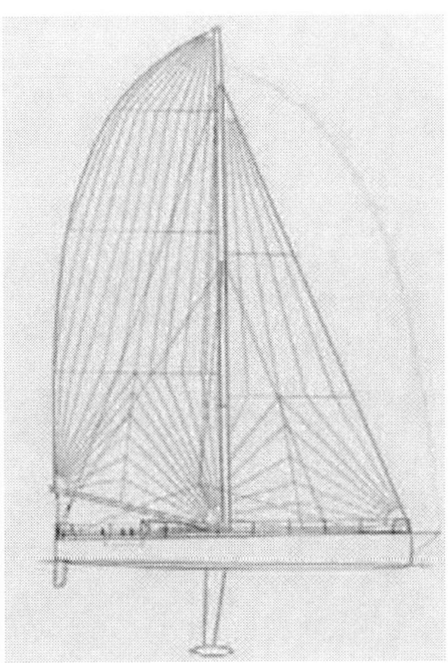

Fig. 1

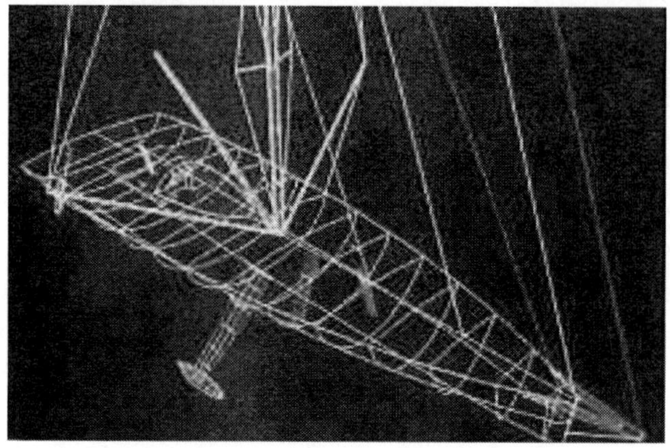

Fig. 2

In order to fully optimize this component, a number of targets had to be taken into account, namely:

• Weight reduction (lighter structure)
• The lowest Centre of Gravity position
• The stiffest structure in the considered loading conditions for both the original bulb and for a redesigned bulb (the latter with a remarkable mass increase).

2.2 Project Development and Strategy

The starting point for the project development was the original keel (Fig. 3), which had proven its structural stability in its numerous voy-ages.

Fig. 3 Old Keel

Starting with the CAD data for the original keel, a finite element model was created using Altair HyperMesh (Fig. 4).

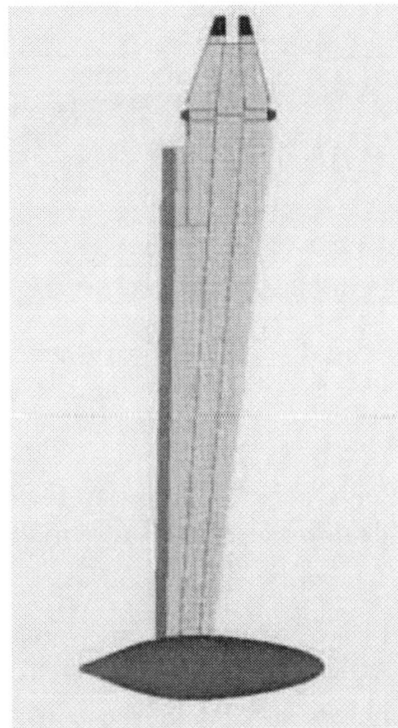

Fig. 4 Old Keel FEM Model

This model was used as a reference, comparing its numerical results with the real response of the structure (Table 1). The load-case considered was the deflection due to the keel's own weight, with the structure in a horizontal position and with the lower end, where the bulb is attached, free to bend (Fig. 5). The exercise showed the accuracy of the discretization used.

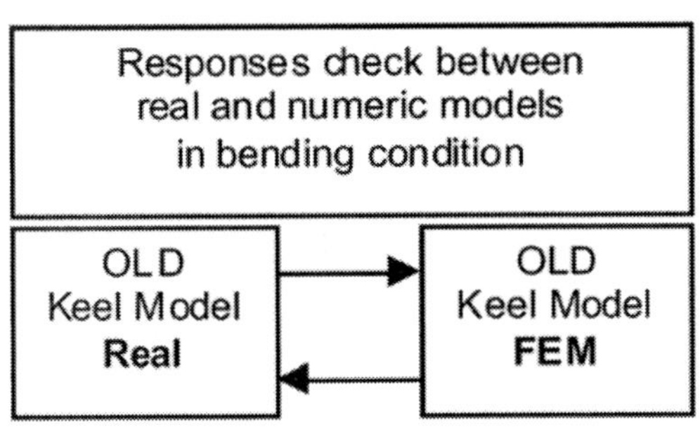

Table 1

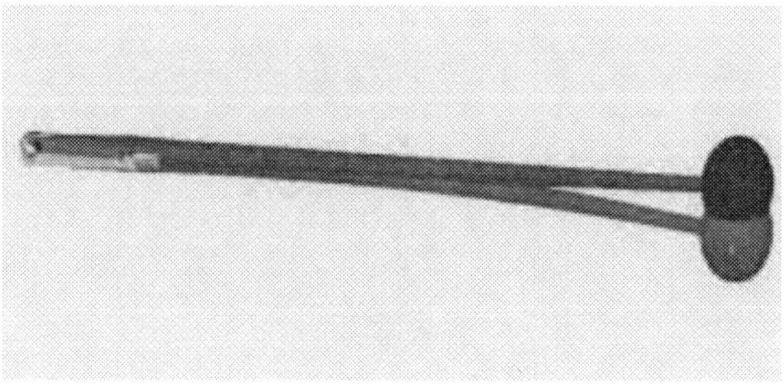

Fig. 5 Lower end free to bend

Once the accuracy of the FE model was verified, a workflow plan for the optimization was developed, considering possible design variables and which boundary conditions to take into account:

a) Materials played a strategic role. New ultra high strength steels, produced by SSAB Swedish Steel, allowed the designers to consider use of materials whose minimum yield strength was between 700Mpa to 1100Mpa, but still maintained a great deal of manufacturability (Table 2).

UHS Steel used (Unit of Measure: Mpa)		
Type	Yield Strength	Ultimate Strength
Docol 1000	800	1100
Domex 700MC	750	850
Weldox 1100	1100	1300

Table 2

b) Since the external profile of the keel had to remain the same, the possible design space available was the volume enclosed by the keel's outer skin. Therefore the topology optimization technique was the only applicable approach (Table 3).

c) Construction technology determined that, for the materials chosen, a combination of operations such as milling, bending and welding was only possible for thicknesses of 2mm to 12mm.

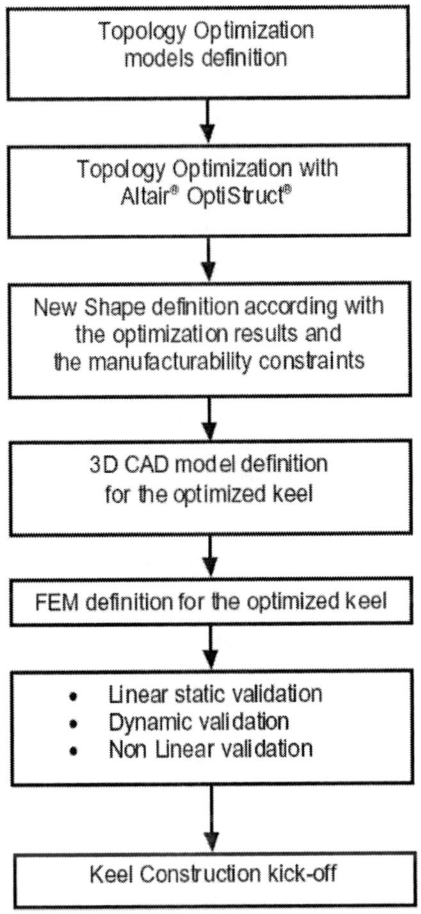

Table 3 Topology Optimization and Validation

2.3 Topology Optimization

The load-cases considered in this phase of the project were the following:

- Bending loading conditions;
- Frontal impact loading conditions;
- Torsional loading conditions;
- First frequency value.

In reality the keel will see a combination of these load-cases, therefore generous safety coefficients were used in order to follow a "Safe Life" philosophy. Due to the behavior of the structure the optimization of the model was split into two parts: the first involves the major part of the keel below its rotation axis and is named "FINN", the second involves the part of the keel above the rotation axis and is named "HEAD" (Fig.6).

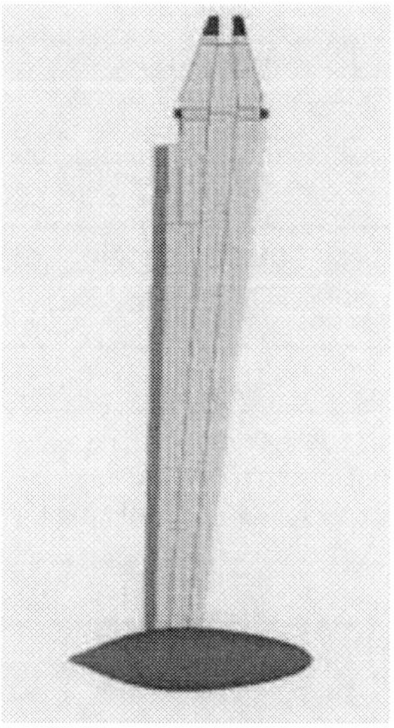

Fig. 6 Keel Composition

For the FINN optimization two models were created, one with the same skin thickness as the original reference model (Fig. 7) and a second with the skin thickness reduced by 50% - allowing an increased design space for the optimizer (Fig. 8). The objective for both models was to minimize mass whilst meeting certain stiffness constraints. The results of the optimization were translated into 3D CAD data by using OSSmooth, a utility included in the OptiStruct package.

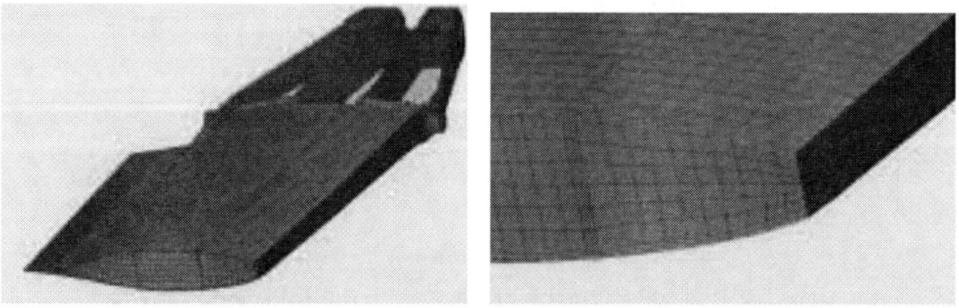

Fig. 7 First model for the FINN optimization (complete and detail)

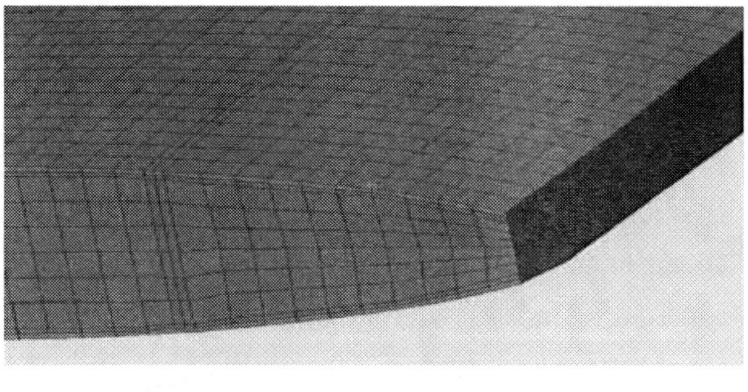

Fig. 8 Second model for the FINN optimization

In order to establish a possible manufacturable layout for the internal structure of the FINN, the old sections of the keel (Fig. 9 in blue) were compared with the ones suggested by OptiStruct (Fig. 9 in red) and a new concept for the optimized design emerged, which took into account the commercial range of steel blanks available for the chosen materials (Fig. 10).

The new design for the FINN is not only advantageous from a structural point of view, but also with regard to construction and durability. The new design removes the need for a horizontal weldline behind the rotation axis of the keel.

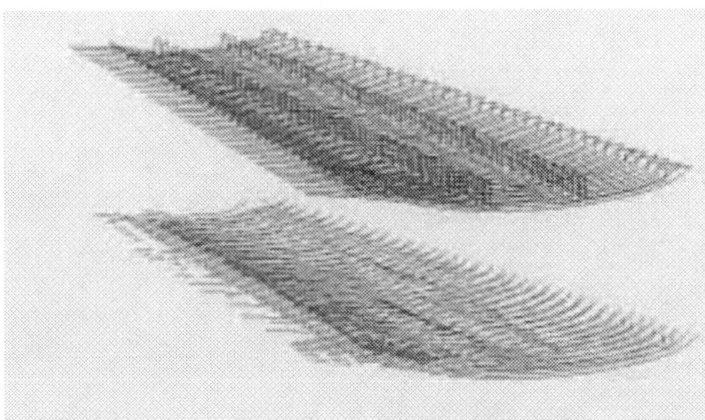

Fig. 9 Sections comparison

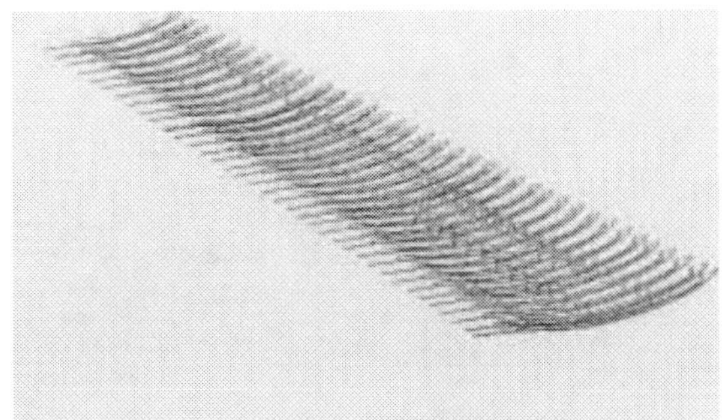

Fig. 10 Optimized section

Unlike the FINN, the HEAD is out of the water and does not have any hydrodynamic constraints, therefore there was greater scope to investigate new shapes in this optimization.

Comparing the reference model of this part (Fig. 11) with the model used for optimization (Fig. 12), one can see that the connection points between the keel and the hydraulic pistons that rotate it were not included in the design space (grayed out).

As with the FINN, manufacturability requirements were considered before a first concept was realized for the new HEAD design (Fig. 13 and 14). Finally the connection between the FINN and HEAD parts was investigated for possible improvements.

Fig. 11 Original HEAD

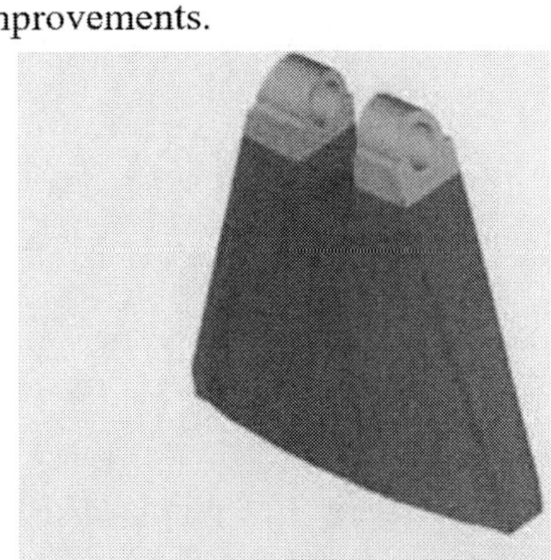

Fig. 12 Model for the HEAD optimization

After checking for manufacturability, the optimized designs for both parts were reviewed in a 3D CAD system in order to produce a full optimized CAD model in accordance with the workflow illustrated above.

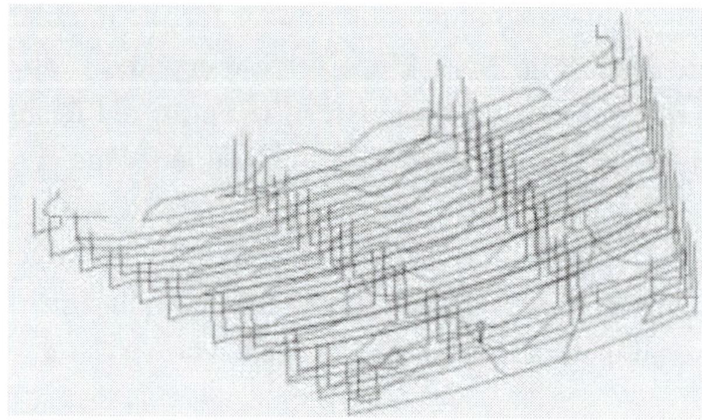

Fig. 13 HEAD sections comparison

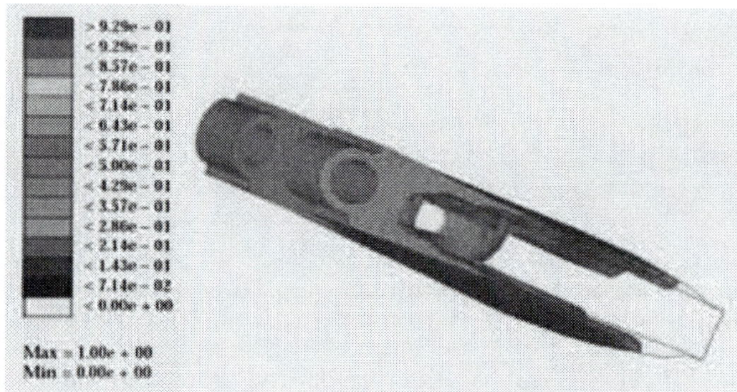

Fig. 14 HEAD optimized topology: iso view

From the optimized CAD data a new FE model was created. The model was then analyzed under the 4 load-cases referred to above. Linear static analyses were performed to simulate the working conditions for the keel (bending, torsion) as well as accidental conditions (frontal impact - Fig. 15).

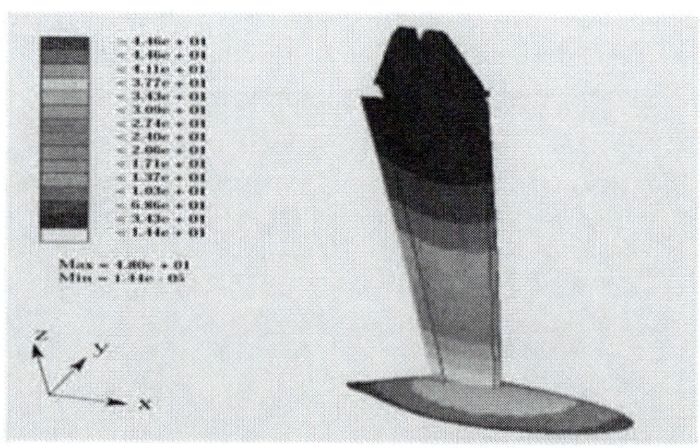

Fig. 15 Optimized keel in frontal impact conditions

Dynamic analyses were also performed in order to investigate modal behavior of the structure and to determine its natural frequency with both the old and new bulbs attached. Non-linear static analyses were performed in order to investigate the sections stability in extreme conditions like the one represented by the mixed torsional-bending load of 5g, which the keel is supposed to experience on the downhill section of big waves at boat speeds of over 25 knots.

2.4 Conclusions

The final optimized design satisfied all the mission profiles. A weight saving of 15% was achieved, whilst lowering the keel's center of gravity by 6%. These results were achieved even with the bulb mass increasing by 11.5%.

The Structural Design Approach for CCS "Guidelines of Small Waterplane Area Twin Hull Craft"

Hong Ying

China Classification Society
Shanghai Rules and Research Institute
Shanghai 200135

Abstract: Including the newest technical standards for both high speed and low speed SWATH with large scale, and adopting the state of the art of ship structural design techniques, CCS latest "Guidelines of Small Waterplane Area Twin Hull Craft" has been established with many new concepts of structural rules based on load criteria system. The new features in this guidelines, such as the solution of transverse bending moment, which is critical for SWATH, direct calculation practice including the FEM methods both for global ship model and local fine mesh, as well as hot spot based fatigue evaluation, etc are integrated and is the best solution for SWATH structural design in the world.

Keywords: **Swath; load criteria; transverse bending moment; scantling; direct calculation**

Supported by China Ship Scientific Research Center and other Marine Institutes, China Classification Society published the latest "Guidelines of Small Waterplane Area Twin Hull Craft" in last year.

The new features for the structural approach in this guideline are:

1. Load criteria based ship structural strength principle forming the new rule system
2. Special design consideration and construction requirement for critical areas of structural details such as the connection between the cross structure and adjacent parts.
3. The formulae of wave induced loads are calibrated based on serially model tests, and the load combinations are proposed.
4. The modern new techniques, such as direct calculation of ship structural and relevant strength criteria of yielding, buckling and fatigue are introduced.

1. New Rule System

This guideline has been established with load criteria based ship structural strength principle forming the new rule system. The load condition and combination should be determined at first, then for the elementary member, prescriptive requirement is given for determine the initial scantling of the member, therefore adopt the direct calculation to verify the strength effect using the requirement of yielding, buckling and fatigue criteria. These concepts and procedures are quite new for the SWATH technical standards and called as "rational" structural rules.

2. Design Loads

In the guidelines, the design loads are generally given as global loads and local loads. For SWATH global loads, the transverse bending moment and torsion moment in horizontal plane are more critical than others. For local loads, besides the general requirements according to the Ship Rules, such as the requirements for deep tank, the special loads, e.g. the wave induced slamming pressure and structural responses are also for special consideration and should be checked.

Generally, when using global FEM analysis to investigate the overall strength of ship structure including the cross structure, it needs to be taken into the global loads only. The local loads are especially for the determination of member initial scantling. The load characteristics of SWATH are described below:

1) The global loads

Theoretically, the wave induced loads are obtained from the hydrodynamic studies for irregular type of vessel, and then verified by means of model tests. But actually, since the hydrodynamic studies and model tests are very sophisticated and exhausted for some design companies, the simplified load evaluation formulae are given in this guidelines, the transverse bending moment and torsion moment are listed at the top for significances. The global loads are:

(1) Transverse bending moment

The transverse bending moment is most critical load for SWATH, see Figure 1,and determines the scantling of cross structure and its connection. The transverse split force is relatively for draught, displacement and the length of struts. For low speed SWATH, the determination for transverse bending moment is based on 17 serially model tests ranging from 200 to 3500 tones and calibrated considering the effect of sensitive scaling factors for accuracy improvement (more close to the result of model tests) and simplify. For low speed SWATH, the determination for transverse bending moment is based on hydrodynamic studies and some design experience. The transverse split forces are shown as follows:

(a) apply to low speed SWATH:

$$F_y = \pm 9.81 D_F T L_F \Delta \qquad kN$$

where:

$$D_F = 3.24 - 0.55\log_{10}\Delta ;$$
$$T = 1.754d/\Delta^{1/3}$$
$$L_F = 0.75 + 0.35\tanh(1.65l_s/\Delta^{1/3} - 6.0)$$

(b) apply to high speed SWATH:

$$F_y = 57C_1 d\Delta^3 \alpha_1 L_F \qquad kN \qquad \text{for } L \le 50\text{m}$$

where:

C_1= factor of sea state

$$\alpha_1 = 1.55 - 0.75\tanh(\Delta/11000)$$

(2) Horizontal torsion moment

The horizontal torsion moment is generated due to quartering sea or hull form difference along the ship length, and act as "yaw split". The load distribution along the waterlines are specified as triangle or trapezia shown in Figure 2. This specification can consider the combination of transverse bending moment and horizontal torsion moment.

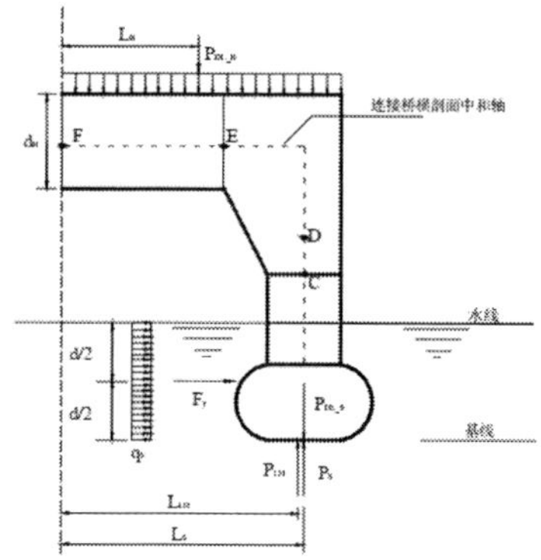

Figure 1

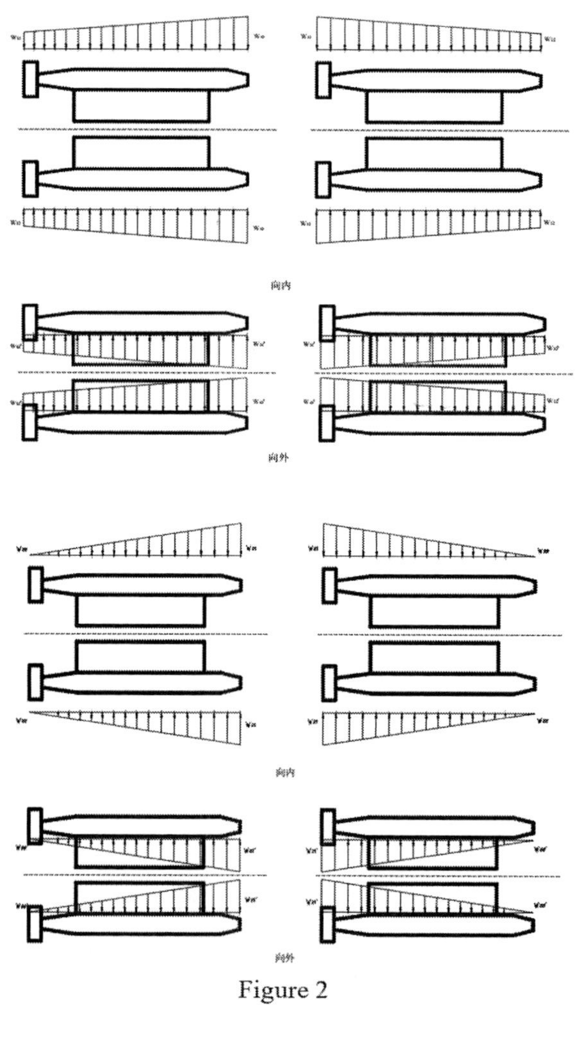

Figure 2

(3) un-synchronous pitch moment

The un-synchronous pitch moment is induced by un-synchronous pitch movement between two lower hulls. The value is taken as the greater of the following equations, see Figure 3：

$$M_{P1} = 0.125\Delta a_{cg} l_{lh}$$
$$M_{P2} = 0.25\Delta a_{cg} b$$

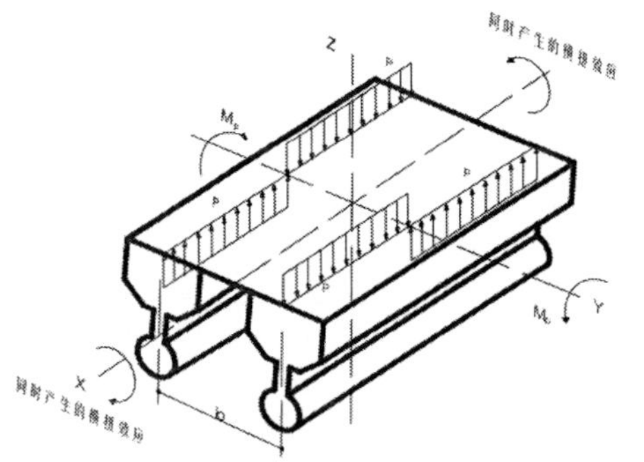

Figure 3

(4) Vertical shear force at centerline

The vertical shear force is to represent the un-synchronous heave movement between two lower hulls or the weight difference from the superstructure including the inertial effects, see Figure 4. In FE model, the vertical shear force can be applied at the bottom of the lower hull for the purpose of investigating the strength of longitudinal section at centerline of the ship.

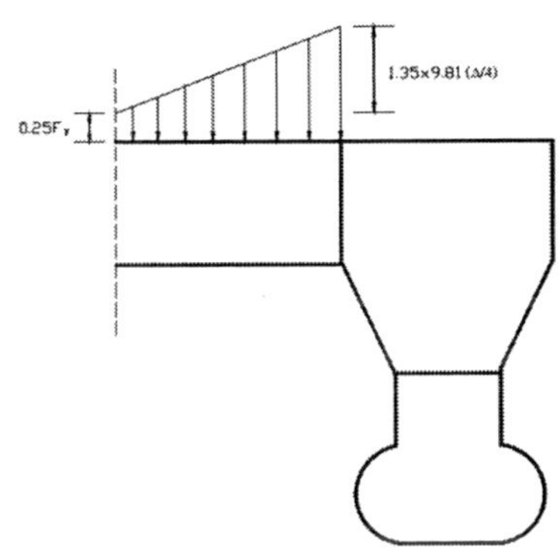

Figure 4

(5) Longitudinal bending moment

For lower speed SWATH, the longitudinal bending moment is assumed as follows:

Wave induced bending moment: $M_W = 0.082(C_{WP}/0.75)^{2.5} l_s^{2.5} B_s$

Still bending moment: (approximately): $0.2M_W$

2) Load combination

The load combination should cover the maximum load acting onto the ship at each wave attack angle, e.g. **90°** for maximum transverse bending moment and vertical shear force, **0°/180°** for longitudinal bending moment and **45°/135°** for severe torsion combination condition, etc.

According to the above, 11 load combinations, including 5 for checking transverse strength, 4 for torsion moment and 2 for longitudinal bending moment, are given in the guidelines applying in the global FE model analysis.

3) Slamming impact

For high speed SWATH, the local loads including the slamming pressure are taken complied with the requirements similar as the general high speed craft in the relevant CCS high speed craft rules. Except above, for the low speed SWATH, the index n in equations of slamming pressure are specified more accurately based on 10^8 exceed probability(20 years lifetime) shown as:

$$P_{max} = nL_P\ (L_b)^{1.08} \qquad kN/m^2$$

where: $n=6.78$

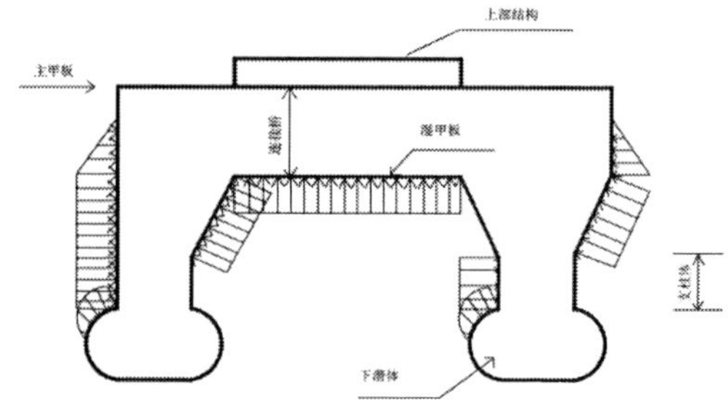

Figure 5

3. Scantling

The requirements for member scantlings e.g. cross area, section module, aspect ratio and minimum plate thickness, length, etc., are corresponding to the local strength requirements for resist-bending/shear and buckling, corrosion, fabrication and inspection, and sound engineering practices. Therefore the requirements for member scantlings in the rule should be considered as the "deadline" of the minimum requirements. In general, during the design stage, the initial scantling can be determined based on "scantling requirements", and then perform the strength verification using another ways, such as direct calculation.

For the scantling, the called "prescriptive" requirements are given, but where design head in the load calculation are relevant to the above "design load", e.g. the slamming pressure and deck local pressure, etc. should be taken according to the requirements of "design load".

4. direct calculation

The main points of direct calculation are:

(a) Determination for application of global and local analysis and relevant requirements

(b) Determination for FE modeling, including the element, mesh size and boundary conditions. The fine meshing method is introduced the newest concepts in IACS CSR for Tankers

(c) Determination for allowable stresses, especially for fine mesh analysis, which is introduced in IACS CSR for Tankers.

(d) Determination for buckling requirements applying to direct calculation.

The Figure 6 shows the boundary conditions applying to the analysis of globe longitudinal/transverse bending moments and torsion moment, longitudinal pitch moment and transverse bending moment, as well as vertical shear force, respectively.

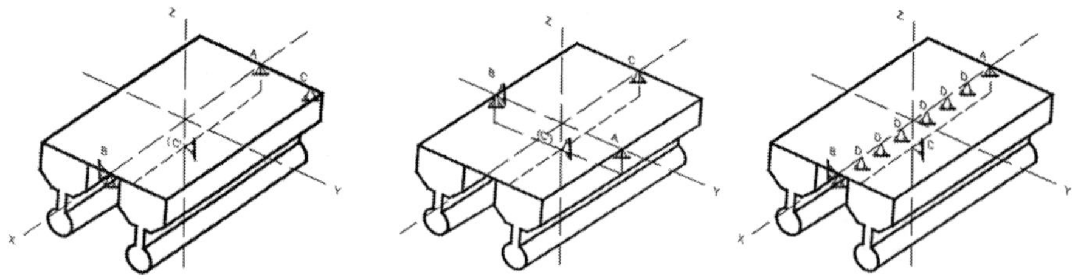

Figure 6

5. Fatigue assessment

Based on the real cases that the crack damages are easily found within the area of the high stress level on the cross structure, the fatigue requirements are given in this guidelines according to the structural responses of SWATH, and critical areas including the connection of stiffeners through the bulkhead and the haunch structure to the strut and to the wet deck, etc. The typical critical areas are shown in Figure 7.

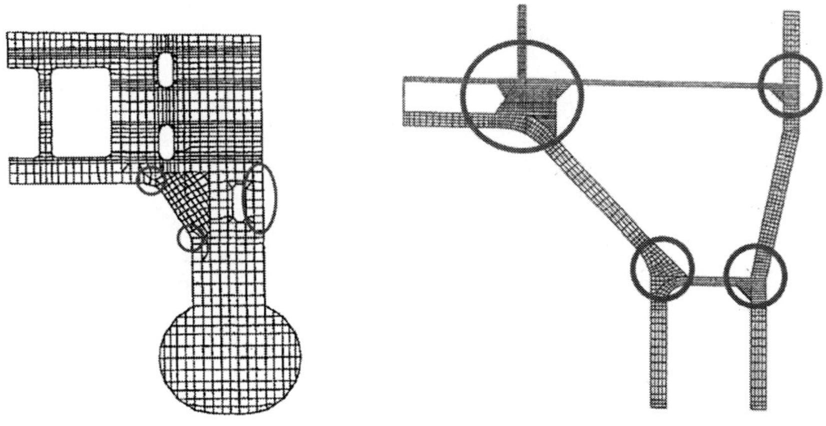

Figure 7

The stress range of long term statistics can be taken the bi-parameter Weibull distribution, with the form parameter $\xi=1$.

The damage cumulation D is taken as:

$$D=\sum_{i=1}^{3} D_i \quad (i=1,2,3 \text{ for wave direction of heading, beam and quartering sea, respectively})$$

where:
$$D_i = \frac{0.6\alpha_i \mu K_1 S_{L_i}^3}{18.42^{3/\xi} f_1 K} \times 10^8$$

in addition, **S-N** curve can be taken from the CSS "Guidelines of fatigue assessment of ship structure".

Study on Synthesis Optimization of Speed and Seakeeping PerformanceBased on On-Line Simulation of High-speed Single-hull Ship

Wang Yi， Yang Song-lin
(School of Naval Architecture and Ocean Engineering,
Jiangsu University of Science and Technology,
Zhenjiang , Jiangsu 212003)

ABSTRACT: In this paper, the mathematic model of high-speed single-hull ship's propulsion system (adjustable-pitch screw propulsion) is firstly established. Then, based on MATLAB, the simulation model of the propulsion system is built with the help of simulation software—Simulink. And the intelligent control on the propulsion system is carried out by using fuzzy controller of Look-up Table.

After that, the mathematic model of synthesis optimization of high-speed single-hull ship is established by considering control, speed and seakeeping performances. Based on forementioned established simulation model and applying GAOT— MATLAB's optimization toolbox for genetic algorithm (GA), the authors carry out the optimization design of the system.

At last, the authors have given an example and validated the accuracy and reliability of the simulation and optimization models.

KEYWORDS: high-speed single-hull ship; modeling and simulating; speed performance; seakeeping performance; control performance; synthesis optimization

0 INTRODUCTION

The navigation performances of a ship mainly contain speediness, maneuverability and seakeeping. The synthesis optimization of the ship navigation performances is a complex problem that has many objectives, variables and constraint conditions. It has relationship with many factors. With the help of the definition of speediness we can see that it depends on resistance and propulsion performances, and resistance performance relates ship shape parameters, velocity and environment; while propulsion performance relates engine rotational speed, propeller parameters, ship shape parameters and environment.

Maneuverability mainly depends on shape and rudder parameters of the ship. Navigation environment also partly affects maneuverability. Seakeeping mainly depends on sea conditions, ship shape parameters, velocity, course and etc. In this paper, the authors do study on synthesis optimization of speed and seakeeping performance based on on-line simulation of high-speed single-hull ship on condition that the shape parameters and principal dimensions of the ship have been fixed and the sea conditions has been appointed.

1 MATHEMATC MODEL OF ADJUSTABLE-PITCH SCREW PROPULSION

1.1 Resistance

The resistance of single-hull ship consists of frictional and rest resistance. Using Mother Type Ship method, we adopt **Maxsurf** to calculate the resistance R_T. We find the relationship between R_T and velocity so that each resistance corresponding to a certain velocity can be gotten.

1.2 Adjustable-pitch screw propulsion system

Thrust of propeller $T = K_T \rho n^2 D^4$; Torque of propeller $Q = K_Q \rho n^2 D^5$;

Propeller efficiency in the open $\eta_0 = K_T J / 2\pi K_Q$; Effective thrust $T_p = T(1-t)$;

Entrance velocity $V_A = (1-\omega)V$; Advance coefficient $J = V_A / nD$

where K_T is the thrust coefficient of propeller, K_Q is the torque coefficient of propeller, w is the wake fraction of propeller, ρ is the water density, D is the diameter of propeller, n is the rotational speed of propeller, t is the thrust deduction fraction.

Use the following expressions to receive K_T and K_Q:

$$K_T = \sum_{i=0}^{n1} \sum_{j=0}^{n2} \sum_{k=0}^{n3} A_{ijk} (P/D)^i (J)^j (A_E/A_O)^k$$

$$10K_Q = \sum_{i=0}^{n1} \sum_{j=0}^{n2} \sum_{k=0}^{n3} B_{ijk} (P/D)^i (J)^j (A_E/A_O)^k$$

where, A_{ijk} and B_{ijk} are the regression coefficients of the propeller.

2 MATHEMATIC MODEL OF SIMULATION

The model of simulation is built based on two dynamic equations: transplant movement and rotational movement.

2.1 Transplant movement dynamic equation of propulsion system

$$m \frac{dV_S}{dt} = T - R_T$$

where, m is the mass of the ship (including accessional mass) .

2.2 Rotational movement dynamic equation of propulsion system

$$2\pi I\frac{dn}{dt}=M_d-M_c$$

where, M_d is the torque received from engine, M_c is the hydrodynamic torque of the propeller and I is moment of inertia of the rotational part (including accessional inertia moment).

3 LOOK-UP-TABLE FUZZY CONTROLLER

In this paper, the authors use 2-D fuzzy controller and its inputs are velocity gap E and acceleration ΔE, output is control value U. For adjustable screw propulsion which employs two fuzzy controllers respectively for n and P, U respectively represents the adjustive amount of n and P.

$$E=V_O-V_S$$

$$\Delta E=\frac{\mathrm{d}V_S}{\mathrm{dt}}$$

These two fuzzy controls are realized by the form of look-up-table. The table is given as Table 1.

Tab.1 Fuzzy rules of the ship propulsion system control

ΔE	U						
	E=-1.2	E=-0.8	E=-0.4	E=0	E=0.4	E=0.8	E=1.2
-1.2	12	10	8	6	4	2	0
-0.8	10	8	6	4	2	0	-2
-0.4	8	6	4	2	0	-2	-4
0	6	4	2	0	-2	-4	-6
0.4	4	2	0	-2	-4	-6	-8
0.8	2	0	-2	-4	-6	-8	-10
1.2	0	-2	-4	-6	-8	-10	-12

4 MATHEMATIC MODEL OF OPTIMIZATION

4.1 Design variables

Nine parameters are selected as design variables: objective velocity V_O, D, A_E/A_O, three coefficients of the fuzzy controller for n : K_1, K_2, K_3, three coefficients of the fuzzy controller for P: K_4, K_5, K_6. For the sake of simplicity, we use a vector X to express these nine design variables:

$$X^{\mathrm{T}}=\{V_O,\ D,\ A_E/A_O,\ K_1,\ K_2,\ K_3,\ K_4,\ K_5,\ K_6\}$$

4.2 Objective function

4.2.1 Speed

We define a criteria of speed that is similar to Admiralty Coefficient. Its expression is:

$$C_{sp}=P_E\Delta^{-2/3}V_S^{-3}/(\eta_0\eta_H\eta_R\eta_S)=R_T\Delta^{-2/3}V_S^{-2}/(\eta_0\eta_H\eta_R\eta_S)$$

where, Δ is the displacement of the ship, P_E is the effective power,η_0 is propeller efficiency in the open,η_H is the hull efficiency;η_R is the relative rotation efficiency andη_S is transmission efficiency of shafting.

We choose speed objective function in this form: integrate speed criteria in the time domain and then get its mean value. The expression is

$$F_1(X)=\frac{1}{T}\int_0^T C_{sp}dt$$

where, T is the simulation time span. As far as this sub-objective function is concerned, its minimal value is wanted.

4.2.2 Seakeeping Performance

In the paper, the authors choose effective pitching L_{ra} and heaving S_{rs} to be the index of ship seakeeping performance. $S_v = 0.5\left|L_{ra(0-1)}\right| + 0.5\left|S_{rs(0-1)}\right|$, where (0-1) means each of the index must be transformed to the bounds of 0 and 1. S_V is the function about design variables X and environment variables (wind, wave, flow and etc.).In the paper, the sea conditions is appointed. For this is a whole-voyage optimization problem, we choose seakeeping objective function in this form:

$$F_2(X)=\frac{1}{T}\int_0^T S_V dt=\frac{1}{2T}\int_0^T\left(\left|L_{ra(0-1)}\right|+\left|S_{rs(0-1)}\right|\right)dt$$

About the calculation of L_{ra} and S_{rs} :

Effective pitching value means the pitching angle of the ship, its unit is degree (°). The expression of L_{ra} is:

$$L_{ra}=A_0+A_1C_{wp}+A_2C_B+A_3\frac{L}{B}+A_4\frac{L}{T}+A_5X_b+A_6\frac{K_{yy}}{L}+A_7\frac{0.552V_S}{\sqrt{L}}+A_8\left(\frac{0.552V_S}{\sqrt{L}}\right)^2$$

where K_{yy} is the longitudinal inertia radius, and $K_{yy}=\sqrt{0.07C_{wp}L^2}$, V_S is velocity of the ship and its unit is kn. Coefficients $A_0 \sim A_8$ can be received with the help of **Tab.2**

Tab. 2

Wind level	L_{BP}(m)	A_0	A_1	A_2	A_3	A_4	A_5	A_6	A_7	A_8
5	122	3.65	-16.19	9.10	0.222	-0.0510	-0.069	14.8	7.60	-5.13
	152	3.96	-12.39	6.97	0.203	-0.0329	-0.068	5.0	5.76	-4.24
	183	3.38	-8.21	4.43	0.165	-0.0167	-0.057	0.2	3.76	-2.9
	244	2.03	-3.59	1.86	0.078	-0.0037	-0.032	-1.7	1.40	-1.28
	305	1.17	-1.61	0.85	0.038	-0.0001	-0.018	-1.7	0.40	-0.47

Effective heaving value means the range when the ship does heaving motion in the waves. And its unit is meter (m).

$$S_{rs}=0.3048\times\left[A_0+A_1C_{wp}+A_2C_B+A_3\frac{L}{B}+A_4\left(\frac{L}{T}\right)^{-1}+A_5X_b+A_6\frac{K_{yy}}{L}+A_7\frac{V_S}{\sqrt{L}}\right]$$

where K_{yy} is the longitudinal inertia radius, and $K_{yy} = \sqrt{0.07C_{wp}L^2}$, V_S is velocity of the ship and its unit is kn. Coefficients $A_0 \sim A_7$ can be received with the help of **Tab.3**

Tab. 3

Wind leve	L_{BP}(m)	A_0	A_1	A_2	A_3	A_4	A_5	A_6	A_7
	122	-0.18	-13.30	4.62	0.10	78.6	0.205	23.5	3.83
	152	0.9	-12.54	3.63	0.175	65.4	0.204	19.2	2.50
5	183	2.06	-9.66	2.01	0.191	41.8	0.163	13.6	1.25
	244	2.38	-5.29	0.63	0.146	13.6	0.083	6.5	0.02
	305	2.18	-2.67	0.04	0.097	0.3	0.035	2.4	-0.49

We can see that when the principal dimensions and major parameters of the ship are certain, the ship seakeeping is the function of sea conditions and velocity.

4.2.3 Quality of control

We want to get the minimal control overshoot and the fastest speed stabilization time, so the sub-objective function of control quality is introduced as:

$$J = F_3(X) = \frac{1}{T}\int_0^T |V_O - V_S| dt$$

The integrand reflects the gap between V_O and the real-time velocity V_S.

4.2.4 General objective function

Considering genetic algorithm (GA) is chosen to be the optimization method in this paper and GA is to get the maximal value of the fitness function, we add a minus sign in front of the general objective function as fitness function when programming the GA. The general objective function is of the following expression:

$$F(X) = aF_1(X)_{01} + bF_2(X)_{01} + cF_3(X)_{01}$$

The sign 01 means each sub-objective function has been put into the bounds between 0 to 1, and $0 \le a,b,c \le 1$, a+b+c=1.

4.3 Constraint conditions

4.3.1 Inequation constraint

(1) Bounds of design variables' value. Design variables' value must be within the lower and upper bounds. These nine variables are: V_O, D, A_E/A_O, K_1, K_2, K_3, K_4, K_5 and K_6.

(2) Cavitation requirement for propeller.

According to Keller formula, the inequation constraint is as follow:

$$\frac{1.3+0.3Z}{(P_0 - P_v)D^2}T - \frac{A_E}{A_0} \le 0$$

where, P_O is the static pressure of propeller shaft center , P_v is the gasification pressure of water at 15℃ and T is the thrust of the propeller.

4.3.2 Equation constraint

(1) buoyancy. The displacement keeps a constant, Δ(displacement)= ρkLBTC$_B$.

(2) Transplant movement dynamic equation of propulsion system

$$m\frac{dV_S}{dt}=T-R_T$$

(3) Rotational movement dynamic equation of propulsion system

$$2\pi I\frac{dn}{dt}=M_d-M_c$$

5 OPTIMIZATION METHOD

In the paper, Genetic algorithm optimization toolbox **GAOT** is chosen to be the optimization method.

The corresponding relationship between **GAOT** and mathematic optimization model is:

Genes: V_O, D, A_E/A_O, Ke_1, Ke_2, Ke_3, A_1, A_2, A_3, which are corresponding to the design variables.

Chromosome: character string of the real number code of vector $\{V_O, D, A_E/A_0, Ke_1, Ke_2, Ke_3, A_1, A_2, A_3\}$.

Population size: when the number of design variables is large, the bigger the population size is, the better, because the probability of falling into local optimization is reduced. However, the size shouldn't be too large, either, or it will waste the computer memory resources.

Generation number: in theory, the bigger the number is, the better; however, saving time is ought to be taken into consideration.

Fitness function: $f(X)=-\left[aF_1(X)_{01}+bF_2(X)_{01}+cF_3(X)_{01}\right]$

In the paper, a=0.35, b=0.35, c=0.30.

6 EXAMPLE

In this paper, a 1000t high-speed single-hull ship is taken for example. Its principal dimensions and major parameters are given in Table 4:

Tab. 4 Principal dimensions and major parameters of a 1000 t high-speed single-hull ship

Length **L**/m	Breadth **B**/m	Draught **T**/m	C_{WP}	X_b/m	C_B
75.2	11.86	2.68	0.71	-1.16	0.407

In the paper, we evaluate w=0.1, t=0.06 and Z = 5.

The bounds of every design variable are given in **Table 5**:

Tab.5 Extent of every design variable

Design variable	V_O/kn	D /*m*	A_E/A_O	K_1	K_2	K_3	K_4	K_5	K_6
Lower bound	30	1.6	0.7	0.1	0.1	0.01	0.005	0.005	0.025
Upper bound	31	1.9	0.9	0.15	0.4	0.012	0.02	0.02	0.04

The propulsion and control system model can be seen in **Fig.1**, and the whole simulation model can be seen in **Fig.2.**

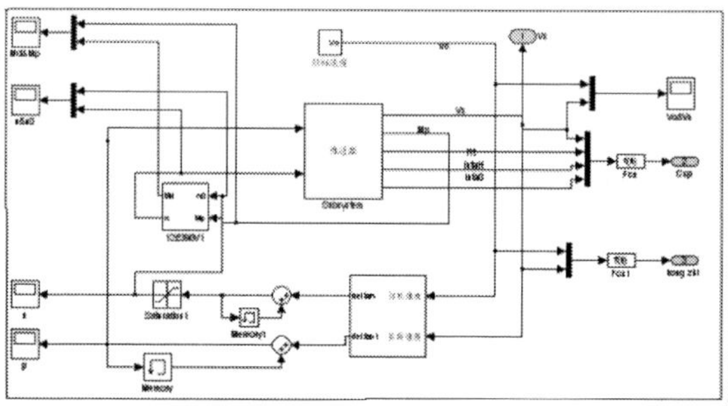

Fig.1 Propulsion and control system model

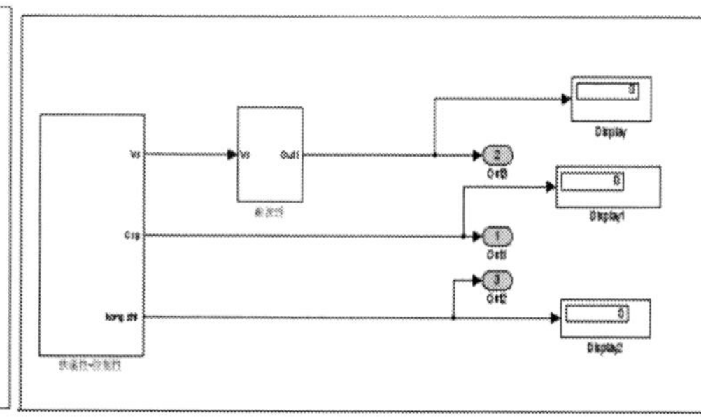

Fig.2 Whole simulation model

The simulation starts at velocity of 28kn and rotational speed of 20r/s. The population size is 20, the generation number is 150 and the simulation time span is 100s. The resistance increases by 5% because of waves.

The result is shown in Table 6:

Tab.6 Optimization result

Design variable	V_O/ kn	D /*m*	A_E/A_O	K_1	K_2	K_3	K_4	K_5	K_6
Result	30.178	1.6094	0.7248	0.1030	0.2348	0.0108	0.0186	0.0102	0.0350

The maximal value of fitness function $f(X)$= -0.7224. **Fig.3** expresses the relationship between fitness function value and generation. We can see that from the 1st to the 80th generation the value of fitness function increases very quickly while after 80th generation it changes evidently gently and becomes a horizontal line.

Fig.4 shows the change of V_O and V_S. From **Fig.4** we can find out that the maximum velocity overshoot is about 1.4kn. After the simulation time T=50s, the real-time velocity Vs moves up and down around the objective velocity Vo. It is because of the waves, and the simulation accords with the factual conditions. Thus, **Fig.4** has testified the rationality of the simulation model.

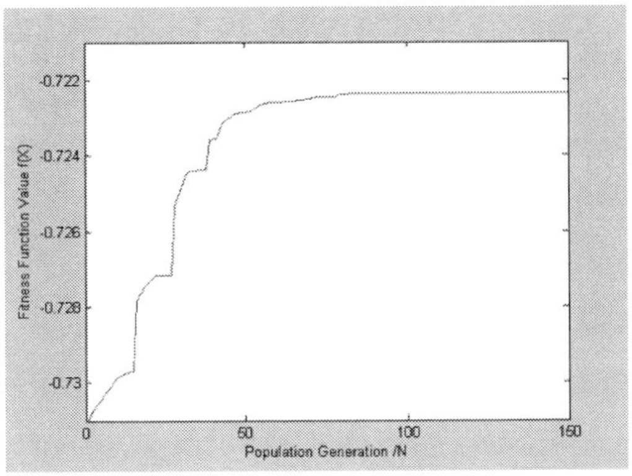

Fig.3 Fitness function value and generation

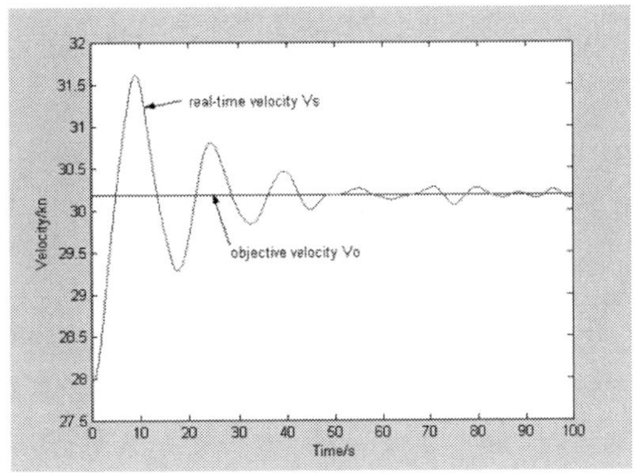

Fig.4 V_O–t & V_S–t curves

7 CONCLUSION

In conclusion, the method of High-speed Single-hull Ship's speed and seakeeping performance synthesis optimization based on on-line simulation has taken resistance, propulsion, speed, seakeeping and control quality into account. Seakeeping performance has close relationship with shape parameters and principal dimensions of the ship. It will have very important value if we make shape parameters and principal dimensions also be design variables and take speed and seakeeping and maneuverability performances into consideration to do overall optimization of ship navigation performances based on on-line simulation.

Authors' Brief Introduction:

Wang Yi (1981-), Male, Postgraduate Student of Jiangsu University of Science and Technology, China.

Yang Song-lin (1956-), Male, Professor of Jiangsu University of Science and Technology, China.

REFERENCES

[1] Yang Song-lin. Method of Engineering Fuzzy and Its Application. Beijing: National Defense Industry Press, 1996

[2] Zhang Zhi-yong. Master MATLAB 6.5 Edition. Beijing: Beihang University Press, 2003

[3] Wang Guo-qiang ,Sheng Zhen-bang. Ship Propulsion. Shanghai: Shanghai Jiaotong University Press, 1993

[4] Shao Shi-ming, Zhao Lian-chang, Zhu Nian-chang. Ship Resistance. Beijing: National Defense Industry Press, 1992

[5] Gao Shang. Optimization Based on MATLAB's Genetic Algorithm Toolbox. Microcomputer Applications, 2002,18(8)

[6] Yang Song-lin, Zhu Ren-qing, Wang Zhi-dong, Zhang Huo-ming. On the Overall Optimization of Speediness and Maneuverability of Large-Sized Medium-Speed Ships. Ship.2003, 10(5)

[7] Wu Xiu-heng. Ship Maneuverability and Seakeeping Performance. Beijing: People's Communication Press, 1999

[8] Zhang Huo-ming, Yang Song-lin, Zhu Ren-qing, Wang Zhi-dong. Fuzzy-genetic Algorithm of Ship Navigation Performance Optimization. Shipbuilding Of China, 2002,43(3)

[9] Tao Yao-sen, Feng Tie-cheng, Ding Hao. A Discussion on Seakeeping Prediction Method in Ship Optimization Design. Journal of Ship Mechanics, 2000, 4(2)

Discussion on high speed boat run up on rocks and its impact force

Chen Guoyu

Shanghai Marine Steel and Structure Research Institute

Abstract: "Boat run up on rocks" means boat and rock collision. This refers to a kind of boat collision with fixed things underwater, and is similar to the crash of a boat with pier. There are many researches at home and abroad on ship colliding with pier. In his summary at 1991, a European author found 150 papers on the subject published in the West, while the number of such special documents in China reached 120 in the year of 2005 alone, showing the great attention paid by China in this field. All these can provide references to those who study the collision of high speed boats with fixed things.

The research of "ship run up on rocks" and collision has been included in engineering mechanics, requiring the application of theoretical analysis, value calculation and test to ensure successes. For instance, a certain paper, containing a formula inconsistent with Newton's basic principle on mechanics, was thus cancelled at an international meeting on "ship collides with pier".

This paper put forwards a formula on the calculation of the impact force of "boat run up on rocks". The formula includes the parameter such as angle, speed, rigidity and qualities of the "impact pair" and may provide help with the designing of the bow of high speed boats.

Key words: "ship run up on rocks", force of ship collide with fixed thing, high speed boat.

1. Preface

1)"Run up on rocks"have a common definition:

"Run up on rocks"have a common definition. That is :'Boat in navigating bump against rock, underwater thing or ice cube etc. to cause ship shell damage, leakage and hull to be stranded or sink accident.'[7]. It also has a professional definition:'run up on rocks': 'Ship is since losing control and the wave influence of gale unsuitable or other reasons the boat bump against the rock.' [6]。The professional definition of 'Grounding' is: 'The phenomenon of the ship draft to be bigger than the depth of water that had been put and can not float [6] '. So ,the boat hull touch the rock in sport, that phenomenon should belong to the 'run up on rocks', this paper from the above-mentioned ordinary concept known , discuss the mechanics problem of run up on rocks concerning high speed ship. So, it has common language with boat professional publication.

According to above-mentioned definition, grounding is usually occur at the parallel middle body of the ship, it occur the ship bottom put on the sand base and mud base, or other sea bed .The method of relieve the grounding is : Wait tide, go astern, turning the rudder, and towage. Run up on rock can not go astern at once. It can leave contact only after analyses the condition of contact. Otherwise, it can brew larger average accident.

On world, some insurance company will be grounding and run up on rock put to manage together to settlement of claims, but when making mechanics analysis, should distinguish.

2) The mode of ship collies fixed things in water [4, 8]

About the mode of ship collies with the fix thing in water, in 1981, there are Japanese have a ship contact bridge pier mode, with 3 kinds in planes: Hull bumps against the pier, the side surface of ship bow against the pier and sharp angle of the pier against the ship. These 3 kinds of mode are not typical. Author given out a mode of ship against pier in 1995: the pier against the bow, the parallel middle body and the stern, besides extremes mode condition of drift, this dividing can be summarized all condition [8]. This dividing is through 10 years, in same occupation, getting to affirm [9]. Therefore we can use this dividing to discuss monomer ship run up on rocks. When studying catamaran, from this principle, extend to go out , add the characteristic of catamaran: One body run up on rock to arouse hull to incline; One body run up on rocks to arouse violently bridge load to increase ----- - etc.. We can get practical mode. (There are some mode is very few occurred, we need not parallel carry out research.)

2. On history, have negated two approximate formulas

1), The Woisin formula of ship collision force, with only one variable

$$P_{max} = 0.88(DWT)^{\frac{1}{2}} \pm 50\%$$

In which: Pmax is the ship collision force; DWT is ship loading capacity.

At 1983, Knott M. A . suggest that for Woisin - Saul formula must be revised, replace with two variable formula.

$$P_{max} = 0.88(DWT)^{\frac{1}{2}}(V/g)^{\frac{2}{3}}(D_{aot}/D_{max})^{\frac{1}{3}}$$

In which: V is ship speed; Dact is the actual displacement when ship against pier; Dmax is the biggest displacement of ship; DWT is ship loading capacity. g is gravity acceleration.

The reason is very simple, according to impulse formula FT = mV; so, F = (mV)/T .The force is influenced by the quality of impact objects, speed and impact course time. Some bridge designer, let the course time of certain scope (size similar) ships as a constant, and enters the constant of formula, and then they have the formula of two variables M and V.

We can express Woisin formula as:'one ship mass have one collision force'(speed is unconcerned). This is inconsistent with mechanics principle obviously, is too rough, therefore, now, used person is already very few.

2) Use the kinetic energy as a variable to calculate the collision force [3, 5]

At 1983 , Amadahl. J makes a formula [3]:

$$P_{bow} = P_0(5 \bullet E \bullet L)$$ In which:

P_0 is reference load; Pbow is ship bow biggest collision load; L is length of the ship; E is the kinetic energy of ship.

You may think: In the E of kinetic energy, has m and V, but the difference group of m_1, V_1; and m_2 , V_2 can get a same E。(The formula give the same impact force) Actually they has a different impulse , corresponding different impact forces。

So, though, this formula is listed in the working summary of April 1991 that the LABSE first committee of annual meeting approves [3], (in the 4th of LABSE structural document , ship collision force item 14). But in September 1991 the final version of annual meeting [2] (only 13 items, no this item) it do not place.

From this, the person who knows mechanics principle will not be with variable energy E , when going to calculate the impact force with approximate formula.

3. China and foreign countries common approximate (half experience) formula and recommend formula

The International association of bridge and structure engineering (IABSE) in 1991 [2] recommends the 13 results of calculations of the force of ship collision with pier。Within than, there are 6 formulas。

In the foreign country formula of above-mentioned, there are 3 formulas can carry out speed together as quality variable acting income and calculation。Add the formula in highway standard of our country [11] and railway standard [10], there are 5 formulas have quality and speed as variable。 The formula in the highway standard of our country source from impulse theory, it is impact time to as a variable; The formula in the railway standard of our country source from energy theory, some variable for appearance in each object just the coefficient of rigidity level.

Before, the calculation result of different formula have differ time, let the designer difficult to application. Actually, so long as doing , ask the mechanism of extant formula on world , make clear the discrepancy between them , take its reasonable part , put together the place that abandons it to be below; Follow the design standard of our country, make statistics as in formula, carry out analysis and experiment for the specific value of each variable entered to collect , coordinate for the discrepancy between each formula, can use the already experience formula of our country and on world , preliminary estimation the force of the ship collision the fixed things in the water。

According to Newton mechanics law, fix thing in water give a force to contact personally normal direction to ship to turn over the bow; boat go forward force decompose two vector, one equilibrium the force turn over the bow, another maintain boat to go forward.

Ship collision the fixed thing in water, and ship is mounted do not move , this force is one theoretically biggest value, the kinetic energy exchange completely。If ship collision with the fix thing in water, the head of the ship does not be mounted , force is little , ship is continued go forward., partial kinetic energy can be retained on board。

The 5 calculation formula of theoretically biggest force of ship collision fix thing in water is as follows:

1) The highway standard of China [11] 4. 4 sections

"In navigational river the force of the ship collision with the bridge pier, when doing not have actual information, can adopt according to the t subordinate able 4 ------. The force of drift thing can press as the formula:

$$P = WV \ / (gT) \qquad \text{------(1)}$$

In which: P-- The force of drifts thing, (kN);

W-- drifts thing gravity (kN), should drift about in river thing condition, determine according to actual inquiry;

V-- current speed(m / s);

T-- time(s), should use 1 s normally according to actual information estimate, when doing not have actual information;

G-- gravity acceleration 9.81(m / s^2).

2) the railway standard of China [10] Sub clause 3.4.6

"The force of the ship or raft collision with the pier to calculate according to:

$$F = \gamma V \sin\alpha[w/(C_1+C_2)]^{0.5} \qquad \text{-----(2)}$$

In which: F—collision force, kN;

γ - - kinetic energy fold to reduce coefficient, $s / m^{0.5}$. work as ship or raft can adopt 0.2 (point at ship or raft to run near direction and ram place the normal direction is not consistent), straight to ram (ship or raft run near direction with ram point the direction is consistent) can use 0.3;

V-- ram speed of ship or raft, m/s. This speed can adopt that ship adopting shipping department offers for raft adopt the water issue speed;

α - - the angle of ship or queue up raft to ram place, become by tangent, should determine according to specific condition, if difficult, can adopt $\alpha = 20^0$;

W-- ship heavy or raft values, kN;

C_1, C_2 - - the flexible deformed coefficient of ship or raft and masonry. When lacking information, can assume $C_1 + C_2 = 0.0005$ m / kN . "

Author suggestion: when having anti-impact gear, increase C_3, is representative anti-impact gear just the average flexible coefficient of gentle level.

Within the guide of International association of bridge and structure of engineering (IABSE) calculation the force of ship collision the pier, have 3 formula can enter two variables M and V :

3) Formula of Minorsky, Gerlach, Woision:

" $P = 0.024(VDmax)^{2/3}$ ------(3)

In which: P—force (MN); V-- ship speed(m / s); Dmax--the fully loaded displacement of ship (t);

4) The formula of Knott, Saul-Svelsson, Greiner:

" $Pmax = 0.88(DWT)^{1/2}(V/8)^{2/3}(Dact / Dmax)^{1/3}$ ------(4)

In which: P_{max}—biggest collision force; Dact is the actual displacement when ship against pier; Dmax is the biggest displacement of ship; DWT is ship loading capacity. g is gravity acceleration."

5) American state highway transportation worker association (AASHTO) formula, by 1994 [11] revise is:

" "$Ps = 1.2\chi 10^5 V(DWT)^{1/2}$ ------(5)

In which: Ps—ship collision force (N); DWT-- the loading capacity of ship(t); V-- ship collision speed(m / s). "

Formula 5 is the special case of formula 2 , apply to the condition of certain ship and speed range only, it is included in the formula 2. after studying and comparing these approximate formulas, recommend a formula [12]. the formula of recommendation make normalizing formula for railway make to extend modification at two point : first point is the flexible deformed coefficient of ship bow not worth with assumption, get with the method of calculating ; second point is to add C_3, C_3 is the deformed coefficient of extra thing around the fix thing in water (anti-impact gear , timber by rush , the broken ice besides iceberg , ------)

Now ship bump against to artificial facility (gate, pier and stake column etc. for instance.)go up than the proportion that bumps against natural rock.

6). The formula that recommends to use

Now author recommendation the force of ship run on rock to calculation with the following formula:

$F = \gamma V \sin\alpha[w / (C_1+C_2+C_3)]^{0.5}$ --------(6)

In which: F—collision force , kN;

γ—kinetic energy fold to reduce coefficient, s / $m^{0.5}$. work as ship or raft oblique can adopt 0.2 (point at ship or raft to run near direction and ram place is not consistent), straight to ram (ship or raft run near direction with ram point is consistent) can use 0.3;

V-- the speed of ship or queue up raft, m/s. This speed for ship and queuing up raft to transport can adopt the data that shipping department offers, for the drift raft adopt current speed of issue;

α -- ship or queue up raft to run near direction and ram place the angle become by tangent, should determine according to specific condition, if difficult, can adopt $\alpha = 200$;

W-- ship weight or queue up raft weight, kN;

C_1 -- the flexible deformed coefficient of ship or row raft. m / kN

C_2—masonry work deformed coefficient. m / kN

C_3--deformed coefficient of anti-impact gear or other outside bullet thing. m / kN

※ : Ship weight is ship displacement and even water the sum attached

This formula come from Newton mechanics, [13] in have discussed it push course. Each parameter each physical meaning and γ impact dynamics meaning. [12] in have expounded will this formula extend the course of joining C_3.

For C_1 and C_2 is listed after carrying out series calculation into subordinate list, to let many mechanics workers check use. In which, C_3 can offer from anti-impact gear designer calculation.

4. The discussion of studying approximate formula

1) Study high speed ship to run up on rocks to reflect speed influence certainly, but in above approximate formula as speed V is entered , there is one time and 2 /3 time, two Chinese standards is directly chosen from Newton mechanics formula to use one time. (within approximate formula by quality M influence for force also useful one time ,1/2 and 2 /3times , is the formula directly chosen from Newton mechanics use one time).

2) Use this recommendation formula, in order reflect the specific condition of example variable more.

3) When you prepare to carry out value calculation, if the factors that program can consider lose the formula of recommendation, however consider the method of rich this kind of value calculation , or choose calculation additionally program.

4) When you finished a value calculation wanting to sum up an approximate formula: may be consider how many variable do you contain? The approximate formula and existing appearance s what superiority comparatively, have any existent meaning, and to consider that made formula is conform to mechanics principle.

5) Now, if there are already existent one compares good practical formula , urgent matter is the parameter needed for how to collect this application formula, instead of go to seek some parallel formulas.

6) When some one use approximates formula. A high speed ship design initial stage, usually with approximate formula to evaluate the force of collision. For the later stage of important boat in design, use the dynamics calculation is desirable. If the data needed for calculating is not ample or when need to verify the calculate result, need to still carry out experiment. Therefore, normalize to be chosen to select a kind of fine approximate (half experience) formula, look very necessary.

5. Methodology of research

Since, the approximate (half experience) formula is look very necessary. Then, the approximate formula needs to verify need to development. And what is the research technique of the approximate formula ?

Anti collision research and design, on principle, is what belongs to engineering mechanics this big category. Since 20 century, experiment, analysis and value calculation are the 3 major means that engineering mechanics research adopts. Sometimes 3 though each have stress , but actually the solution of a engineering mechanics problem, the close cooperation that often needs this 3 promotes mutually, talent is deepened further the course of knowing of " experiment—analysis-- value calculation". If have no firsthand test result offer build mould basis and binary calculate the verification of result, do not imagine so called "theory" and "calculation" will look so pale and powerless; Conversely, if having no the guidance of scientific theory , we possibility return vast bump against groundlessly in dark to grope. And value is imitated in project mechanics development recover to recover important position, no matter from reason knowledge, is still from practice, it is generally recognized to have gotten generally.

As money study QIAN Xueseng sir give a letter of clear China university project mechanics department in 1997 in "Look forward 21 Age" , have emphasized to point out "Mechanics add computer are the major means of engineering project ".

So 21 century, the methodology of research on engineering Mechanics should adopt: experiment, analysis and value calculation are the 3 major means. Also, we say :experiment, analysis and value calculating 3 combinations.

6. Conclusion and suggest the consideration for bow structure of high speed boat

1) Collision force and the ship speed

People use the theory such as material mechanics, structural mechanics and flexible mechanics when they go to value calculated, explain that in the boat object and scope that collides research, Newton mechanics is suitable. In Newton mechanics, collision force to be directly related the quality of object, speed and the course time. Therefore estimation the formula of collision force, should be directly quality and speed with the course time (or reflection the course time that rigidity) 3 variable.

2) hull sharp department will not design very hard

As now calculate from approximate formula and value can explanation: Ship bow the more hard the collision force more bigly, so in design ship bow, it is hard to need not pursue. The author suggests before the collision bulkhead it is hard to need not design . Little collision t a little hollow, so, the collision force is little , return to repair is not difficult.

3) The 3 coordination's example of hardness, strength and rigid in ship bow

Hardness is the ability of resisting part to invade, strength is the ability of resistance fracture and rigidity is a measure of deformed out size , 3 will pursue respectively, will satisfy use again mutually to ask.

Lift example, go up century 80 decade, Beihai fishing port is imported from America the 10 fast fishing vessels of glass reinforced plastics. Since these ships are more rapid than original flexible fishing vessel, reach fishing area and first net bumper harvest quickly. Add again to catch up public attention to rush wind end, fish obtains quantity larger. In our country that is to use the f the fast fishing vessel of glass reinforced plastics on first. His hull sharp department uses the sandwich glass reinforced plastics with foam, strength enough and has little rigidity, promise warps big, collision force is not large, conform to the anti-collision principle. However, because of the surface layer of sandwich comparatively thin(3-5 mm), anti-invades ability to be not enough. Fisherman for increase hull anti-invade ability -- hardness, the surface increase of glass reinforced plastics covered a copper skin of 3mm, accomplish do not increase rigidity can at the same time guard against invading again, it is harmonious to accomplish strength, rigidity and hardness 3.

Study of Longitudinal Motion Control System on SWATH

ChenZhengchao, LiTieli

Department of Naval Architecture and Ocean Engineering
Dalian University of Technology

Abstract: Based on ship linear strip theory the state control system model on longitudinal motion for SWATH was established. And then the LQR optimum control theory was employed to verify this model. After several numerical simulations, the basic laws between the system response performance and the matrix Q and R have been found and some pre research in LQG was taken. The experimental results showed the conclusion is right.

Key Words: SWATH; State-space control system model; LQR; LQG; Optimum control theory.

Preface

State-space model is the latest and the most scientific method used in control systems. All states and capabilities of the system can be completely expressed by the model. Compared with transfer functions of classical control theory which can only express outside relationship between single input and single output state-space model is obviously more comprehensive. As well as, this matrix method makes the expression of mathematical model of control system more distinct. It can be considered as a powerful tool for the analyzing and researching of multivariable systems.

An idea in establishing of SWATH was provided in this paper. And analyzing and design of a real system were carried out based on LQR theory.

1 Mathematical model design of SWATH longitudinal motion control system

SWATH has a good seakeeping performance. But the trim restoring moment is always reduced, especially when ships are in high speed caused by the small waterline areas. Furthermore, the 'MUNK' moment makes it easy to lose the stability. The aim of designing the controller is to make a specific control to the fin stabilizers in order to ensure longitudinal stability of the ship as well as improve movement capabilities when sailing.

Motion of a launch vehicle consists of longitudinal motion in which heave motion and pitch are included and lateral motion in which sway motion, rolling and bow rolling are included. Only longitudinal motions are considered in this paper.

1.1 State-space equation of SWATH longitudinal motion control system

Suppose fluid medium in calm water is even and a little disturbance in vertical plane is present, then, state-space equation of SWATH longitudinal motion control system can be concluded as

$$\begin{cases} \dot{X} = AX + BU \\ U = -KX \end{cases} \quad , \tag{1}$$

where $X = (x_1 \quad x_2 \quad x_3 \quad x_4)$,

in the formula

$x_1 = \xi_3$ (Displacement of heave motion), $x_2 = \dot{\xi}_3$ (Velocity of heave motion), $x_3 = \xi_5$ (Displacement of pitch motion) , $x_4 = \dot{\xi}_5$ (Velocity of pitch motion)

$$A = \begin{bmatrix} 0 & 1 & 0 & 0 \\ -\dfrac{C_{33}}{M + A_{33}} & -\dfrac{B_{33}}{M + A_{33}} & -\dfrac{C_{35}}{M + A_{33}} & -\dfrac{B_{35}}{M + A_{33}} \\ 0 & 0 & 0 & 1 \\ -\dfrac{C_{53}}{I_5 + A_{55}} & -\dfrac{B_{53}}{I_5 + A_{55}} & -\dfrac{C_{55}}{I_5 + A_{55}} & -\dfrac{B_{55}}{I_5 + A_{55}} \end{bmatrix}$$

$$B = \begin{bmatrix} 0 & 0 \\ \dfrac{1}{M + A_{33}} \cdot \rho V^2 A_f \cdot \dfrac{\partial C_f}{\partial \alpha} & \dfrac{1}{m + A_{33}} \cdot \rho V^2 A_a \cdot \dfrac{\partial C_a}{\partial \alpha} \\ 0 & 0 \\ \dfrac{1}{I_5 + A_{55}} \cdot \rho V^2 A_f \cdot \dfrac{\partial C_f}{\partial \alpha} \cdot l_f & -\dfrac{1}{I_5 + A_{55}} \cdot \rho V^2 A_a \cdot \dfrac{\partial C_a}{\partial \alpha} \cdot l_a \end{bmatrix}$$

$$U = \begin{bmatrix} \alpha_f \\ \alpha_a \end{bmatrix} \quad \text{(Control vector)}$$

$$K = \begin{bmatrix} k_1 & k_2 & k_3 & k_4 \\ k_5 & k_6 & k_7 & k_8 \end{bmatrix} \quad \text{(Feedback gain matrix)}$$

In these formulas, M and I_{55} are mass of ship and inertia moment in lateral axis y through the gravity of ship respectively. $A_{i,k}$, $B_{i,k}$ and $C_{i,k}$ are additional mass, coefficient of damping and restoring force respectively. Subscript i, k are motion mode labels, i(or k)=3 is heave motion,5 is pitch motion. α_f α_a are angles of attack generated by fore and aft fins, l_f, l_a are distance between fore and aft fins axis lines to axis y, $\dfrac{\partial C_f}{\partial \alpha}$, $\dfrac{\partial C_a}{\partial \alpha}$ are differential of power-angle that corresponding lift coefficient of fore and aft fins respectively.

Following Fig 1 is flow chart of longitudinal motion control system on SWATH.

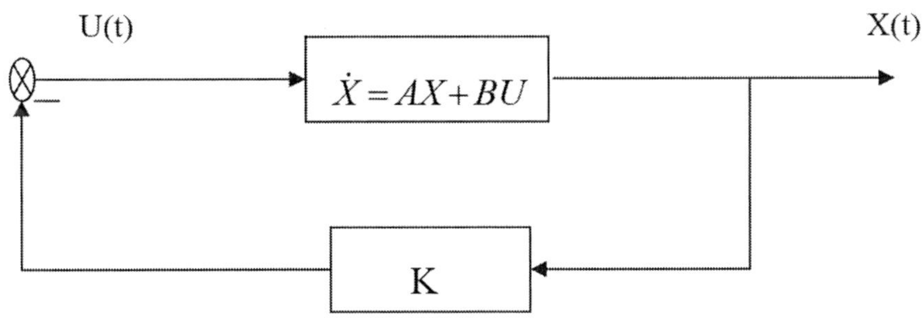

Fig 1 Flow chart of longitudinal motion control system on SWATH

Obviously, system above is negative feedback proportional plus derivative (PD) controller. The estimated performance of differential control is exactly what the gain system needs. At the second hand, this multiple-degree-of-freedom system is the reason for using state-space model[1].

1.2 Parameter confirmation on SWATH longitudinal motion control system

In the paper, the date root in theoretical calculation and test date on SWATH "New century 1"[2].When Vs=19kn, Fn=0.58(only aft fin),corresponding parameters are as follows:

Tab.1.Figure of ship parameter when Vs=19kn, Fn=0.58

Parameter of hull	Value	Parameter of fin	value
B(m)	14.0000	C（m）	1.5260
Lb(m)	28.0000	S(m)	2.0000
d（m）	3.5000	R/Lb	0.1637
2b（m）	11.0000	T（m）	0.0600
V0（m^3）	330.0400		

Additional mass($A_{i,k}$), coefficient of damping and restoring force($B_{i,k}$ and $C_{i,k}$) based on the longitudinal motion linear strip theory have been proved by the test. Then matrix A and B can be determined through processing corresponding parameter above.

Here

$$A = \begin{bmatrix} 0 & 1 & 0 & 0 \\ -2.7170 & -1.5600 & 3.1450 & 0.5608 \\ 0 & 0 & 0 & 1 \\ -0.4063 & 0.4197 & 4.8144 & 7.6745 \end{bmatrix}$$

In this processing, $\frac{\partial C_i}{\partial \alpha}$ in matrix B and corresponding coefficient in matrix U have been combined to be updatable B1 and U1, as follows,

$$B \cdot U = B1 \cdot U1$$

$$B1 = \begin{bmatrix} 0 & 0 \\ 0 & 165.9262 \\ 0 & 0 \\ 0 & -44.0489 \end{bmatrix}$$

$$U1=\begin{bmatrix}\dfrac{\partial C_f}{\partial \alpha}\cdot\alpha_f\\ \dfrac{\partial C_a}{\partial \alpha}\cdot\alpha_a\end{bmatrix}=\begin{bmatrix}C_{Lf}\\ C_{La}\end{bmatrix}$$

In matrixes above, C_{Li} （i=f,a） are lift coefficients generated by fore and aft fins.

2 Design of SWATH longitudinal motion control system

Optimum control method can be finally ascertained for linear control system exactly after determining stability, controllability and observability. Fig 2 shows the design process above[3].

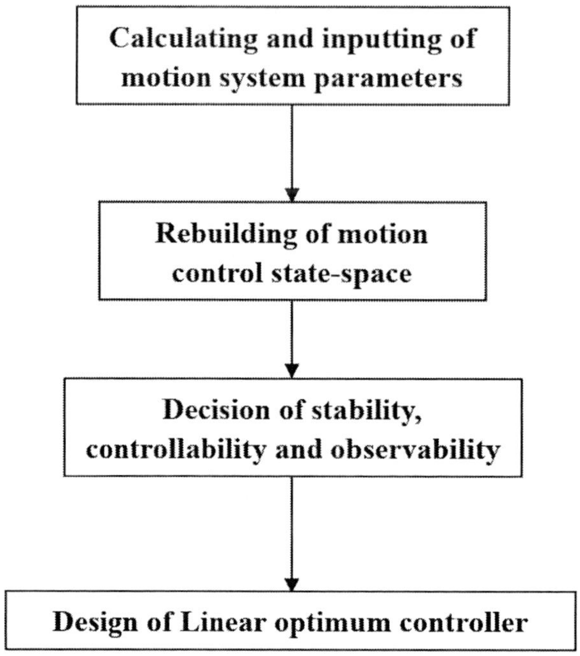

Fig.2. Designing of longitudinal motion control system

2.1 Calculating and inputting of motion system parameters.

This part consists of Calculating and inputting of corresponding parameters on SWATH longitudinal motion equation.

2.2 Rebuilding of motion control state-space equation

Rebuilding of motion control state-space equation is actually the conversion from motion system equation to state-space equation.

2.3 Decision of stability, controllability and observability of linear control system

Stability of linear time-invariant system can be determined by calculating characteristic root of matrix A. The system is stable when characteristic root is located in left half plane in complex plane.

Characteristic root of matrix A calculatedly is as follows,

[8.2662，-0.82043+1.4566i，-0.82043-1.4566i,-0.51087]

Obviously, this system is unstable control system and need external controller.

Necessary and sufficient condition of linear system controllability is as follows,

$$rank[B \quad AB \quad \cdots \quad A^{n-1}B]=n$$

This system is controllable after calculating.

Necessary and sufficient condition of linear system observability is as follows

$$rank\begin{bmatrix} C \\ CA \\ \vdots \\ CA^{n-1} \end{bmatrix} = n$$

In this paper, the main object concerned is longitudinal trim. So, matrix C can be fixed as follows

$$C = [0 \quad 0 \quad 1 \quad 0]$$

Finally, control system equation is as follows

$$\begin{cases} \dot{X} = AX + B1 \cdot U1 \\ U1 = -KX \\ Y = CX \end{cases}$$

the system is judged as observable by calculating.

As a result, this linear control system is unstable, observable and controllable, but can be stabled by appending external controller.

2.4 Design of Linear quadratic optimum controller (LQ) and pre research of simulation on Matlab with Gauss optimum controller (LQG)

2.4.1 Brief introduction of LQ and LQG

State equation of linear fixed system $\dot{x}(t) = Ax(t) + Bu(t)$, $x(0) = x_0$,

Where $x(t)$ is a n-dimension vector, $u(t)$ is a p-dimension system input control vector which is restrictedless. A is a $n \times n$ constant matrix, B is a $n \times p$ constant control matrix.

Quadratic performance index is $J = \frac{1}{2}\int_0^{\infty}(x^T Qx + u^T Ru)dt$,

Where terminal time t_f is infinity, Q and R is weighted matrix used to balance state vector and input vector. Q is a $n \times n$ positive semidefinite constant matrix, R is a $p \times p$ positive definite symmetric matrix.

If $u^*(t)$ can help system back to it's zero condition and minimize J when system departure it's condition due to outside disturbance, $u^*(t)$ is called optimum control. According to optimum control theory, when one of the following conditions being satisfied some conclusions will be achieved.

(1) $Q > 0, \quad Q = Q^T$, and Q is complete controllable to [A,B].

(2) $Q \geq 0, \quad Q = Q^T$, and Q is complete controllable to [A,D], $DD^T = Q$ where D is a random matrix.

The conclusions are as follows.

a. Optimum feedback matrix $K = R^{-1}B^T P$.

b. Unique optimum control $u^*(t) = -Kx(t) = -R^{-1}B^TPx(t)$.

c. Optimum Quadratic performance index $J^* = \frac{1}{2}x^T(0)Px(0)$,

where P is positive semidefinite constant matrix and it is the unique root of equation

$$PA + A^TP - PBR^{-1}B^TP + Q = 0.$$

d. Closed loop system $\dot{x}(t) = (A - BR^{-1}B^TP)x(t), x(0) = x_0$ is progressive stable and has a optimum trace $x^*(t)$.

Take random input noise and random measuring noise into consideration, optimum quadratic control became LQG.

According to separation principle LQG can be considered as two types of problems quadratic adjuster and optimum controller.

(1) According to statement above optimum feedback matrix $K = R^{-1}B^TP$,

optimum control $u*(t) = -Kx(t) = -R^{-1}B^TPx(t)$

where P is fixed to $PA + A^TP - PBR^{-1}B^TP + Q = 0$.

(2) Optimal estimation $\hat{x}(t)$ can be got applying Kalman filter theory. And $\hat{x}(t)$ can minimize variance of estimation error, that is $J = E\left[\tilde{x}^T(t)\tilde{x}(t)\right] = \min$.

So, optimal estimator is $\dot{\hat{x}}(t) = (A - LC)\hat{x}(t) + Bu(t) + Ly(t)$,

where L is Kalman filter gain $L = P_0C^TR_0^{-1}$,

P_0 is root of formula $AP_0 + P_0A^T + GQ_0G^T - P_0C^TR_0^{-1}CP_0 = 0$.

Optimum solution can be got by combining the optimum controller and optimum estimator of LQ model.

2.4.2 Researching of optimum control of SWATH LQR

The premise and difficulty of designing control laws is the selection of weighted matrix Q and R. Bad selections can fail the design or generate abnormal laws. The influences of different matrix to control laws are compared based on dynamic capabilities of state variables and a method to determine weighted matrixes is provided as well as.

In the process of designing SWATH optimum control system by the LQR method diagonal matrixes are vested to Q and R. Responses will be got after applying a step pulse to the system. The method of trial and error was used when determining Q and R. Principles of this method can be expressed as following. First, Initialize Q and R by primary analyzing. Second, check simulating results to make whether the matrixes are reasonable or not. Current Q and R will be hold if the answer is positive and the optimum gain matrix K can be easily determined by computer. On the contrary, Q and R will be reinitialized if the answer is negative until a positive answer coming out. The design of the controller is finished when K is assigned to controller variable practically.

In the test, sampling rate is 2 per second and the procedure lasts 30 seconds. By analyzing response curses whether Q and R are reasonable can be judged. After a series of trial and error, Q=[55 0 0 0; 0 1 0 0; 0 0 1 0; 0 0 0 1], R=[1 0; 0 1] are determined.

Some conclusions are achieved through the process of Q and R determining.

(1) A bigger q_{11} means punishment to any continuous transient movement in heave motion. So a bigger Q is selected. In addition, adjusting time t_p and state-feedback control vectors are directly influenced by q_{11}.When increasing $q_{11,}$ t_p will be decreased and amplitude of vectors increased.

(2)The feedback gain matrix K and state variables are directly influenced by R. When increasing R, K decreased and amplitude of variables increased. So it is not wise to choose a bigger R.

(3)Determining of Q and R can influence capability index and feedback gain matrix to a big extend. So determination of Q and R should follow the rules mentioned above and be adjust to the real situation in practical work.

By the analyzing of weighted matrix, R=1 and Q=diag(55,1,1,1) are determined. The optimum feedback matrix K of SWATH longitudinal kinetic control system is obtained under this situation.

$$K = \begin{bmatrix} 0 & 0 & 0 & 0 \\ -5.7398 & -2.2164 & -8.4754 & -12.749 \end{bmatrix}$$

In order to process datum easily the following assumption was made at the beginning of this test.

$$BU = B1U1$$

so the final optimum control matrix should be converted as follows.

$$U1^* = -KX$$

and

$$U = -B^{-1} \cdot B1 \cdot K \cdot X(t)$$

$$K_{final} = -B^{-1} \cdot B1 \cdot K$$

2.4.3 Researching of LQG optimum control simulation using Matlab

The premise and difficulty of designing Gauss optimum control system are also determination of weighted matrixes. According to separation principle, in the designing process of secondary regulator of linear system weighted matrixes R=1, Q=diag(55,1,1,1).

By querying mass of material and simulation, a conclusion came out that during the process of selecting and adjusting the model noise covariance matrix Q_0 and measuring noise covariance matrix R_0, the system gives sound responses only when g_3 is not zero. So, by setting

$$G = \begin{bmatrix} 0 \\ 0 \\ 1 \\ 0 \end{bmatrix}$$

and adjusting of Q_0 and R_0, the Unit-step response curses of LQG optimum control system are shown as following figure.

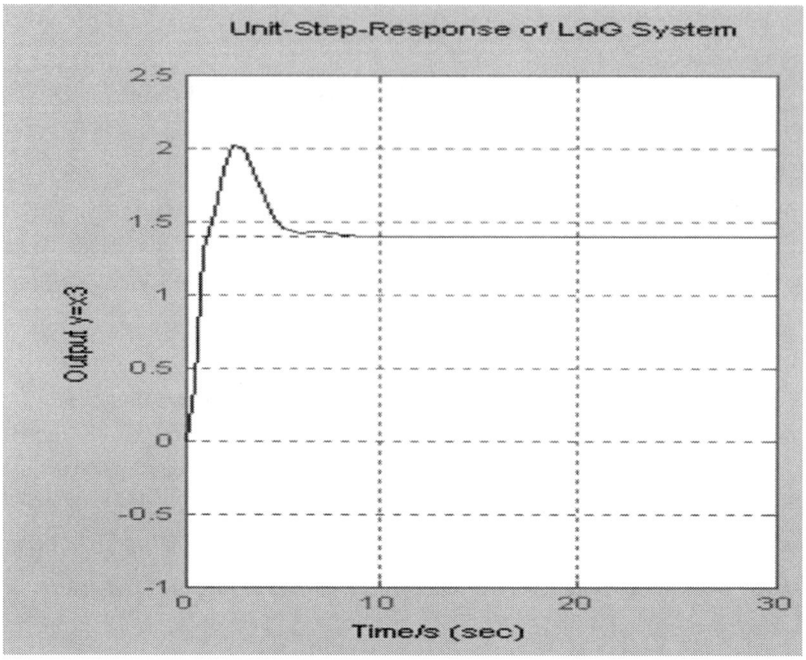

Fig.2.Curve of Unit-step-response of LQG system when Q_0=100, R_0=1

By analyzing the figure above and calculating the Kalman filter gain matrix

$$L = \begin{bmatrix} 1.9764 \\ 19.673 \\ 25.64 \\ 278.69 \end{bmatrix}$$

Some conclusions are achieved through additional simulating test.

(1)Assume R_0=1, when increasing Q_0, system output will be increased slightly, so is the system terminal balanced state. Kalman filter gain will be increased too. But adjusting time t_p keeps steady.

(2)Assume Q_0=100, when increasing R_0, system terminal balanced state will be decreased sharply. Kalman filter gain changed slightly and adjusting time t_p keeps steady.

(3)By further research, a law came out that system step response curves and calculation results seems to be same under situations Q_0/R_0=100 or Q_0=100, R_0=1. So, system can get optimum unit-step response when Q_0 and R_0 are related by this multiple.

3 Conclusions

(1) SWATH longitudinal motion control model is based on SWATH longitudinal motion strip theory.

(2) The idea of LQR quadratic optimum control is lead in SWATH longitudinal motion control system and proved to fulfilled basic demands of designing a system theoretically. This method is also valuable for referencing when determining Q and R of some other control systems.

(3) In the part of LQR theory, weighted matrixes Q and R selected fulfill demands of system designing. The relationship between system dynamic response and matrixes are found by analyzing. As well as, the relationship between system dynamic response and weighted matrixes Q_0 and R_0 are found though further researching of LQG quadratic Gauss optimum control algorithm. Simulation results show that plan mentioned in this paper fulfills the demands of system designing.

Reference

[1] 朱炳泉,畦爱国.小水线面双体船纵向运动控制系统的试验研究[J].中国船舶科学研究中心技术报告,2004.

[2]Tieli Li,Yan Lin,Zhuoshang Ji.Optimized SWATH From Design with High Speed and Seakeeping Performance[J]. Proceeding of The Fifteenth(2005)International Offshore and Polar Engineering Conference, ISBN 1-880653-64-8(Set):ISSN 1098-6189(Set)Vol.4:785~792.

[3] 钱伟康，方宗达.基于 MATLAB 的二级倒立摆控制系统设计方法[J].上海理工大学学报，2004，Vol.26:589~592.

[4] 黄忠霖.控制系统 MATLAB 计算与仿真[M].北京：国防工业出版社，2004：428—430.

Development and Application of Fuel Cell

Zhang wen yu

Luoyang Ship Material Research Institute, Luoyang 471039

Abstract: Fuel cell is high efficiency ,safety, convenient cleaning energy. This paper mainly relate the main type of fuel cell, present situation both at home and abroad ,development and application etc. Through analysis and discussion, there are all understand about fuel cell.

Keywords: fuel cell hydrogen cell

0 Introduction

Fuel cell is as high efficiency , cleaning, no pollution energy change device, is not new thing. Early160 year ago human found common cell. Fuel cell have existed .But ,Fuel cell car by fuel Cell as power is new development in last year car industry. America, Japan, Germany, Canada etc Advance country have walk in front. Fuel cell said on all benefit, other energy technology are not compared with it, and at present research prove fuel cell feasibility.[1]

1. Present situation both at home and abroad of fuel cell

1.1 External fuel cell present situation

Advance country all will be large fuel cell development as key research projects, business circles are engaged in research development on fuel cell by huge sum, now have get many important achievement, so that fuel cell will displace traditional generator and combustion engine to widely use in generate electricity and motor . Fuel cell high efficiency ,no pollution, building period short, easy repair and low cost proficiency will ignite 21 century new energy and environmental protection green revolution.

America

In America ,regard energy ministry as first, on a big scale prepare and popularize to use fuel cell, from 1992 year to 2001year , only from America energy ministry to appropriate 7 hundred million U.S.dollar fund, use in fuel cell prepare and development, recently America energy ministry declared to appropriate 1380 ten thousand U.S.dollar to assist America General Motors Corporation ,using research and development high polymer fuel cell and its application in motor. First hydrogen cell motor combined to prepare by America environmental protection bureau and DaiMule-klsg company, American combined postal service company, have put into field use inspection period, this fuel cell first provide power motor vehicle drive on all America road.

Canada

In research and development PEMFC technology, at present Balad company in lead position, development completely depend on Canada government aid financially, in this field occupy first place in the world. In 1995 february, Balad company second PEMFC motor by Canada government and America Ja state south seaboard air quality control government office etc unit common invest 440 ten thousand U.S.dollar have used, its power is 275 horsepower, carry passengers 17 people.

Germany

In 1999 march Benc company showed Necar-4 type PEMFC electric vehicle , its use liquid hydrogen as fuel, one time fill hydrogen can drive 450 kilometre. At present, Benc company have developed Necar-4 type small car and NEBUS big public bus. Using carbinol reforming hydrogen
NECAR5 will come out.

Fuel cell have volume big ,specific energy high ,power scope widely, noise small etc merit, from a long-term looking, it is a promising AIP technology. Germany in this fuel cell AIP technology study field occupy first place in the world.. In 1995, Siemens company have completed 212 grade submarine to use all measurement PEM fuel cell prepare work. Germany navy have decide to adopt PEM fuel cell system in nearing start 212 submarine. 212 grade first boat estimate to start sea test in 2001,is on active service in 2003.[2]

Italy

Italy in order to rapid fuel cell technology research and development ,promote scientific and technological circles and industry world to cooperate, they hope to reduce fuel cell production costs through cooperate study, from 3000~5000 U.S.dollar per kw at present ,to 250~300 U.S.dollar (2005) and 50~100 U.S.dollar (2010) per kw.

Japan

From 1981 to 1990 time, Japan government set up a "moonlight engineering" .At present ,a new fuel cell engineering ---"new sunlight engineering"is going on .In Japan government trade industrial province aid financially and big motor company make great efforts, Japan work out a laudably ambitious study ,shown and spread fuel cell motor plan----Japan ` hydrogen energy and fuel cell shown plan. Join this plan main units have Fengt motor company and Bengt motor company etc 14 households , since 2001-2002 year ,have 5 motor company development fuel cell motor to go on drive on road, and build a few filling hydrogen stations.

1.2 Present situation of my country fuel cell

My country fuel cell research started in 1958,the organization electronic industry branch Tianj electric source research institute first carry out MCFC study. In 2000, Dal chemistry physical research institute and china science academy electrotechnics research institute have completed 30KW motor using fuel cell all test work. In 2005, fuel cell made in china will compete with foreign class, seize to equip Beijing the

Olympic Games green environmental protection motor markets. This will initiate industry circles to pay close attention to fuel cell.

My country fuel cell research work have shown that:1.china proton change film fuel cell have come to can entrucking technology level; 2. Dal chemistry physical research institute proton change film fuel cell is to have my country autonomous intellectual property high technology achievements; 3. in fuel cell research ,my country may compete with world advance country ,and in markets share, my country may have ability to possess fair scale. But my country have large gap in PAFC,MCFC,SOFC research, at present still place on prepare stage.

2 Development and application of fuel cell

China

In recent year, china acquire many breakthrough in fuel cell key materials ,key technology innovation, have get in world advance country. In future hydrogen center energy system, hydrogen fuel cell beyond all question become key.. China government very lay stress on research and development of fuel cell, in recent year ,one after another develop out hundred watt grade –30KW grade oxyhydrogen fuel electrode, fuel cell electric motor etc. Fuel cell technology special is proton change film fuel cell technology get rapid development , develop out 60KW, 75KW etc many type proton change film fuel cell group, electric motor using 40KW, city bus using 100KW fuel cell engine, make china fuel cell technology step in world advance country. Magical power hydrogen motive power science and technology come up to advanced world standards. My country autonomous prepare fuel cell come up to advanced world standards. My country fuel cell industry study and development level place precedent in the world. In 2005 January, zero emission middle car by 30KW proton change film fuel cell success to drive in Shiy city of Hubei province. This is my country a first to have autonomous intellectual property fuel cell electric motor. And adopt proton change film fuel cell technology assembling 1KW cell system long –range under water robot to enter the water in recently.

America

America Stf university environment department associate professor Mak·yagbs etc published research report in 24 of June new <<science>>magazine. If all America now motor vehicle change into hydrogen fuel cell motor, will improve people life quality from environment and health sides, may reduce air pollution to dead human number, and save consumer expenses.

America scientist use microbe fuel cell to make hydrogen gas, at same time can clean effluent .America study group scientist use new materials solid acid as electrolyte, this materials electric conduction, use solid acid to produce fuel cell, may give electric to a middle warm electromobile.[3][4]

Japan

Japan SHINANO KENSHI at present develop successfully a new material named Carbon Silk, will use fuel cell electrode to reduce Pt catalyst expenses, can raise maximum power density. This materials have silk texture, and spinning and weaving materials, make powder silk etc materials to sinter molding .A few days ago Hitachi publish towards notebook computer and portable information terminal etc field, study content is fuel cell by carbinol water solution as fuel (DMFC). Developing to adopt" organic inorganic mix electrolyte film" direct carbinol fuel cell (DMFC) function come up to advanced world highest standards. Japan develop out to make small fuel cell new materials. Japan Hi wire company develop a titanium electrode materials to use in making fuel cell, may hope fuel cell to lessen only big like cards clip, apply to notebook computer etc portable electrical equipment .

North Korea

North Korea modern times motor declare a few days ago, with United Technologies to devote fuel cell development subsidiary—America UCT fuel cell company reach an agreement, both sides will unite to develop motor fuel cell starting under zero temperature. Starting under zero temperature is one of motor fuel cell research and development question.

2.1 Fuel cell motor

Fuel cell motor three merits :

(1) Low discharge. Fuel cell through electrochemistry method, hydrogen and oxygen combine, direct make electricity and heat, discharge water , no pollution environment.

(2) Fuel in varied forms. Fuel cell use hydrogen gain from natural gas ,propane, carbinol , gasoline, diesel oil, coal and renewable energy resources. Fuel source many are advantageous to energy supply safe and use now available traffic infrastructure (fuelling station).

(3) High efficient and high property. Owing to fuel cell no plunger or turbine etc machine element and middle link, its efficiency promote ,about 2~3 times than internal-combustion engine. Fuel cell motor obvious good to other cell motor on cost and all property.

At present, many fuel cell motor development merchant attention concentrate on carbinol. Benc,Fengt and Mazd select carbinol as light motor fuel .Fuel cell motor development on the one hand environment pollution pressure, on the other hand fuel cell rapid development. Most suit to use motor power fuel cell is proton change film fuel cell (PEMFC). Its structure compact, work temperature low (only 80℃),start rapid , efficiency density high, work life long. In19 of April, General motor hydrogen fuel cell field most up-to-date achievement----Sequel idea motor first show in Shanghai .

Modern times motor declare the second fuel cell type. Jazou Los Angeles at just open Los Angeles motor exhibition, modern times motor company fuel cell electric vehicle item(FCEV) group publish it's the second fuel cell type ----Tuseng FCEV. This model motor have make marked progress in technology than former generations product Shengfd FCEV.

At present ,in Beijng convene 2008 the Olympic Games new technology, new product recommend meeting, a few ten scientific research units and industry show each new technology, new product . 60Kw fuel cell city bus common prepare by Qingh university and china science academy Dal chemistry physical research institute , be light points product in recommend meeting. Everybody is in agreement on fuel cell bus most have development potential, may full outstanding 2008 year "Green the Olympic Games" theme. Everybody hope that environmental protection means of transportation formed by fuel cell city bus will add honourable to capital green the Olympic Games during 2008 Bejing the Olympic Games.[5]

2.2 Family fuel cell development

Japan is one of depend on energy imports most high country, domestic energy 90% dependent on imports. For this reason, Japan highly pay attention to development alternate energy source, among fuel cell list most important . Japan energy technological renovation mainly round people daily life , most attract the gaze of people achievement is family fuel cell. At present, Japan have 30 industries ,study organ develop fuel cell. Songx etc company 2004 year push oneself development family fuel cell. one after another. These product general electric power generation activity about 1000W, fuel use hydrogen extract from coal gas, can not make air pollution thing. Owing to technology have mature, Tokyo natural gas company decide to provide service to 200 families by lease since February, 2005, this fuel cell first go in Japan family. Tokyo natural gas company plan to use three years time in order to popular family fuel cell collection date. Japan market push family type solid electrolyte fuel cell, main body use a catalyst element , not only raise anti-CO corrosion property, but also electric power generation activity 35%, heat recovery ratio 30%, using life over 4 ten thousand hours . Family small fuel cell electric power generation equipment output power 1KW,may use waste heat to heat water to 60℃, lay in specially heat water container to reserve, and make direct current to alternating current electric power through frequency transformer, offer lighting and family electric use.[6]

2.3 Portable fuel cell development

Recently year, portable fuel cell have great popular demand in Japan market. So far have 20 types product come out. Hi company have push hand machine using fuel cell. Product output value some W ~ some ten W. Suon and NEC etc all make notebook computer and hand machine moving terminal as use aim.

Japan biggest hand machine manufacturer NEC company have prepared successfully fuel cell using carbinol and nanometer technology, its electric energy reserves is ten times than general Li cell, this mean that user using now hand machine need not charge within 1 month, or can last 24 hours to use notebook computer.

England an energy manufacturer prepared a "portable fuel cell" ,this make business electron product to have market foreground. In September of 2003,a company come from England London pushed a fuel cell to make notebook computer to last work 8 hours, now this new type "portable fuel cell" not only volume large reduction, actual cost save over 33% , but also may provide work 12 hours electricity to notebook computer. Other, manufacturer call this fuel cell more environmental protection, environmental pollution reduced to minimum.

3 Conclusion

External fuel cell development quickly, keen competition. My country autonomous prepare fuel cell to come up to world advanced standards. But it have gap in development and application and industry, from now on, under country government support, through more study in a deep going way, surely can catch up with world advanced standards, gain high economic benefits.

References:

(1) Guoxiaoxia ,Solid oxide fuel cell joint materials progress, function materials,2000,31(1),23~25.

(2) Yangqibai, Submarine AIP technology development study, navy equipment and naval vessels new technology progress study,222~223.

(3) Huangchuanyong, Fuel cell YSZ film gel ---flow pattern prepare study, materials engineering,2000,7:13~16.

(4) Liwenfeng, Direct carbinol fuel cell MEA materials research and fabrication progress, materials guide,2000,15(3):54~56.

(5) Zhenghaijin etc, Bear in with future fuel cell motor , External science and technology trends, 2002,1:35~39.

(6) Family fuel cell development, External science and technology trends, 2000,10:5~6.

Author Brief Introduction:

Zhang wen yu : woman, born in the February of 1968; graduated in July of 1990 from metal materials and heat treament speciality of northwest industry university; gained bachelor's degree; senior engineer; mainly engaged in titanium and its alloy task work;now mainly engag in intelligence study work.

Add:023 letter box 5,luoyang henan;

Postal code:471039;

TEL:0379-67256349;

E-mail:ZWY98@mail.china.com.

Investigation on microstructure and property of explosive welding and rolling B30/CCSB composite plate

Hou Fa-chen, Zhang Bao-qi

(China Shipbuilding Industry Corporation LuoYang Ship Material Research Institute, Luo Yang, China, 471039)

Abstract: Microstructure and property of explosive welding and rolling B30/CCSB composite plate is studied. The result shows that after explosive welding and rolling, in the B30/CCSB bonding interface, better microstructure and higher bonding strength is got. With good resistance to sea water corrosion, higher strength, plasticity, fatigue resistance, and property of various cold and hot working, it is safe and reliable to use these composite plates to manufacture shipping anti-press under-seawater pipe and main-recycling-water system.

Keywords: explosive welding; rolling; composite plate; B30; property.

1.Introduction

Explosive welding and rolling metal composite plate has become a research hotspot and development direction of large area metal composite plate[1-4], which not only embodies strongpoint of extensive applicability of explosive welding technology, better interface joint and higher combination strength, but also integrates advantage of higher yield efficiency and random prolongation for rolling. B30/CCSB composite plate is first developed by explosive-welding and rolling in China Shipbuilding Industry Corporation Luo Yang Ship Material Research Institute, and its microstructure and property is investigated and appraised by all-round testing, result of which has significance and reference value for engineering personnel understanding technics, microstructure, properties and applications of this kind of composite plate.

2.Interfacial microstructure of composite plate

B30/CCSB interfacial photograph by explosive welding and rolling is seen in Fig.1.

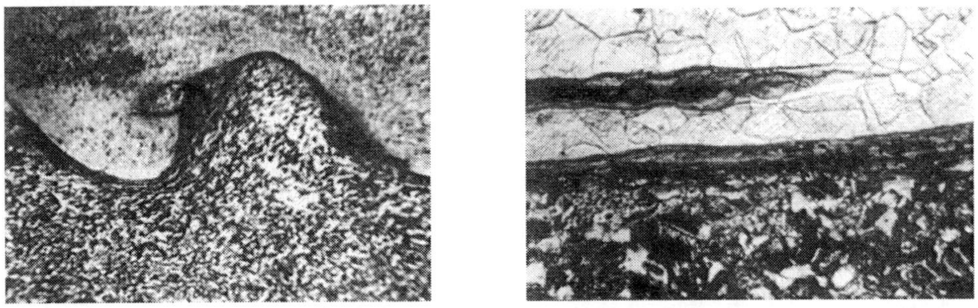

(a) explosive welding ×35 (b) explosive welding and rolling ×110

Fig.1 B30/CCSB interfacial photograph

It is seen in Fig.1a that B30/CCSB composite plate is of sine-wave interface and of front and latter eddies. Violent plastic deformation happens in B30 and CCSB, and grains are elongated。 Micro-hardness testing shows that hardness value in the side of B30 is 1548 for the position far from interface and 2200 in the interface. The corresponding hardness value in the side of CCSB is 1532 for the position far from interface and 2120 in the interface. It is seen from above that obvious work rigidification happens in base materials and cladding materials. It is seen in Fig.1b that after rolling explosive welding interface is made flat, front and latter eddies is melt to strip, and because of heating, B30 and CCSB in the interface is recrystallized in the iso-axial shape. Meanwhile α-phase lies in B30, and pearl and ferrite in CCSB. Microhardness testing value, which is 1520 in CCSB and 1280 in B30, is lower than the explosive-welding-condition one. The above indicates that after hot-rolling, work rigidification is eliminated for either B30 or CCSB.

3. Property of composite plate

3.1 Interfacial bonding strength

Interfacial shear strength and bonding strength is shown in Table 1, according to sample size and testing method of GB6396-1995 criterion.

Table 1 Test result of bonding strength near the interface of B30/CCSB composite plate

State	Shear strength (Mpa)		Bonding strength (Mpa)	
	Testing data	Average	Testing data	Average
Explosive	285 290 322 276 304	295	485 545 474 470 506	496
Explosive-rolling	246 258 248 277 262	258	319 331 326 338 351	333

Interfacial shear strength is usually used as important index of interfacial bonding property of composite plate, and interfacial bonding strength is only for referrence. It is seen in Table 1 for both shear strength and bonding strength that there is the same characteristic that value in explosive state is higher than the one in explosive-rolling state. It is considered by analyzing that its cause is that softening happens in composite interface and parent materials on both sides of interface, because of heating effect in the process of rolling, which is testified by the above micro-hardness testing result. Meanwhile, it is possibly related with the information of ferrite strip in the side of base layer, as a result of diffusing action of carbon atom in two sides of interface. Notwithstanding, shear strength of two states is still higher than state criterion GB13238-91 (≥100MPa)and other state criterion such as ASTM A264 and JISG3604(≥98MPa). It is shown that bonding strength B30/CCSB composite plate in any state of explosive welding or explosive-rolling is dependable.

3.2Tensile property

The samples of clad layer are used for testing, where composite plate is 2+14mm thin, and the result is shown in Table 2.

Table 2 Tensile property of B30/CCSB composite plate

State	σ_s (MPa)	σ_b (MPa)	δ_5 (%)
Explosive-rolling	264 283 276 271 257	430 452 421 407 445	30 31 36 34 31
Explosive	375 381 369 390 387	506 497 520 512 517	18 17 17 15 11

Tensile property is most important parameter demanded for frame materials such as pressure vessel. It is asked in State criterion GB13238 and JISG3604 that tensile strength σ_b of copper or copper alloy-steel composite plate is not lower than calculated value made as:

$$\sigma_b = \frac{t_1\sigma_{b1} + t_2\sigma_{b2}}{t_1 + t_2}$$

whereσ_{b1} and σ_{b2} are lower limit of tensile strength of base layer and above layer, t_1 and t_2 are thickness of base layer and above layer, elongating rate is not lower than base and higher than smaller standard value of base or above layer, yield strength is not given in above two criterion, and, in U.S. criterion ASTM A264 yield strength, tensile strength and elongating rate of composite plate are not lower than these of parent materials.

For B30/CCSB composite plate, calculated value of tensile strength by above equation is σ_b=375 MPa. It is seen in Table 2 that testing value in explosive-rolling state is higher and lies in the range of 375~460 MPa. Meanwhile, elongating rate is also higher than given value by Chinese criterion and other criterion such as U.S. or Japan — where not lower than 26% is for Chinese criterion, not lower than 26% is for U.S. criterion, and not lower than 23% is Japanese criterion.

In addition, B30/CCSB composite plate by explosive-rolling can endure violent plastic deformation, because, for whole elongating sample including shrink zone and cracked surface, delamination and crack are not found and bonding interface is good.

In term of design, the highest pressure is 18.15MPa in site of B30/CCSB composite plate used at shipping anti-press under-seawater pipe, rounding stress of cylinder is 25.47MPa, and axial stress is 12.74MPa. Meanwhile, with increasing of temperature, hot stress exists in cylinder because of difference in line expansion coefficient between B30 and CCSB, and by estimating simply, hot stress of CCSB is 7.85MPa at the running temperature of 100℃. When medium pressure and hot stress are together taken into account, rounding stress on cylinder CCSB is about 33.32MPa and axial stress is about 20.59MPa. As for design of pressure vessel, allowing stress of materials is based on tensile strength at room temperature and its safety factor is not bigger than 4 given in U.S. and Japan criterion. Allowing stress, which is ratio between testing tensile strength of B30/CCSB composite plate and safety factor, is far higher than the biggest stress of cylinder, from which cylinder can run on the safe side.

3.3 Impact property

The test samples with or not with clad layer are from explosive-rolling B30/CCSB composite plate, where standard Charpy V notch sample are taken and the notch is perpendicular to the surface of composite plate. The testing results are shown in Table 3.

Table 3 Impact property of B30/CCSB composite plate

Material	Testing temperature (℃)	A_{kv} (J)		Judge of bonding interface
		Testing data	Average	
B30/CCSB sample with clad layer	20	106 116 108 110 106	109	fine
	0	90 86 95 89 75	87	fine
Samples with no clad layer	20	121 115 98 109 95	108	fine
	0	65 90 75 70 70	74	fine

Impact toughness is important technology index of pressure vessel. As to composite plate between copper or its alloy and steel, despite of no regulation by nation criterion and most other foreign criterion, impact toughness must be insured when pressure vessel being manufactured. In Table 3, as to composite plate with or not with clad layer, testing value of composite plate at temperature of 0℃ or 20℃ is far higher than standard value of CCSB steel plate(not lower than 27J at 20℃), and delaminating phenomenon is not found near bonding interface. It shows that B30/CCSB explosive-rolling composite plate is of better impact toughness. Besides, it also shows that A_{kv} value of B30/CCSB composite plate is higher evidently than that of composite plate whose clad layer is wiped off, maybe because B30 is of no low temperature brittleness. Therefore, impact toughness of B30/CCSB composite plate is better than single CCSB steel plate.

3.4 Cold bend property

Cold bend testing results of explosive-rolling B30/CCSBcomposite plate are shown in Table 4. And photographs of cold bend testing samples are shown in Fig.2 and Fig.3.

Table 4 Cold bend testing results of B30/CCSBcomposite plate

Bend method	Sample number	Testing result
Inner bend (B30 in)	5	d=1.5a, 180° fine
Outside bend (B30 out)	5	d=2a, 180° fine
Side bend	5	d=2a, 180° fine

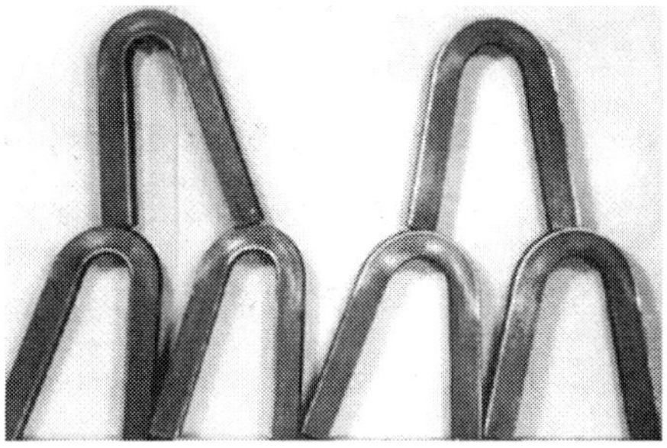

Fig.2 Photograph of inner or outside bend test samples

Fig.3 Photograph of side bend test samples

Cold deformation of material is examined by cold bend test. Inner bend and outside bend is needed in China criterion, U.S. and Japan criterion, and in German criterion, side bend test is needed only if customs ask. In all criterions, bend angle is 180°, but bend heads differ in diameter. It is stated in China and U.S that bend test is done according to request of base steel, while, In Japan, inner bend is done according to request of base steel and outside bend done according to clad layer. If it is not regulated in criterion of clad material, cold bend test can not be done. As to inner bend or outside bend, diameter of 2a is taken if d value is lower than 2a in criterion of base material. It is seen in Table 4 that inner or outside bend test is done according to strict regulation in explosive-rolling B30/CCSB composite plate, while side bend test is done, and the fine results are taken. At the same time, it is seen from the appearance of cold bend test sample that delamination is still not found near the interface after violent bend deformation, from which the firm combination in interface is proved.

3.6 Fatigue property

Elongation fatigue test is carried out on explosive-rolling B30/CCSB composite plate. Invariable load is kept during test, and when the nominal stress loaded is equal to ±257.5MPa, the results show in Table 5.

Table 5 Fatigue life of B30/CCSBcomposite plate

Sample	Stress (MPa)	Fatigue lifetime		Site of crack source
		Single value	Average	
1#	±257.5	208956	223874	1, at the outside of CCSB
2#	±257.5	235682		1, at the edges and corners of CCSB
3#	±257.5	226465		2, at the outside of CCSB
4#	±257.5	230584		2, at the outside or edges and corners of CCSB
5#	±257.5	217683		1, at the edges and corners of CCSB

Fatigue property of composite plate is not regulated in worldwide criterion, but in view of the use of equipment made from composite plate, one time of stress cycle is corresponding to one time of running and shutting down, while cycle times are not more than 10^4 times. It is shown in Table 5 that fatigue property of composite plate is better, and better even under higher stress. Under the condition of pulling and compressing, fatigue crack source comes firstly at the outside or edges and corners of the samples, and not does at the inside interface, which shows that there are, in the inside interface, not evident flaws that can develop to microcrack, because evident flaws, where fatigue source comes first, is sensitive to fatigue. Testing results show that fatigue can not lead to delamination at the interface of composite plate. This testifies that cylinder can run on the safe side.

3.6 Wring property

The size of wring sample is (2+14)×16×240mm, wring length is 170mm, and the test results show in Table 6. The wring test samples are seen in Fig. 4.

Table 6 Wring property of B30/CCSB composite plate

Wring type	No.	Wring angle (°)	Results
Wrung on the single direction	1#	940	Fine interface
	2#	1080	Fine interface
	3#	1120	Fine interface
Wrung to and fro	4#	0-360-0	Fine interface
	5#	0-540-0	Twist off on one of collets, and interface is fine
	6#	0-360-0	Fine interface

Fig. 4 Photographs of the test samples after wrung

The test results show that by explosive welding and rolling B30/CCSB, composite plate is wrung on the condition of large angle or wrung to and fro, no delamination is found in the bonding interface, from which the firm combination in interface is proved more.

3.7 Erosion property

With good resistance to seawater corrosion, B30 serves as clad material. During explosive welding, the surface and interface of B30 clad material undergo instantaneous high temperature, and other part does not undergo violent and lasting heating outside, because of which it is impossible that elements of B30 diffused evidently and separate locally out. At the same time, there is no effect on erosion property of B30.

In order to understand how elements of B30 diffuse during hot-rolling, qualitative and quantitative analysis of main elements (Cu, Ni, Fe, Mn and Mg) in B30 of B30/CCSB explosive-rolling composite plate is by means of SEM done, and the results show in Table 7.

Table 7 SEM composition of B30

State	Distance from interface (μm)	Chemical composition (wt%)				
		Cu	Ni	Fe	Mn	Mg
Original B30 board		67.34	30.51	0.78	0.91	0.008
Explosive-rolling	25	66.89	30.48	0.92	0.90	0.005
	50	66.78	30.75	0.85	0.90	0.005
	100	66.59	30.91	0.76	0.90	0.005
	200	66.32	31.64	0.68	0.92	0.005
	300	65.97	31.75	0.62	0.92	0.005
	Outside surface	66.74	31.31	0.65	0.92	0.005

It is seen from Table 7 that elements don't diffuse evidently during heating and rolling of B30, and content of any element almost doesn't change at the range of from 25μm distance to outside surface, which is in accord with the request of B30 for chemical composition. Therefore, there is no effect on erosion property of B30 after heating and rolling.

4. Conclusion

(1) After explosive welding and rolling, better microstructure and higher bonding strength is got in the B30/CCSB bonding interface.

(2)Explosive welding and rolling B30/CCSB composite plates have better integrated property, with good resistance to sea water corrosion, higher strength, plasticity, fatigue resistance, and better impact property.

(3) It is safe and reliable using explosive welding and rolling B30/CCSB composite plates to manufacture shipping anti-press under-seawater pipe and main-recycling-water system.

References

[1] 稻垣道夫. Press Technology, 19(5): 1-16.

[2]Li Qing-an, Wang Ting-fu et al. Producing of stainless steel by explosive welding and rolling composite plate[J]. Iron & Steel,1986, 21(3): 20-14.

[3]Li Zheng-hua. Development trend of composite plate[J]. Rare metal materials and engineering, 1989, 4: 56-59

[4]Zheng Yuan-mou, Zhang Sheng-jun et al. Explosive welding and rolling of large and thick stainless steel-carbon steel composite slab [J]. Journal of iron and steel research, 1996, 8(4): 14-19

Brief introduction of author:

Hou Fa-chen, Born in 1963, high engineer, research direction: explosive working and metal composite material, about 20 articles published.

E-mail:dlzbq_001@163.com, Tel: 13937908281

Hull Form And Resistance Performance Of WPC Research

Dalian University Of Technology Navy Architecture Department Zhang Xiaoyang Li tieli

Wave Piercing Catamaran (WPC) which works as a new kind of high-performance ship, has got more and more applied and developed. In this paper, the theoretic calculation of wave making resistance has been researched, including the hull form optimization and the relation between the principal dimensions and the resistance.

Keywords: Wave Piercing Catamaran; Hull form optimization; Wave making resistance; Speed and powering of ship.

1 Foreword

By comparison to airlift and drayage, water carriage not only has large cargo capacity but has less investment as well, it should necessarily been developed firstly. This years, high speed ships have got flourish development in the world. Because of the best rapidity, navigability and comfort, WPC has been developed especially fast. With all kinds of high performance ship growing up, it provide a new rational high speed sea transportation mode, madding it possess competitive ability with the other mode of transport.

2 Passenger transport vessel's requirement

2.1 Wave piercing passenger transport vessel's keynote of design

2.1.1Hull form

The main forms of passenger transport vessel are monomaran, catamaran and wave piercing catamaran and so on.

2.1.2Material and hull form

The passenger transport vessels are mainly made of high tensile steel, the superstructure is made of duraluminum. Steel ship has good intensity but its empty ship weight is so large that its speed and carrying capacity will reduce. Steel ship also has advantage, that is low maintenance cost and high residue.

To decrease the rock of ship most passenger transport vessels will install bilge keel, some also have stabilizing fin and self-control system.

2.1.3Mainframe

The mainframe using for passenger transport vessels has two kinds: oil engine and gas turbine. In a broad sense, oil engine has large weight and it is not good for advancing way, but its fuel expenditure and maintenance cost is very low; gas turbine has small weight and it is good for advancing way, its fuel expenditure and maintenance cost is very high on the contrary.

2.1.4Passenger deck

Besides meeting the structural and fireproofing requirement, the arrangement of passenger salon should make the passengers comfort and convenience firstly. And the manager hope to won the most return here even more. As for the choice of fitment material must take account of low maintenance cost and cleanness besides the hue and texture.

2.1.5Cardeck

The shape of driveway has two kinds: simplex driveway and U-mode turnback driveway. Simplex driveway makes it needn't to reversal when unloading, so it expedite unloading speed. Accordingly, the passenger transport vessels with simplex driveway must be able to docking in the beginning and end, that is to say gangways must be setup in the beginning and end.

2.1.6Safety precautions

Passenger transport vessels carry necessary survival equipment usually, so that when emergency occurs passengers and shipmen can transfer quickly in the stated time. All passenger region, cardecks and engine rooms have been installed fire detector, fire-extinguishing system and taking lens. At stress region fireproofing material must be covered on the surface.

2.1.7Weight optimization

Ulterior measures in weight optimization should be taken such as using more fiberglass reinforced plastics or other composites. New material is not only cheaper than traditional material but also easier to installed.

2.2Performance feature of WPC

Because of the specific change of WPC's geometry this hull form has some features as follow:

(1)Rawboned half-hull shape, its length coefficient of fullness of displacement is somehow big belonging to a fractional loaded form, which is good to reduce wave making resistance. Generally speaking $L/b = 12\sim20$, $b/T = 1.7\sim2.0$。

(2)The span between the half-hulls is larger than the average catamaran, this makes the main hull rose to a higher place to prevent main hull flooding and decrease slamming; advances the transverse stability and make it very safe in the big wave; the wave making disturb between two half-hulls decreases sharply; vertical and lateral acceleration improves.

(3)The half-hulls adopt deep V-mode form which has nice seaworthiness, the freeboard at the chine and ship is very small linking with the streamline crossstructure, possessing good piercing ability and buffering the pitching motion.

(4)The section of crossstructure is extenuative streamline, this makes the buoyancy change little in the wave and decrease the heaven movement when steering in the sea.

(5)The freeboard of half-hull is very low, and reduce reserve buoyancy, especially in the beginning so that heaving, pitch and vertical acceleration reduce effectively when steering in the head sea.

(6)WPC is able to steer safely in the poor wave conditions, besides transverse stability is large, it is ease to satisfy damage stability specifications when two cabins damaged, that means maximum heeling angle is small when dissymmetry damage.

(7)This kind of hull form is suitable to high way, because of its nice seaworthiness it stalls low in the wave and can keep high speed.

(8)As the span between the two half-hull is large, this hull form has nice maneuverability. The radius of gyration is about double length in the middle and high speed. Its directional stability and full speed braking performance is nice as wall.

(9)In the end the geometry of the half-hull must satisfy mainframe fixing condition and is ease to construct, operate, service and maintain.

3 WPC's resistance performance theoretical analysis

3.1 Hull form parameter of WPC influence to the resistance performance

3.1.1Half-hull's length drain volume ratio and aspect ratio

In contrast to mono-hull ship, because of the spacious deck area the assess of the least length dose not depend on general arrangement and other requirement. Besides particular cases, it usually decides the best length from the point of least resistance, so the choice of the principal dimensions must base on the F_r or F_{r_∇} in the design speed according to rapidity. When $F_{r_\nabla} < 3.0$, the bigger length drain volume ratio the nicer resistance performance, and the bigger aspect ratio L/b_s. But when L/b_s is too big, wetted area increasing is bad for frictional resistance and viscous pressure resistance. So from the point of choosing length, it must consider the length drain volume ratio Ψ seriously so that we can get the best length.[1]

3.1.2Stern shape

WPC has high speed and it's usually installed hydraulic propulsion system in the end, so its end is square. The principle for major parameter determining is coincident with the mono-hull ship. High speed WPC's deadrise angle of hull bottom β can be determined according to resistance performance and seaworthiness. Usually using small β, when high speed it can attain more dynamic lift and improve rapidity. At the same time it is good for adopting hydraulic propulsion. But this conclusion is just right to the high speed and small displacement WPC. Because the displacement is proportional to the cube of contour scale, and lift is proportional to the square of the contour scale. So, to the low speed and large displacement WPC, too flat end is bad for resistance even degrading seaworthiness and nice directional stability, increasing vertical acceleration at the end.

3.1.3Stem shape

Stem shape of the half-hulls usually use V-mode section, keel goes down to the base line to increase V-mode degree. This can increase pitch damping, avoid the beginning go out of water and reduce wave impacting.

3.1.4Center of buoyancy lengthwise position

Because of the high speed of WPC, center of buoyancy lengthwise position will influent seaworthiness nonnegligibly. Center of buoyancy lengthwise position depend on design speed, the smaller F_r lend to forward position. As to those large F_r WPC, its center of buoyancy lengthwise position may be close to the end. So to the different F_r, there is one only perfect center of buoyancy lengthwise position.

3.1.5Shape of the crossstructure and main hull

The shape of the crossstruture has something to do with the distribution of displacement margin, so it influences navigating-control and seaworthiness of WPC. The cross section's shape of crossstructure have two kinds: straight and arch. The straight crossstructure can takes full use of the breadth to extent deck area, this kind crossstructrue is used in the WPCs which steer in ease wave and with narrow breadth. The arch crossstructure is good for deducing impacting and enhancing transverse strength and hiding.

The main hull's cross section is sharp V-mode, it can soothe wave impacting in the bottom, at the same time supplying additional reserve buoyancy. Usually the main hull dosen't touch with wave. Only in the huge wave its additional reserve buoyancy can prevent nose diving because of lack of reserve buoyancy.

3.1.6Spacing of half-hull's influence

The same with high speed ship, when the spacing of half-hull grows the wavemaking viscosity disturbing affection will reduce and it is good for calmwaterresistance and seaworthiness. In 1988 Australian catamaran company researched the relation between spacing of half-hull and vertical and lateral acceleration in the different wave through model test. It is proofed that spacing of half-hull is the main factor which influence seaworthiness, the larger spacing the better moving performance and samller vertical and lateral acceleration. But it is showed that it is bad for lateral intensity and add the structural weight when the spacing is more than 10 times of breath. [2]

3.2WPC wave making resistance theoretical calculation

WPC is a rawboned kind of hull, using the Michell-mode as the wave making resistance design formula can obtain very actual result, and it is one popular computational algorithm both abroad and home. But the most literatures using this mode did not dispose the influence of square tail and knuckle, and the influence is essential to the wave making resistance calculating.

3.2.1WPC wave making resistance computing method

WPC's Michell-mode wave making resistance design formula:

$$R_w = \frac{4\rho g^2}{\pi V^2}\int_1^{\infty}\left(I^2+J^2\right)\left[1+\cos\left(2bk_0\lambda\sqrt{\lambda^2+1}\right)\right]\frac{\lambda^2 d\lambda}{\sqrt{\lambda^2+1}} \qquad (1)$$

$$I=\int_{-T}^{0}\int_{-L/2}^{L/2} f_x(x,z)\exp\left(k_0\lambda^2 z\right)\sin\left(k_0\lambda x\right)dxdz$$

$$J=\int_{-T}^{0}\int_{-L/2}^{L/2} f_x(x,z)\exp\left(k_0\lambda^2 z\right)\cos\left(k_0\lambda x\right)dxdz$$

$$k_0=\frac{g}{V^2}$$

$$\lambda=\sec\theta$$

R_w — wave making resistance;

I, J — Kochin function;

θ — wave spreading angle;

V — speed;

2b — spacing of half-hull;

x, z — distributive source' s portrait and vertical coordinates;

f — hull function.

Integral formula deform:

$$R_w=\frac{4\rho V^2 k_0^2}{\pi}\int_1^{\infty}\left|A(\lambda)\right|^2\frac{\lambda^2}{\sqrt{\lambda^2-1}}\left[1+\cos\left(2bk_0\lambda\sqrt{\lambda^2-1}\right)\right]d\lambda \qquad (2)$$

$$A(\lambda)=-ik_0\lambda\iint_S e^{k_0\lambda^2 z+i\mu k_0 x} f_x(x,z)dxdz$$

When taking numerical integration, this formula is singularity integral function, $\lambda=1$ is the singularity:

$$\int_1^2 |A(\lambda)|^2 \varpi \frac{\lambda^2}{\sqrt{\lambda^2-1}} d\lambda = |A(1)|^2 \int_1^2 \frac{\varpi}{\sqrt{\lambda^2-1}} d\lambda + \int_1^2 \frac{\lambda^2|A(\lambda)|^2 - |A(1)|^2}{\sqrt{\lambda^2-1}} \varpi d\lambda \qquad (3)$$

$$\varpi = 1 + \cos\left(2bk_0\lambda\sqrt{\lambda^2-1}\right)$$

Deform to nonsingularity integral function:

$$R_w = \frac{4\rho V^2 k_0^2}{\pi}\left[|A(1)|^2 \int_1^2 \frac{\varpi}{\sqrt{\lambda^2-1}} d\lambda + \int_1^2 \frac{\lambda^2|A(\lambda)|^2 - |A(1)|^2}{\sqrt{\lambda^2-1}} \varpi d\lambda + \sum_{n=2}^{\infty}\int_{2^n}^{2^{n+1}} \frac{\lambda^2|A(\lambda)|^2}{\sqrt{\lambda^2-1}} \varpi d\lambda\right]$$

$$\varpi = 1 + \cos\left(2bk_0\lambda\sqrt{\lambda^2-1}\right) \qquad (4)$$

About $A(\lambda)$ use the formula as follow in integral operation:

$$\int_{x_1}^{x_1+2h} e^{kx} f(x)dx \approx \frac{e^{kx_1}}{k}\left[e^{2hk}\left(-\frac{0.5f_1 - 2f_2 + 1.5f_3}{hk} + f_3 + \frac{\Delta^2 f}{(hk)^2}\right) - \frac{1.5f_1 - 2f_2 + 0.5f_3}{hk} - f_1 - \frac{\Delta^2 f}{(hk)^2}\right]$$

$$\begin{cases} f_1 = f(x_1) \\ f_2 = f(x_1+h) \\ f_3 = f(x_1+2h) \\ \Delta^2 f = f_1 - 2f_2 + f_3 \end{cases} \qquad (5)$$

$f(x)$'s approximate result is a quadratic polynomial, k can be real number or plurality.

3.2.2Square tail influencing correction approach

We can add a virtual length to the square tail to calculate wave making resistance at the end, this is the so called "virtual length approach". According to this approach the respective length should include hull surface and the virtual length in wave making resistance calculating. As projection approximation, this approach just get little deviation. [3]

3.2.3Numerical calculation approach and knuckle influence

Firstly we use Simpson numerical calculation approach, and add the waterline density regularly to calculate the influence of knuckle. Then we add state lines at the position where WPC's cross section area varying curvature is big, so the data used in the theoretical calculation can fully reflect the influence caused by the knuckle. This approach is very simple, but it needs more water lines and state lines, so its calculated amount is large. [4]

Simpson numerical calculation formula:

$$\int_a^b f(x)dx = \frac{h}{3}[f(x_1) + 4f(x_2) + f(x_3)] \qquad (6)$$

In this formula a、b are real numbers, $f(x)$ is real number function. As deviation evaluation formula, we need two or three vertical coordinates when using trapezoid theorem to integral. If all the following conditions are satisfied we believe the evaluative deviation satisfies the requirement:

(1)The deviation must be smaller than the given limitation;

(2)The time the deviation satisfies the requirement must more than twice;

(3)The real difference between a、b must less than twice the maximal interval allowed.

Interval calculating: $DX = \dfrac{X(L-1)+Y(L+1)}{2}$

Deviation calculating: $ADD = \left| \left(Y(L) - \dfrac{Y(L+1)+Y(L-1)}{2} \right) \times DX \right|$

Reference:

[1]Liu dongwei, Zhu meiqi, Dong yuansheng, High-speeding Catamaran Resistance Calculating, 1998 (2), 19-23

[2]Xu tongquan, Xie kezhen, Yang huizhong, Zhou deli, High-speeding Catamaran Resistance and Wave-making Performance Experimental Research, Ministry of transport Shanghai watercraft transport scientific institute transaction,VOL17, NO.2, 1994.10

[3]T.H.Havelock.WaveResistance-The Mutual Action of Two Bodies. Proc.Roy.Soc.A 155.1936

[4]Edward.D.Fry, Timothy Grual. Design and Application of Modern High-speed Catamarant.Marine Techology, VOL19,NO.2, July 1972

The feature of the hull form and performance of 2500t grade SWATH Oceanography Survey Vessel

Ge Wei-Zhen Sui Ai-Guo Yang Dong Song Guo-hua

(China Ship Scientific Research Center)

Abstract

According to the requirements of the item of 'New type synthetic Scientific Research Vessel' submitted by Chinese Academy of Science, this vessel must be satisfied to operate on unrestricted navigating area except ice-zone and can be able to engage in the synthetic scientific research works on sea and offshore, such as underwater acoustics, oceanography physics and chemistry, oceanography geology, marine biology and etc.

SWATH has been chosen by Chinese Academy of Science as hull form of oceanography survey vessel and CSSRC undertook the job of design. The synthetic navigating feature of the vessel is perfect. The performance of the vessel is excellent as used for scientific research and can be compared favorably with AGOR26 built in 2002.

1. Introduction

21st century is the era of Ocean. Exploitation of Ocean is one of significant strategy demands for an ocean country. Chinese Academy of Science is responsible to the discovery and research of Ocean. Seeing that some of the available vessels are too old and the others are out of data so the most urgent task on hand is to build an advanced Oceanography Survey Vessel for Chinese Academy of Science (OSV for CAS).

According to the requirement of design, the OSV for CAS must fulfill the demands of unrestricted navigation area except ice-zone and can be able to engage in the synthetic scientific research works on sea and offshore such as under water acoustics, oceanography physics and chemistry, oceanography geology, marine biology and etc. and can undertake the establishment, regulation and supervision of the global and the large scale networks of survey.

About the unrestricted navigation, it requires not only the safe sailing but also the suitable environment for scientific survey. It is well know that if the vertical acceleration exceeds 0.2g one can't read the data on the apparatus. Nothing is more important than the correct choice of vehicle type for survey vessel. SWATH is one of high performance vehicles with unique feature. Good seakeeping for SWATH means that motions of SWATH are long in period of roll, pitch and heave and low in vertical acceleration even on rough sea and the deck area of SWATH is spacious offering a comfortable working space. The submerged bodies are suitable for installation of underwater acoustic instruments. It can't be more right to select SWATH, so did Chinese Academy of Science and CSSRC undertook the job of design.

In order to develop SWATH technique lots of experts of all relevant discipline were organized together in the later period of 20th century and superior force was mustered to surmount the technical keys at CSSRC. Part of the results was applied to the design of supervise craft of National Custom the code of which are HJ201 and HJ202. Navigating records of two crafts for near about five years fully proved their superior performance. The ride quality of two crafts, HJ201 and HJ202, has been won extensive praise.

In this turn of the design of OSV for CAS (Oceanography Survey Vessel for Chinese Academy of Science) lots of specialist of design joined in forming a combine of research and design with tremendous force to overcome various difficulties.

Now the contract design has been approved and the detailed design is in progress.

2.The technical feature of OSV for CAS

2.1 Principal dimensions and particulars

- Length Overall Loa=60.90m Length of water line L_{WL}=50.79m

  Depth, Molded H=10.50m Breadth, Molded B=26.00m

  Draft, Design Td=6.5m Draft for pier Tm≯5.0m

  Displacement △=2569t General tonnage GT=3060

- Speed

  max Vd=15.0kn economical Ve=11.0kn

  min Vc<2.0kn

- Cruising range/endurance 8000 nmile/40day

- Seakeeping

  safe sailing in 11 Beaufort wind scale

  normal operation in 6 sea state

  roll angle ≯5°   pitch angle ≯3.5°   vertical acceleration ≯0.10g

  nature periods of motion   roll 20.8s / pitch 15.6s / heave 9.5s

2.2 Technical feature

• Technology to obtain SWATH hull form in CSSRC has been received the authoritative right of national patent in China.

• Hull form of SWATH provides enjoyable voyages and offers a quiet and comfortable environment so it is very applicable for research works on sea and offshore.

• Good seakeeping ensures the vessel safe sailing in 11 Beaufort wind scale and normal operation in 6 sea state. Long nature periods mean that it will prevent the resonance from sea waves in a great extent. The rate of sorties would be raised from 55% of conventional monohull to 85% and the amplitude of motions would be reduced to 1/3~1/5 of that of conventional monohull.

• Propulsion System is composed of Diesel-Electric System with alternating current frequency modulation control device and directly driving without reduction gear. It is an optimal matched power system of high operational reliability and economy.

• Utilizing the advantages of twin hull, twin propellers and twin rudders, combining with the active bow propulsive unit, the automatic remote controlled dynamic positioning system is established optimistically.

• During Sailing, a pair of the movable fins makes the trim angle controllable, it results in the decrease of ship resistance and fuel consumption too.

3. Difficulties in design

In the course of research and design a lot of difficulties in technique were surmounted and it is not easy to describe in short words at this paper. Here we'll only discuss something directly about general arrangement and hull form.

3.1 The requirement of 'Double Draught'.

Owing to the special demands of anchorage area and waterway, the draft must less than 5.0m in fresh water area and the draught should be 6.5m on sealing in sea- water. It is a strict requirement causing great difficulties in ship design, because the load and draft are very sensible to SWATH. Dealing with alternative design approaches, painstaking investigation and coordinated design, the sticky problems have been solved successively within the work team.

3.2 'Twin Speed' for optimum object

Based on the 'Contract Design' the max speed of OSV for CAS is 15kn and the economical speed is 11kn. In the step of approval design some equipment was changed and some equipment was added, it resulted in the increase of the displacement about $57m^3$. It is the demand that the shipowner requested to keep the installed power and the maximum speed both unchanged although displacement was raised. From the view of hydrodynamic's point it is impossible or to say the least it is very difficult. In order to overcome the difficulty, a special kind of optimization technology for hull form and lines were developed and a series of model tests were performed. It is marvelous at last that not only the displacement is raised near 2.2% and the max speed remains unchanged, but also the propulsive power is reduced from 991.7kW to 967.8kW and fuel consumption is decreased by 2.4% at economical speed.

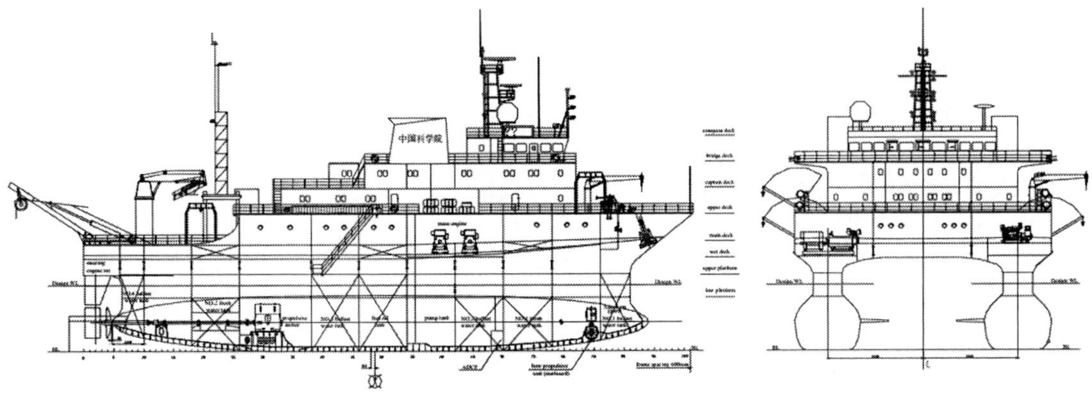

Fig1 A part of overall layout of Oceanography Survey Vessel for Chinese Academy of Science

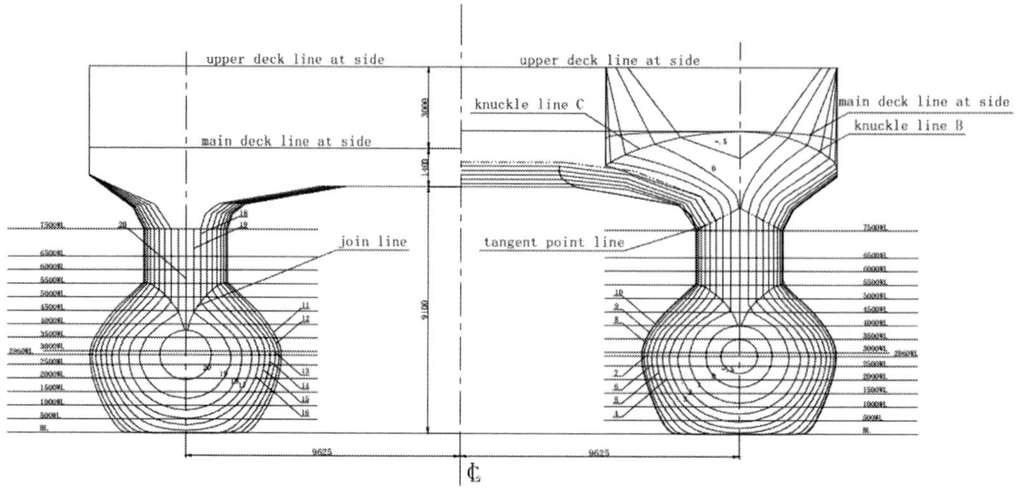

Fig2 Body plan of Oceanography Survey Vessel for Chinese Academy of

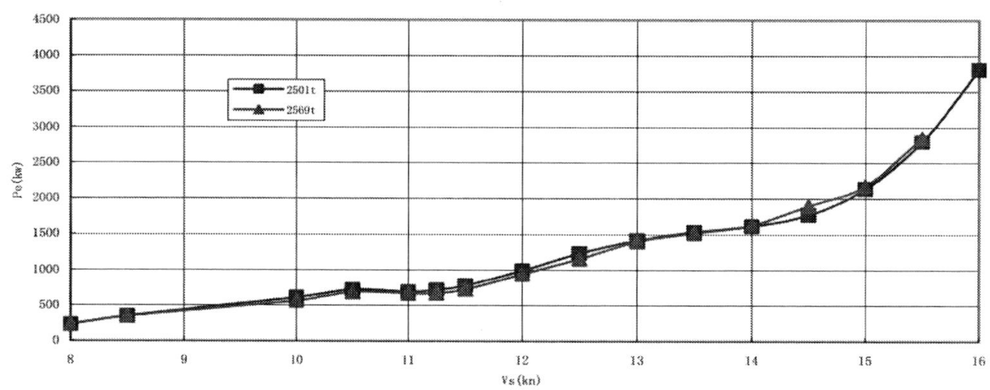

Science

Fig3 Effective power curves of Oceanography Survey Vessel for Chinese Academy of Science

3.3 Habitability

As mentioned above, this vessel will provide a comfortable working and living platform because of her superior performance. Comparing with similar type of SWATH in the world, AGOR26, OSV for CAS would be rather outstanding.

4. Remarks

4.1 Innovation and advancement

The mission and performance of OSV for CAS are similar to that of AGOR26 built by US in 2002. She will be a research vessel of high quality and economy.

The completion of this vessel would promote the undertakings of ocean exploration and ship building significantly and would enhance the ability of competition of national science and technology.

Table 1

speed (kn)	wave direction (°)	encounter wave		Single significant Amplitude (SSA)				
		$H_{1/3}$ (m)	Sea state	pitch (°)	roll (°)	vertical acceleration at midship		heave (m)
						m/s	g	
0	180	3.25	5	2.39		0.51	0.05	1.35
	135	3.25	5	2.13	2.0	0.56	0.06	1.44
	90	3.25	5	1.05	5.0	0.71	0.07	1.58
	0	3.25	5	3.31		0.45	0.04	1.23
10	180	4.0	5~6	2.40		0.76	0.07	1.45
12	180	4.0	5~6	2.32		0.73	0.07	1.28
15	180	4.0	5~6	2.25		0.73	0.07	1.30
6	180	5.0	6	3.46		0.75	0.08	2.42
8	180	5.0	6	3.75		0.77	0.08	2.45
10	180	5.0	6	3.62		0.78	0.08	2.51

Table 2 The seakeeping compared with monohull(18000t) and similar SWATH

Speed (kn)	wave direction (°)	motion	monohull (18000t)	2500t grade SWATH	
				AGOR26	OSV for CAS
0	90	heave (m)	2.27	1.40	1.65
	180	pitch (°)	2.51	2.00	2.30
	90	roll (°)	9.31	2.00	5.00
	180	vertical acceleration at midship (g)	0.15	0.06	0.07

Table 3

No			AGOR26	OSV for CAS	OSV for CAS Compared with AGOR26
1	Principal Dimension	Loa (m)	56.56	60.9	longer by 4.34m
		B (m)	26.82	26.0	more narrow by 0.82m
		T (m)	7.62	6.5	more shallow by 1.02m
2	Crew & scientist		17+31=48	27+45=72	1.33 times
3	Habitable cabin		Chief scientist 1set Capt dayroom/bedroom1set Single bedroom for crew and scientist 6set Double bedroom 20set	Single dayroom/bedroom with W.R 3set Single bedroom with W.R 13set double bedroom with W.R 8set trio bedroom with W.R 4set Four-some bedroom with W.R 7set	Rooms are bigger Amount is greater
			Total 28 set	Total 35 set	
4	Laboratory area(m^2)		279	407.43	1.59 times
5	Poopdeck area(m^2)		205	405.6	1.98 times
6	Dining-room (m^2)		28	54	1.93 times
7	Mess room(m^2)		55	96	1.75 times

4.2 The method of wave induced loading calculation and the basis in construction design will be described in another papers, say, 'The determination of wave induce loads for 2500t grade SWATH vessel'.

Key words SWATH Optimizing design

Reference

1. 'Technical requirement and task in ship design of OSV for CAS', Management team of OSV for CAS, 2003 (in Chinese)

2. 'Specification of OSV for CAS', CSSRC 913A-100-002SM, CSSRC, 2005 (in Chinese)

3. 'Overall layout of OSV for CAS', CSSRC 913A-100-003, CSSRC, 2005 (in Chinese)

4. 'Lines and Offsets of OSV for CAS', CSSRC 913A-100-004, CSSRC, 2005 (in Chinese)

5. 'Report of Model test for seakeeping of OSV for CAS', CSSRC 913-060-002SY, CSSRC, 2005 (in Chinese)

6. 'Report of Model test for resistance and self-propulsion of OSV for CAS', Towing Tank of SJTU, No.20040525 (in Chinese)

7. 'Report of Model towing test of OSV for CAS', CSSRC 913A-060-001SY, CSSRC, 2005 (in Chinese)

8. 'Skewed propeller design and Performance prediction of OSV for CAS', CSSRC Report, 2005.8 (in Chinese)

RESEARCH ON RESISTANCE PERFORMANCE OF HYSWATH

Liu Ya-dong, Feng Yong-jun
(SHANGHAI JIAOTONG UNIVERSITY)

Author: Liu Ya-dong, female, SHANGHAI JIAOTONG UNIVERSITY, lecturer
Feng Yong-jun, male, SHANGHAI JIAOTONG UNIVERSITY, Engineer
SHANGHAI JIAOTONG UNIVERSITY, School of Naval Architecture and Ocean Engineering Design & Research Institute of Ship and Ocean Engineering

abstract:

Based on the analysis of resistance changing principles of the HYSWATH, the resistance calculation of HYSWATH ship in foil borne state, including wave-making resistance, viscous resistance, hydrofoil resistance, disturbing resistance between the hull and hydrofoil and so on is discussed and a draft and trim angle calculating method of HYSWATH in foil borne state is gained.

1. Introduction

Hybridizing the different high-speed ship, learning from others' strong points to offset one's weakness and obtaining the better performance, i.e. taking the procedure of "hybrid" ship type, is a trend in development on high performance ship at present. This concepts combined the three existing sustention source: unpowered static lift (buoyancy), powered static lift (the lift of air cushion) and dynamic lift onto a new ship in different proportions.

Hydrofoil will be sated up in fore part and after part between the two slice body and that the distance between the fore hydrofoil and the after one will be carefully adjusted, thereby when the ship is sailing with high-speed, the hydrofoil can bear on 50%~60% of buoyancy, at this time, the displacement of slice bodies accordingly is decreased. In theory, the area of wetted surface will be reduced by 35%~45, and the frictional resistance will be reduced by 45%~55%. Thus, the speed of SWATH will be advanced. The moment induced by the two hydrofoil can balance the Mank moment so that this type ship can sailing with safety in high speed. So this new type ship that focus advantages of SWATH and Hovercraft is a HYSWATH with perfect performance about speediness, seakeeping and stability.

The resistance of HYSWATH mainly consists of hull resistance, hydrofoils resistance and interacting resistance between hull and hydrofoil. This article mainly studies the calculation methods of HYSWATH's resistance and its motion of carriage, when it enters into the stage of foil borne, comparing the results with them of model test, in order to verify reliability of the methods, and meanwhile, initially discuses the regularity how Principal Dimensions of HYSWATH impact on resistance.

2. Resistance calculation of HYSWATH ship

The resistance of hydrofoils themselves varies with the lifting force of hydrofoil, which also changes the hull's draught and trim angle, and meanwhile, variety in carriage on foil borne is to result in variety in the lifting force of hydrofoils. Therefore, calculating hull resistance of HYSWATH, its carriage of foil borne should be taken account.

The hull resistance of HYSWATH is comparatively complicated, as includes upper body resistance、strut resistance、main body resistance and hydrofoil resistance, and that, each part of which could be subdivided into wave resistance, viscous resistance, and spray resistance. When in low speed, the lifting force of hydrofoils is comparatively low; upper part of hull still doesn't lift off water surface; at the moment, HYSWATH is in the stage of body borne; wave resistance induced by strut and main body only count for a small proportion of total, as thus could be omitted; hull resistance is similar to regular ship of displacement type, as can be calculated in traditional method, thereby this article doesn't lay stress on it.

Along with speed increasing, the lifting force of hydrofoils also increases, upper part of hull lifts off water surface, and waterline is located at strut. When HYSWATH enters into the stage of foil borne, hull resistance consists of main body resistance and strut resistance, at the moment, the mathematical model of HYSWATH's resistance is similar to the Small Water Surface Twin-Hull ship, however the draught on foil borne is still a unknown variable, as also the trim angle needs to be gotten through solving equilibrium equation of force and moment acted on hull, after that, calculation methods applied for the Small Water Surface Twin-Hull ship can be adopted to get wave resistance of HYSWATH.

The resistance of hydrofoils is composed of aerofoil resistance, inducement resistance and wave resistance. Aerofoil resistance resulted from water viscosity includes friction resistance and shape resistance. [6]; Inducement resistance is mainly made up of additional resistance resulted from limited wingspan. The calculation methods of hydrofoil's resistance are comparatively mature; when calculating, some methods for hydrofoil boat can be referred

2.1 Equilibrium equation of HYSWATH on Foil Borne

Take $o_1\xi\eta\zeta$ as reference frame with ship's parallel movement, original o_1 is situated at center of main body plane, take $o_1\xi$ axis to parallel water surface, and take $oxyz$ as frame system fixed with ship, original o is situated at center of design waterline of main body, ox axis in longitudinal section parallel baseline. (see fig.1)。

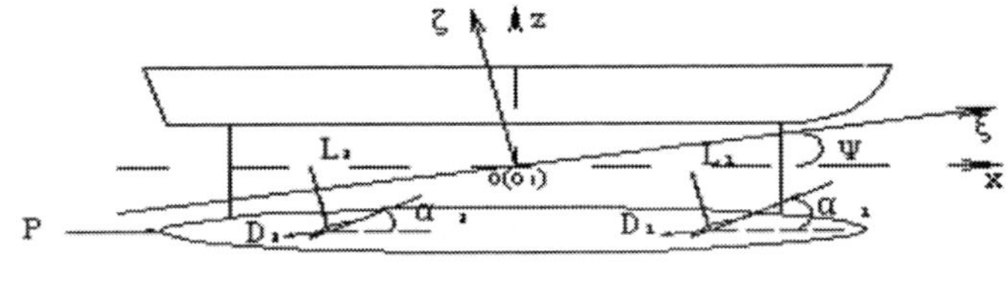

Fig.1 the Sketch Map of HYSWATH Carriage on Foil Borne

In view of outside forces acted on HYSWATH on foil borne (fig. 1), equilibrium equation of force and moment can be obtained as follows:

$$\begin{cases} P\cos\psi - R_T = 0 \\ \sum_{i=1}^{2} L_i + \rho g V - P\sin\psi - mg = 0 \\ P\cdot z_p - \sum_{i=1}^{2} L_i(x_i + z_i \sin\psi) - \sum_{i=1}^{2} D_i(z_i - x_i\sin\psi) - R_H\cdot z_H \\ \qquad - \rho g\nabla_h[x_c + (r_\psi - z_c)\sin\psi] + mg(x_g + z_g\sin\psi) = 0 \end{cases} \tag{1}$$

Signification:

P ----- thrust

ψ ----- trim angle

R_T and R_H ----- total resistance and hull resistance on foil borne

L_i and D_i ----- No. i lifting force and resistance of hydrofoils

∇_h ----- displacement on foil borne

x_i and z_i ----- x axis coordinate and z axis coordinate of installation position of hydrofoil ($i=1$ express fore. hydrofoil, $i=2$ express after hydrofoil)

x_c and z_c ----- x axis coordinate and z axis coordinate of center of buoyancy on foil borne

x_g and z_g ----- x axis coordinate and z axis coordinate of center of gravity

z_p and z_H ----- z axis coordinate of acting point of thrust and hull resistance

r_ψ ----- longitudinal metacenter radius

Suppose trim angle is very little, non-dimensional equation group can be obtained as follows:

$$\begin{cases} \sum_{i=1}^{2} C_{Zi}\bar{S}_i - \sum_{i=1}^{2} C_{Xi}\bar{S}_i\cdot\psi + \dfrac{2g\nabla_h}{V^2 S_0} - C_H\bar{S}\cdot\psi = \dfrac{2mg}{\rho V^2 S_0} \\ (\sum_{i=1}^{2} C_{Xi}\bar{S}_i + C_H\bar{S})\cdot\bar{z}_p - \sum_{i=1}^{2} C_{Zi}\bar{S}_i\cdot(\bar{x}_i + \bar{z}_i\cdot\psi) - C_{Xi}\bar{S}_i\cdot(\bar{z}_i - \bar{x}_i\cdot\psi) \\ \qquad - C_H\bar{S}\cdot\bar{z}_H - \dfrac{2g\nabla_h[\bar{x}_c + (\bar{r}_\psi - \bar{z}_c)\cdot\psi]}{V^2 S_0} + \dfrac{2mg\cdot(\bar{x}_g + \bar{z}_g\psi)}{\rho V^2 S_0} = 0 \end{cases} \tag{2}$$

$$\overline{S}_i = \frac{S_i}{S_0},\quad \overline{S} = \frac{S}{S_0},\quad \overline{x}_i = \frac{x_i}{L_0},\quad \overline{z}_i = \frac{z_i}{L_0},\quad \overline{z}_p = \frac{z_p}{L_0},\quad \overline{z}_H = \frac{z_H}{L_0}$$

Denotation:

S_0 and L_0-----specific area and specific length of HYSWATH

C_H 和 S ----- resistance and wet surface area coefficient of HYSWATH

2.2 Calculation of the draft and trim angle in foil borne state

Expression (2) can be regarded as a non-linear equation group in which draught and trim angle are taken as independent variable and speed is taken as parameter, solving with Graphic Method.

Recompose expression (2) as follows:

$$\begin{cases} p_1(z,\psi;V) = g_1(V) \\ p_2(z,\psi;V) = 0 \end{cases} \tag{3}$$

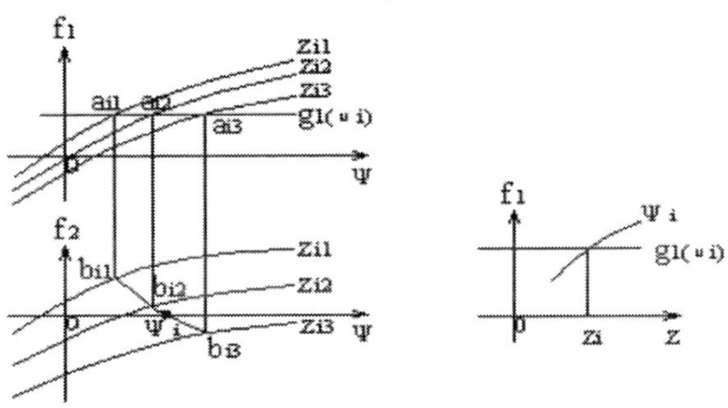

Fig.2 the Sketch Map of Solving Equilibrium Equation with Graphic Method

Within the range of foil borne, a couple of speeds are to be chosen; V_1, V_2, ... V_n.To a fixed speed V_i, a couple of draught are also to be chosen; z_{i1}, z_{i2}, ... z_{im}.To data group $(V_i, z_{ij}); j = 1,2,\cdots m$, solve left part of equation group (3) $f_{1j}(\psi) = p_1(z_{ij}, \psi; V_i)$ and $f_{2j}(\psi) = p_2(z_{ij}, \psi; V_i)$, and plot two curve groups (see fig. 2).Right part of equation group shows as a group of strait line in fig.2. (As shows as parallel in the fig. above and shows as axis $o\psi$ in the fig. below).

In the fig. above, solve intersection $a_{ij}(j = 1,2,\cdots m)$ of parallel and curve group $f_1(z_{ij}, \psi; v_i)$. Then get intersection $b_{ij}(j = 1,2,\cdots m)$, which corresponds to a_{ij}, at corresponding draught curve in the fig. below. Intersection ψ_i of the curve going through point serials b_{ij} and $o\psi$ axis is the trim angle at speed V_i. In the case trim angle ψ_i is confirmed, another data group (V_i, ψ_i) can be taken the same method as data group $(V_i, z_{ij}); j=1,2,\cdots m$, to get draught z_i. To each speed $V_i (i=1,2,\cdots n)$, same method is to be repeatedly utilized to get draught and trim angle $(z_i, \psi_i), i=1,2,\cdots n$ of HYSWATH on foil borne. Then, according to the draught and trim angle which have been solved, utilizing calculation method of the Small Water Surface Twin-Hull ship, calculate the total resistance of HYSWATH on foil borne.

3. Compared between calculation and model test of total resistance

HYSWATH is a brand new type of ship. The theoretical calculation methods for analyzing its capability of many kinds, are still need verification through the corresponding model tests, thereby disclosing its essence, further revising and perfecting these methods. The model test of HYSWATH is applied with Froude similarity code. Considering the capacity of experimental apparatus and varied conditions, scale between actual ship and model is to be taken with λ=20。Tab.1 actual dimensions of HYSWATH to be calculated.

Tab.1 Main Dimensions of HYSWATH

name data	name data	name data
Hydrostatic displacement Δ /ton 400 Displacement on foil borne Δ_B/ton 240 Main body length LB/m 32.4 Strut length LS/m 25.92 Strut thickness 2t0/m 1.02 Center distance of two Slice bodies 2b/m 10.8 Main body diameter D/m 2.02 draught(on foil borne)	------ Fore. hydrofoil ----- Fore. hydrofoil chord length c1/m 1.16 Fore. Hydrofoil span length s1/m 8.848 Fore. Hydrofoil thickness t1/m 0.105 Fore. Hydrofoil installation angle α_1/deg. 2 Fore. Hydrofoil position lf/m 12.96 Fore. Hydrofoil lifting force derivative CL1 3.12 Fore. Hydrofoil strut	------Aft. hydrofoil ----- Aft. Hydrofoil chord length c2/m 1.38 Aft. Hydrofoil span length s2/m 9.626 Aft hydrofoil thickness t2/m 0.124 Aft. Hydrofoil installation angle α_2/deg. 2 Aft. Hydrofoil position la/m 13.76 Aft. Hydrofoil lifting force derivative CL2 3.03 Aft hydrofoil strut chord length b2/m 1.38 Aft hydrofoil strut span

d/m 3.3 longitudinal position of gravity Xg /m 0.74 height of gravity KG/m 4.29	chord length b1/m 1.16 Fore. Hydrofoil strut span lengthd1/m 4.81 Fore. Hydrofoil strut thickness e1/m 1.21	length d2/m 4.81 Aft. Hydrofoil strut thickness e2/m 0.25

Data in tab. 2 and fig. 3 are hull resistance, which are gotten by theoretical calculation method as before discussed, utilizing data of the HYSWATH model with scale 1:20, which are listed in tab. 1 and gotten from corresponding model test results.

Resistance coefficient in tab. 2 and fig.3 is to be expressed as Ri/ρΔ2/3V2, amid which, Ri express resistance composition of all parts.

Tab.2 The Theoretical Calculation and Test Results of Hull Resistance of HYSWATH

speed		theoretical calculation (resistance coefficient ×103)				Model test	deviation (%)
Fn	(m/s)	Viscous resistance	Wave resistance	Hydrofoil resistance	Total resistance	Resistance coeffcient	
0.462	1.839	26.172	20.449	5.476	52.096	68.242	-23.7
0.519	2.069	25.324	19.988	8.502	53.814	67.628	-20.4
0.577	2.299	24.242	16.385	10.687	51.314	59.220	-13.4
0.635	2.529	22.249	12.126	11.566	45.941	48.163	-4.6
0.692	2.758	20.157	8.779	10.589	39.525	38.788	1.9
0.750	2.988	19.177	6.798	6.921	32.896	33.457	-1.7
0.808	3.218	18.172	5.333	4.819	28.324	29.802	-5.0
0.865	3.448	17.404	4.307	3.526	25.237	26.828	-5.9
0.923	3.678	16.967	3.593	2.972	23.533	24.894	-5.5
0.981	3.908	16.421	3.021	2.731	22.173	23.123	-4.1
1.038	4.138	16.044	2.586	2.790	21.421	21.601	-0.8
1.096	4.367	15.794	2.244	2.970	21.008	20.561	2.2
1.154	4.597	15.696	1.965	3.233	20.894	20.172	3.6
1.211	4.825	15.570	1.730	3.528	20.828	19.624	6.1

We can come to a conclusion through the analysis with comparing with data of the table, when HYSWATH is on body borne, the difference between them is larger than 20 percent, the basic reason for such large deviation is that when HYWSWATH sails at low speed, it doesn't lift off water surface, as result in that bow wave is comparatively high; wave resistance increases, and at same time, because wet surface area of hull enlarge, viscous resistance also increases, but all of these are omitted in theoretical calculation. When it enters into the stage of foil borne, deviations at most of operating speed are quite small, and results of theoretical calculation and model test are quite well coincident. The calculation method presented in this article for solving resistance of HYWSWATH on foil borne is reliable, thereby it can be utilized to estimate resistance and main engine power.

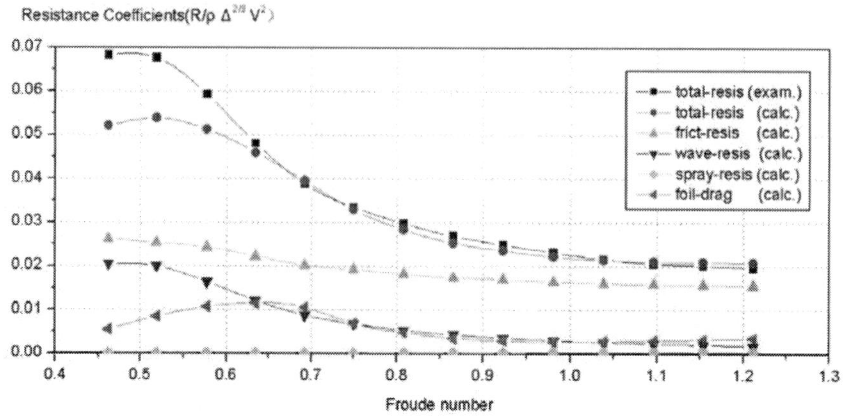

Fig.3 the resistance coefficient curve of HYSWATH model

Reference docments:

[1] Huang Dingliang. Performance theory of small water surface twin-hull ship. Bei Jin: the Defense Industry Publishing House, 1993.

[2] [Russia]Б.А. 柯雷扎耶夫等著. 许百春，丁玮，郭琪英，孙永权译. 水翼船、气垫船设计手册. 北京：国防工业出版社，1985.

[3] Satoru Ishikawa. Study on Hydrodynamic Interaction between Hull and Submerged Foils. Journal of The Society of Naval Architects of Japan, 1991, Vol.169,135-142

[4] Hoppe K.G. Hydrofoil System increases the speed of catamarans. Work Boat World, No.16(1),1997

[5] Li Baiqi. Calculation Method of Hydrostatic performance of Hydrofoil boat on foil borne. Performance study of naval and merchant ship .1980(2): 53-74

[6] Zhang Daxiong. Conversion methods of hydrofoil boat resistance. performance study of naval and merchant ship.1982(4): 1-12

Solar Cell and Its Application in Ship Industry

WEI Guang-Pu

School of Materials Science and Engineering,

Shanghai University, Shanghai, 201800, China

(gpwei@sh163a.sta.net. cn)

ABSTRACT

In China, the Research and Development (R&D) of PV technology were started in 1958. At the beginning of 1970's, c-Si solar cells with efficiency about 13 %(AM0) had been made out and then successfully used in the Chinese satellite launched in 1971. Now, about 50 institutes, universities and companies are engaging in the PV works. Until the end of 2005, about 12 solar cell plants with certain scale had been established. The annual production capacity of solar cells/modules was about 100MW and the total installed capacity of PV system was about 250MW. The highest efficiency of 20.4% for single crystal silicon solar cell ever reported. Some Application fields (including ship industry) of solar cell are described.

Key Words: Solar cell, Photovoltaic, PV system, Ship Industry

1. Introduction

Due to the economical development and population increase, the consumption of energy resource is rapidly increasing, and the global environment is degrading year by year owing to the use of fossil fuels. To solve these problems, developing clean energy resource has become one of the most important tasks assigned to modern science and technology. Solar energy has attracted worldwide attention because it is clean, inexhaustible and available everywhere in the world. Solar cell (PV system) can convert the sunlight energy to electricity energy directly, so PV industry is getting a very rapid development. In the past 5 years, the worldwide PV industry got about 30%~ 60% growth rates yearly. For example, the PV production is 287 MW in 2000, 390MW in 2001, 561MW in 2002, 741MW in 2003 and 1195MW in 2004. The highest efficiency has reached 24.7% for single crystalline Si solar cell and 36.5% for concentrator InGaP/InGaAs/Ge 3-junction stacked cell at 200 suns (AM1.5).

In China, PV industry is rapidly increasing; the PV production is 15MW in 2003, 36MW in 2004 and 100MW in 2005. Up to 2005, the estimated production capacity of China is about 10% of the worldwide capacity. China is very abundant in solar energy resource, and China has a big latent market for PV industry. So the future for PV industry is very bright.

PV system can be used in many fields. Family PV roof top system got great successful in Japan, America and Europe in recent years. Usually, a 3KW PV system can supply the electricity power for a family. Especially PV system can also be used in ship industry.

2. Rich Sunshine Resource

Since China is situated in the area between the 20th and 45th north latitude, the amount of sunshine energy is very rich. On the total 9.6 million km^2 area of China, the average solar radiation is about 140 $Kcal/cm^2$ per year, and the average power is 1.7×10^{12} KW. So the whole China receives about 5.6×10^{22} J solar energy annually, which is equivalent to 1.9×10^{12} tons of standard coal.

According to the annual sunshine time, China can be divided into five kinds of sunshine area, as shown in Fig.1.Two-thirds area of China has a sunshine time more than 2000 hr/year. The total irradiation amount ranges from 80 to 200 Kcal/cm2 per year, and decreases from west China to east China. The peak value of sunshine radiation is at Qinghai-Tibet high land.

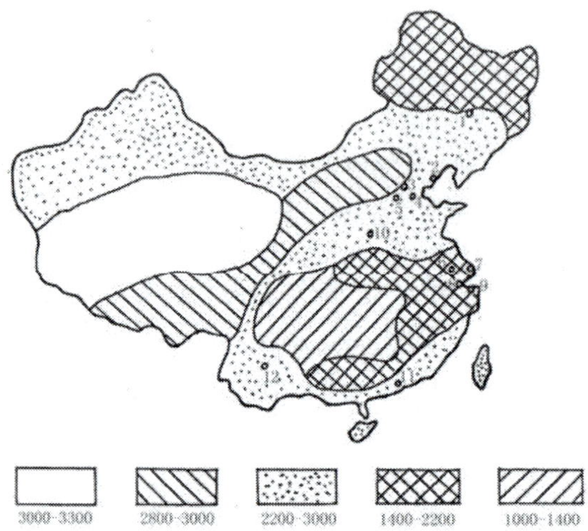

Fig.1 Distribution of solar energy resource in China. The numbers show the cities where solar cell production line was established

1-Harbin, 2-Qinhuangdao, 3-Beijing, 4-Tianjin, 5-Baoding, 6-Wuxi, 7-Shanghai, 8-Hangzhou, 9-Ningbo, 10-Keifeng, 11-Shenzhen, 12-Kunming

3. History & Present Status

In China, the Research and Development (R&D) of PV technology were started in 1958 and got significant progress in recent decades. Crystal Silicon (c-Si) solar cell was developed firstly. At the beginning of 1970's, c-Si solar cells with efficiency about 13 %(AM0) had been made out and then successfully used in the Chinese satellite launched in 1971[5]. During 1970's, some small c-Si solar cell production lines were established. Before the end of 1980's, totally seven PV cell production lines were imported from USA, Canada and other countries (include five c-Si solar cell production lines at Kunming, Ningbo, Kaifeng, Beijing, Shenzhen, and two amorphous Silicon (a-Si) thin film solar cell production lines at Harbin and Shenzhen). Their production capacity is about 4.5 MW/year. From the end of 1990's, some production lines started to make poly-crystal Si solar cells. At the beginning of 21 century, five companies were established at Wuxi, Shanghai, Baoding, Tianjin and Jingsu. They can make single-crystal Si, poly-crystal Si and a-Si(amorphous silicon) thin film.solar cells.

Now, about 50 institutes, universities and companies are engaging in the PV works (solar cells, PV materials, PV devices and systems). Until the end of 2005, about 12 solar cell plants with certain scale had been built. In 2005, the annual production capacity of solar cells/modules was about 100 MW, in which silicon crystal solar cells/modules products 95 MW, a-Si solar cell modules 5 MW. The highest efficiency of 20.4% for single crystal silicon solar cell (area 2cm x 2cm) ever reported. For commercial c-Si solar cells, the efficiency was about 14~16%. By the end of 2005, the total installed capacity of PV system was about 250MW, and the sale prices were about RMB 30~40 Yuan/W. Fig.2 shows the develop statue of annual production and the total installed capacity.

Table 1 Laboratory efficiency and size of some main photovoltaic cells in China.

Cell Type	Highest Efficiency (%)	Maximum Size
Single crystal Si cell	20.4	2cm×2cm
	15.0~17.0	125mm×125mm(applied type)
Poly-crystal Si Cell	16.5	2cm×2cm
	14.0~16.0	125mm×125mm(applied type)
Poly-crystal Si thin film cell	13.6	1cm×1cm (non active silicon substrate)
	11.2 (single junction)	A few mm^2
	11.4 (multi-junction)	
Amorphous Si thin film cell	8.55	10cm×10cm
	7.88	20cm×20cm
	6.17	30cm×30cm
GaAs Cell	20.1	1cm×1cm
Ge substrate GaAs single junction cell	20*	3cm×4cm
Ge substrate GaAs double junction cell	30*	2cm×2cm
CdS/CuS cell	12.0	A few mm^2
$CuInSe_2$ cell	8.57	1cm×1cm
CdTe cell	7.0	3 mm^2
TiO_2 nanometer organic cell	10.0	1cm×1cm (non-active Si substrate)

*: Denoting the efficiency at space condition, while the rest denoting the efficiency in ground condition.

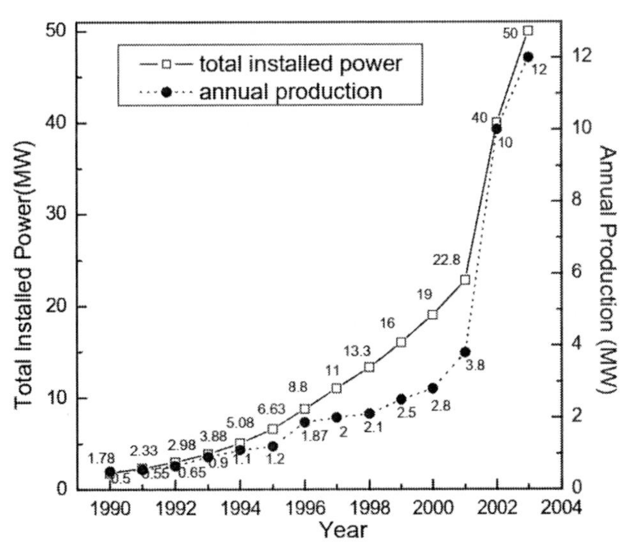

Fig.2 Progress of annual production and total installed capacity in China

Some institutes and universities are engaging the research and development of compound semiconductor solar cells, such as high efficiency GaAs solar cell, hetero-junction solar cell, concentrator solar cells, $CuInSe_2$ solar cells, CdS/CuS solar cells, CdS and SnS solar cells etc. Table 1 shows the laboratory efficiency and size of some main photovoltaic cells in China.

4. Application of Solar Cell(PV system)

In China, the PV system have been applied to many fields, such as home lighting, water pumping station, TV and broadcast system, school and hospital, microwave transmission, mobile communication, railway station signal system, cathode protection for oil/gas transmission piping line, optic fiber telecommunication system, desert virescence, etc. Specially, in order to solve the electric power supply of about 30 million people in the non-electric area of northwest China, Chinese government invested RMB 2,6 billion for the purpose of "Power Electric to Every Township"(be called "Brightness Project" or "Bright Engineering") in recent two years. From 2001 to 2003, more then 600 townships have built PV power station. A 1MW PV power station has been established at the Garden Exhibition of Shenzhen city in 2004, it is the biggest PV power station in China up to now.

The applications of PV power system in architecture are getting attention year by year. Some green houses with PV roof top systems have been built in many cities, such as Beijing, Shanghai, Wuxi and so on.

World Bank, UNGEF (United Nations Global Environment Foundation), and many countries, such as Japan, Germany, gave China a lot of supports to promote the application of PV system.

5. Application of Solar Cell(PVsystem) in Ship Industry

Fig.3 A Solar Cell Ship on the Alster River in Hamburg

Fig.3 shows a solar cell ship on the Alster River in Hamburg (Germany). It is the largest solar boat in the world before 2000, and began operating on the Alster River in Hamburg on May 23, 2000. The almost 27 meter long and 34 ton "RA 82", which was baptized "Alster Sun", glides on the river without any noise at up to 15 kilometers per hour. While cruising in front of Hamburg landmarks, the ship itself has become an attraction. 40 tons of stainless steel and a dainty roof of solar modules glitter in the sun.

The boat can hold up to 120 passengers for excursions and charter tours. The "Alster Sun" reaches a speed of 5 kilometers per hour just from solar power. If it has to go faster, the rest of the energy comes from batteries. The power reserves of the 80 batteries (2340 ampere-hours at an operating voltage of 80 volts) last for at least 12 hours or for a distance of 120-160 kilometers. The boat is powered through two 8-kilowatt marine screw propellers and casts off five times a day.

Fig.4 Solarshuttle "RA 66". (Photo: Kopf Solardesign)

The weatherproof RA 66 ship is fit for service for six to eight hours even when the sun doesn't shine, depending on the whether it is traveling at a speed of 10 kilometers per hour or at its peak speed of 15 kilometers. The generator with 52 modules produces 4.20 kWp; 24 batteries with an operating voltage of 48 volts ensure the needed reserve of 900-ampere hours. Another boat from Kopf Inc. serves as a solar hotel shuttle on Lake Lucerne. Solar boat tours are offered on the Danube in Ulm, and small, emission free ships are used to control the quality of water of reservoirs that supply the population of Bavaria and Baden-Wuerttemberg with drinking water.

Fig.5 Solar ship "Neckarsonne"

Since June 2004 Heidelberg has yet another attraction: a solar-boat made of glass which will cruise the Neckar River driven by pure solar energy. The equipment includes an interior deck seating up to 80 persons, an open air deck for 30 persons and two bistro bar counters. The program provides Neckar Round Trips (50 minutes) as well as trips during special events such as the Heidelberg Castle Illuminations and special charter trips for groups.

In China, a solar ship was made by Sichuan chengdou Travel Company in 2001.Some solar travel boats were also be made in other places such as Zhejiang and Jilin provinces

In Australia and many other countries, many solar cell ships were made.

Solar cell (PV systems) can also be used in big ship, because in such ship, many places and big area can be utilized for assembly of solar cell, and it is significant for conserve energy and ship safety, especially for lifesaving boat.

6. Big Potential Market and Bright Future

According to the statistical data of the end of 2002, there are more than 1061 townships, including about 20000 villages, 9 million households and 30 million people still without electric power supply in the remote areas of China. Photovoltaic (PV) or wind electric generator is the most suitable power supply. On the other hand, PV power system is also suitable for electric power supply in many fields such as telecommunication stations, highway maintenance equipment, oil/gas transmission piping lines, railway stations, and soldier stations in the remote areas of China.

Due to the large population, the requirements of energy resource are very big. For example, about 100 million tons petroleum was imported to China in 2003. The use of fossil fuels brings on serious pollution of environment. So develop the renewable energy is very important for China. Utilization of the rich solar energy resource is a good way. Specially, family PV systems have a bright future. So there is a great application market for photovoltaic industry in China. China government is making detailed plans for development of PV systems.

7.Conclusion

The history and present status of photovoltaic science and engineering were described. PV science and industry have a bright future in the China and all over the world. There is also a bright future for application of PV system in ship industry.

Reference.

1. Yan Luguang: Proceeding of Shanghai environment protection and Renewable Energy International forum. P.50, Shanghai, China, 2003.
2. Zhao Yuwen and Cui Rongqiang: "The 21st Century's New Technology of Solar Energy"-----Proceeding of 2003 Anniversary Solar Energy Conference of China Solar Energy Society, p.9, Shanghai, China, 2003.
3. Li Yi: Technical Digest of the International PVSEC-14, p.537, Bangkok, Thailand, 2004.
4. Liu Zuming etc.: Technical Digest of the International PVSEC-14, p.1055, Bangkok, Thailand, 2004.
5. Xu Honghua etc.: Technical Digest of the International PVSEC-14, p.1081, Bangkok, Thailand, 2004.
6. Arnulf Jager-Waldau: PV Status Report 2003, (European Commission Joint Research Center), 2003.
7. Arnulf Jager-Waldau: PV Status Report 2004, (European Commission Joint Research Center), 2004.
8. Shi Dinghuan: Proceeding of 15th International Photovoltaic Science and Engineering Conference, Shanghai, China,October2005.

Study on optimization of automatic control parameters for Ekranoplan（WIG Ship）

ZHOU Qian, XU Xing-fa, YUAN Chang-hua

(China Ship Scientific Research Center, Wuxi 214082, China)

Abstract: In this paper, a method that based on the space motion equation of ekranoplan for optimization of automatic control parameters is introduced.

First, the space motion equation of 《Ekranoplan + Automatic Control System》 is established. And then **MATLAB** is chosen as the main software support for establishing the mathematical model of optimization system with the help of its non-linear multivariable function **fmincon ()**. Finally the optimum procedure of automatic control parameters is programmed on the platform **MATLAB/GUIDE**, and the optimization calculation of automatic control parameters can be carried out.

The results show that, through optimized calculation of automatic control parameters of ekranoplan, optimal results may be achieved, and for solving the questions of the dynamic stability and maneuverability of ekranoplan, it is unnecessary purely to depend on the aerodynamic design only.

Key words: ekranoplan; the space motion equation; automatic control parameters; optimization

1 Introduction

Usually the questions about the motion stability and maneuverability of ekranoplan are solved using the aerodynamic design, but there exist certain limitations, under some situations it is often difficulty to satisfy the purpose, and when an operating region is quite wide, the limitation is more outstanding. Aerodynamic properties of ekranoplan without automatic control system are completely determined by the aerodynamic design and the structural parameters of ekranoplan, and it is extremely difficult to satisfy the prescriptive technical specifications through modifying the aerodynamic design, because the aerodynamic design often has many times been modified, and certain sacrifices (for example aerodynamic quality or structural weight and so on) must be made for design, even under some situations it is impossible to modify the aerodynamic design. Therefore an ekranoplan needs to be equipped with the automatic control system, and it is possible to satisfy all kinds of technical specifications for motion stability of ekranoplan through reasonable selection of the automatic control parameters, for instance, period of vibration, maximum amplitude of vibration, time of attenuation half amplitude and so on. Compared with modifying the aerodynamic design this should be much easier.

Therefore, using the space motion equation with control parameters, the optimized computation of automatic control parameters for ekranoplan may be carried out by the design variable, the objective function and the constraint condition respectively, until the optimal automatic control parameters which satisfy the purpose are obtained.

2 The space motion equation of ekranoplan with automatic control parameters

Let $X_1 = \Delta\overline{V}$, $X_2 = \Delta\alpha$, $X_3 = \Delta\overline{H}$, $X_4 = \Delta\vartheta$, $X_5 = \Delta\beta$, $X_6 = \Delta\phi$, $X_7 = \Delta\overline{\omega}_x$, $X_8 = \Delta\overline{\omega}_y$, $X_9 = \Delta\overline{\omega}_z$, $X_{10} = \Delta\psi$, $X_{11} = \Delta\delta_{ta}$, $X_{12} = \Delta\delta_{ea}$, $X_{13} = \Delta\delta_{fa}$, $X_{14} = \Delta\delta_{aa}$, $X_{15} = \Delta\delta_{ra}$.

And $K_{\dot{V}} = k_1$, $K_V = k_2$, $K_{\dot{\vartheta}} = k_3$, $K_\vartheta = k_4$, $K_{\dot{H}} = k_5$, $K_H = k_6$, $K_{\dot{\phi}} = k_7$, $K_\phi = k_8$, $K_{\dot{\psi}} = k_9$, $K_\psi = k_{10}$.

$a_{11,1}^0 = \frac{k_1}{\tau}$, $a_{11,1} = k_2$, $a_{12,4}^0 = -\frac{k_3}{\tau}$, $a_{12,4} = -k_4$, $a_{13,3}^0 = \frac{k_5}{\tau}$, $a_{13,3} = k_6$, $a_{14,6}^0 = -\frac{k_7}{\tau}$, $a_{14,6} = -k_8$, $a_{15,10}^0 = -\frac{k_9}{\tau}$, $a_{15,10} = -k_{10}$.

The linear and dimensionless space motion equation of ekranoplan with automatic control parameters is educed as following:

$$
\begin{cases}
a_{11}^0 \frac{dX_1}{d\bar{t}} + a_{11}X_1 + a_{12}^0 \frac{dX_2}{d\bar{t}} + a_{12}X_2 + a_{13}^0 \frac{dX_3}{d\bar{t}} + a_{13}X_3 + a_{14}^0 \frac{dX_4}{d\bar{t}} + a_{14}X_4 + a_{15}^0 \frac{dX_5}{d\bar{t}} + \\
+ a_{15}X_5 + a_{16}X_6 + a_{17}X_7 + a_{18}X_8 + a_{19}X_9 + a_{1,11}X_{11} + a_{1,12}X_{12} + a_{1,13}X_{13} + a_{1,15}X_{15} = \\
= -a_{1,11}\Delta\delta_{tc} - a_{1,12}\Delta\delta_{ec} - a_{1,13}\Delta\delta_{fc} - a_{1,15}\Delta\delta_{rc} + \Delta C_{xp} \\
a_{21}^0 \frac{dX_1}{d\bar{t}} + a_{21}X_1 + a_{22}^0 \frac{dX_2}{d\bar{t}} + a_{22}X_2 + a_{23}^0 \frac{dX_3}{d\bar{t}} + a_{23}X_3 + a_{24}^0 \frac{dX_4}{d\bar{t}} + a_{24}X_4 + a_{25}^0 \frac{dX_5}{d\bar{t}} + \\
+ a_{25}X_5 + a_{26}X_6 + a_{27}X_7 + a_{28}X_8 + a_{29}X_9 + a_{2,11}X_{11} + a_{2,12}X_{12} + a_{2,13}X_{13} + a_{2,15}X_{15} = \\
= -a_{2,11}\Delta\delta_{tc} - a_{2,12}\Delta\delta_{ec} - a_{2,13}\Delta\delta_{fc} - a_{2,15}\Delta\delta_{rc} - \Delta C_{yp} \\
a_{31}^0 \frac{dX_1}{d\bar{t}} + a_{31}X_1 + a_{33}^0 \frac{dX_3}{d\bar{t}} + a_{33}X_3 + a_{34}^0 \frac{dX_4}{d\bar{t}} + a_{34}X_4 + a_{35}^0 \frac{dX_5}{d\bar{t}} + \\
+ a_{35}X_5 + a_{36}X_6 + a_{37}X_7 + a_{38}X_8 + a_{3,12}X_{12} + a_{3,13}X_{13} + a_{3,15}X_{15} = \\
= -a_{3,12}\Delta\delta_{ec} - a_{3,13}\Delta\delta_{fc} - a_{3,15}\Delta\delta_{rc} + \Delta C_{zp} \\
a_{41}X_1 + a_{43}X_3 + a_{44}X_4 + a_{45}X_5 + a_{46}X_6 + \frac{dX_7}{d\bar{t}} + a_{47}X_7 + a_{48}^0 \frac{dX_8}{d\bar{t}} + a_{48}X_8 + \\
+ a_{4,14}X_{14} + a_{4,15}X_{15} = -a_{4,14}\Delta\delta_{ac} - a_{4,15}\Delta\delta_{rc} + \Delta\overline{m}_{xp} \\
a_{51}X_1 + a_{53}X_3 + a_{54}X_4 + a_{55}X_5 + a_{56}X_6 + a_{57}^0 \frac{dX_7}{d\bar{t}} + a_{57}X_7 + \frac{dX_8}{d\bar{t}} + a_{58}X_8 + \\
+ a_{5,14}X_{14} + a_{5,15}X_{15} = -a_{5,14}\Delta\delta_{ac} - a_{5,15}\Delta\delta_{rc} + \Delta\overline{m}_{yp} \\
a_{61}X_1 + a_{63}^{00} \frac{d^2X_3}{d\bar{t}^2} + a_{63}^0 \frac{dX_3}{d\bar{t}} + a_{63}X_3 + a_{64}^0 \frac{dX_4}{d\bar{t}} + a_{64}X_4 + a_{65}X_5 + a_{66}X_6 + \frac{dX_9}{d\bar{t}} + \\
+ a_{6,11}X_{11} + a_{6,12}X_{12} + a_{6,13}X_{13} = -a_{6,11}\Delta\delta_{tc} - a_{6,12}\Delta\delta_{ec} - a_{6,13}\Delta\delta_{fc} + \Delta\overline{m}_{zp}
\end{cases}
$$

$$\begin{cases} a_{71}X_1 + a_{72}X_2 + \dfrac{dX_3}{d\bar{t}} + a_{74}X_4 + a_{75}X_5 + a_{76}X_6 = 0 \\ \dfrac{dX_4}{d\bar{t}} + a_{88}X_8 + a_{89}X_9 = 0 \\ \dfrac{dX_6}{d\bar{t}} + a_{97}X_7 + a_{98}X_8 + a_{99}X_9 = 0 \\ \dfrac{dX_{10}}{d\bar{t}} + a_{10,8}X_8 + a_{10,9}X_9 = 0 \\ a^0_{11,1}\dfrac{dX_1}{d\bar{t}} + a_{11,1}X_1 + a^0_{11,11}\dfrac{dX_{11}}{d\bar{t}} + X_{11} = 0 \\ a^0_{12,4}\dfrac{dX_4}{d\bar{t}} + a_{12,4}X_4 + a^0_{12,12}\dfrac{dX_{12}}{d\bar{t}} + X_{12} = 0 \\ a^0_{13,3}\dfrac{dX_3}{d\bar{t}} + a_{13,3}X_3 + a^0_{13,13}\dfrac{dX_{13}}{d\bar{t}} + X_{13} = 0 \\ a^0_{14,6}\dfrac{dX_6}{d\bar{t}} + a_{14,6}X_6 + a^0_{14,14}\dfrac{dX_{14}}{d\bar{t}} + X_{14} = 0 \\ a^0_{15,10}\dfrac{dX_{10}}{d\bar{t}} + a_{15,10}X_{10} + a^0_{15,15}\dfrac{dX_{15}}{d\bar{t}} + X_{15} = 0 \end{cases}$$

Where $V_0 = \sqrt{\dfrac{2G}{\rho S C_{y_0} \cos\phi_0}}$, $\tau = \dfrac{2G}{g\rho V_0 S}$, $\mu_c = \dfrac{2G}{g\rho S\bar{C}}$, $\mu_b = \dfrac{2G}{g\rho Sb}$, $\Delta\bar{V} = \dfrac{\Delta V}{V_0}$, $\Delta\bar{H} = \dfrac{\Delta H}{\bar{C}}$,

$\Delta\bar{\omega}_x = \dfrac{b}{2V_0}\Delta\omega_x , \Delta\bar{\omega}_y = \dfrac{b}{2V_0}\Delta\omega_y , \Delta\bar{\omega}_z = \dfrac{\bar{C}}{V_0}\Delta\omega_z$.

3 Mathematical model of optimized systems

(1) Design Variables

In this paper, the automatic control parameters in the space motion equation of ekranoplan are taken as the design variable. These automatic control parameters are: automatic control parameter K_V and $K_{\dot{V}}$ corresponding with the velocity; automatic control parameter K_ϑ and $K_{\dot{\vartheta}}$ corresponding with the longitudinal motion pitch; automatic control parameter $K_{\bar{H}}$ and $K_{\dot{\bar{H}}}$ corresponding with the longitudinal motion flying altitude; automatic control parameter K_ϕ and $K_{\dot{\phi}}$ corresponding with transversal motion heel; automatic control parameter K_ψ and $K_{\dot{\psi}}$ corresponding with transversal motion yaw respectively.

(2) Constraint Conditions

In this paper, the constraint conditions are the boundary conditions of automatic control parameters, the lower and upper bounds of these constraint conditions which are based on our experience can be given below:

$K_{\dot{V}} = 0 \sim 0.5$, $K_V = 0 \sim 0.8$, $K_{\dot{\bar{H}}} = 0 \sim 1.1$, $K_{\bar{H}} = 0 \sim 0.6$, $K_{\dot{\vartheta}} = 0 \sim 1.3$, $K_\vartheta = 0 \sim 1.8$,

$K_{\dot{\phi}} = 0 \sim 1.2$, $K_{\phi} = 0 \sim 1$, $K_{\dot{\psi}} = 0 \sim 5$, $K_{\psi} = 0 \sim 1.5$.

lb= [0 0 0 0 0 0 0 0 0 0];

ub=[0.5 0.8 1.1 0.6 1.3 1.8 1.2 1 5 1.5].

Here, lb is the lower bounds of automatic control parameters, ub is the upper bounds of automatic control parameters.

(3) Objective Function

In this paper, the square of attenuated value for initial amplitude of motion parameters is defined as the objective function:

$$J = \sum_{i}^{m} (f_i(X_i) - f_{0i}(X_i))^2$$

Where $f_{0i}(X_i)$ is the initial amplitude of the motion parameter under the condition of initial automatic control parameters; $f_i(X_i)$ is the initial amplitude of the motion parameter under the condition of optimized automatic control parameters; X_i is the motion parameter of ekranoplan, here $i = 1$ represents the velocity V, $i = 3$ represents the flight height $\bar{H}$, $i = 4$ represents the pitch angle ϑ, $i = 5$ represents the sideslip angle β, $i = 6$ represents the bank angle ϕ, $i = 10$ represents the yaw angle ψ. To find the optimal values of design variables and to make the value of the objective function J minimum are our aim.

4 Optimization calculation

In this paper the nonlinear multivariable function **fmincon ()** which is provided by the **MATLAB** optimization toolbox is selected, and the optimization calculation of automatic control parameters for ekranoplan may be carried out by **fmincon ()**.

General format: x = fmincon(fun,x0,A,b,Aeq,beq,lb,ub)

Where x, b, beq, lb and ub are vectors, A and Aeq are matrices respectively.

Based on the nonlinear multivariable function **fmincon ()** and the solver of ordinary differential equation **ode45**, the optimum procedure of automatic control parameters for ekranoplan may be programmed with the help of **MATLAB/GUIDE**.

Then practical format: k=fmincon(@ObjS_Equation,k0,[],[],[],[],lb,ub,[],[])

Where, k0 is the initial design variable, lb and ub are lower and upper bounds of the design variables, and the objective function name is ObjS_Equation, k is the optimal solution of automatic control parameters.

5 Example

In this paper, we take an ekranoplan as example to demonstrate optimization calculation.

Fig.1 illustrates the convergence of curve along with time for the primary motion parameters, and the initial condition is: $\vartheta_0 = 1^0, \bar{H}_0 = 0.3$ and the automatic control parameters are zero, i.e.:

k0=[$K_{\dot{V}}, K_V, K_{\dot{\bar{H}}}, K_{\bar{H}}, K_{\dot{\vartheta}}, K_{\vartheta}, K_{\dot{\phi}}, K_{\phi}, K_{\dot{\psi}}, K_{\psi}$] = [0, 0, 0, 0, 0, 0, 0, 0, 0, 0].

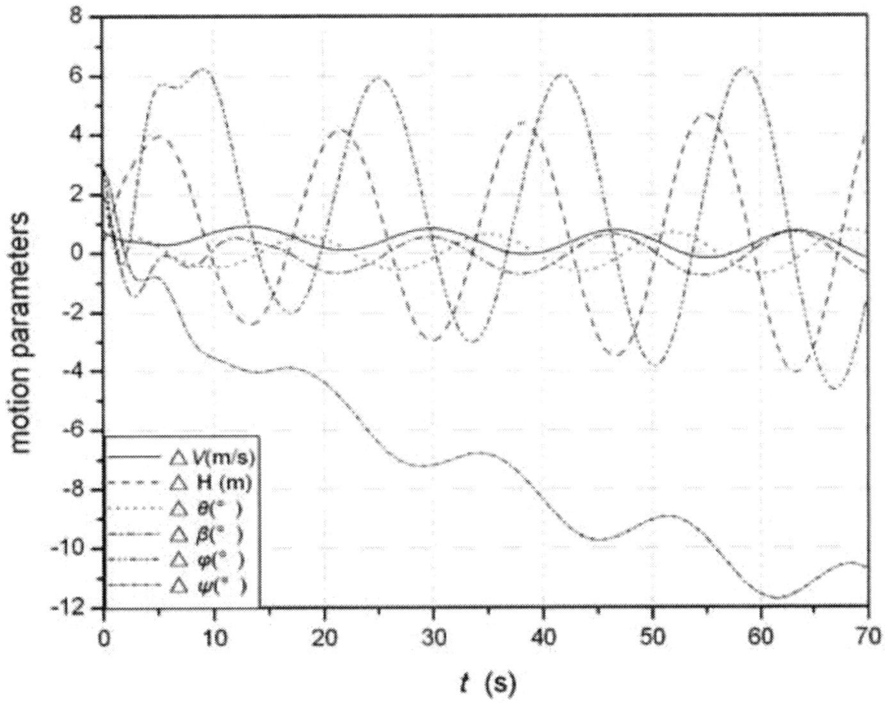

Fig.1 The curve of the primary motion parameters before optimization

Suppose that the initial values of design variables are k0, and the attenuated multiple of initial amplitude for the primary motion parameters is 0.2, the iterative number of optimization calculation is 100, then the optimal results of design variables are obtained, and the optimized automatic control parameters of ekranoplan are k, i.e.:

k=[$K_{\dot{V}}, K_V, K_{\dot{\bar{H}}}, K_{\bar{H}}, K_{\dot{\bar{\vartheta}}}, K_{\vartheta}, K_{\dot{\phi}}, K_{\phi}, K_{\dot{\psi}}, K_{\psi}$]=[0.0019, 0.7796, 1.1000, 0.1896, 1.2716, 1.7969, 1.0270, 0.0225, 4.9846, 0.7446]

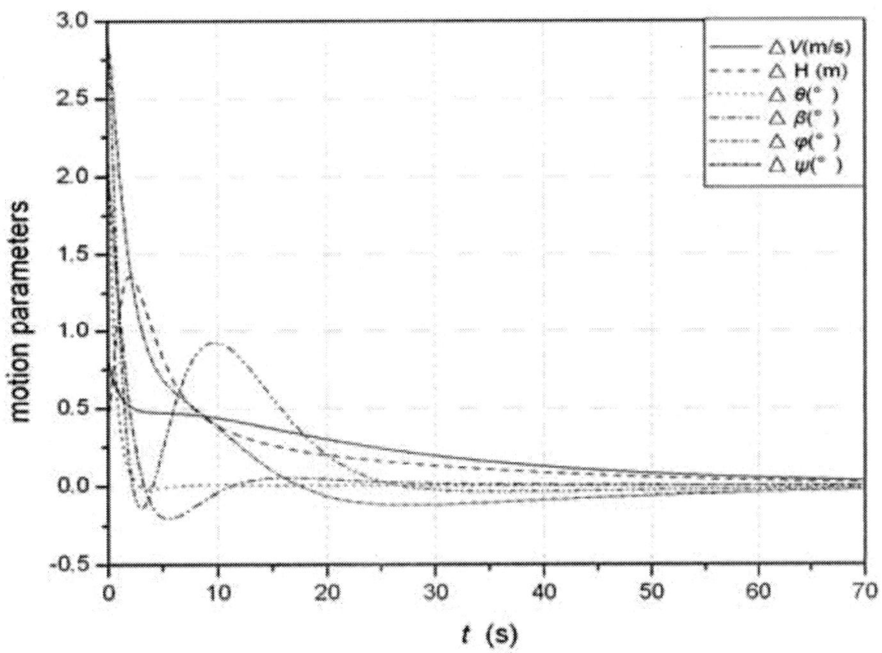

Fig.2 The curve of the primary motion parameters after optimization

Fig.2 illustrates that after optimization the initial amplitudes of all motion parameters have been largely attenuated, and before optimization the flight height and the bank angle are two maximal initial amplitudes, but after optimization the initial amplitude of the flight height has been attenuated from about 3.8 meters to about 1.25 meters, and the initial amplitude of the bank angle has been attenuated from about 5^{0} to about 1^{0}.

6 Conclusions

Large quantity of results which had obtained from other optimization calculations and the above example show that: after the aerodynamic characteristics of ekranoplan have been confirmed and the performance index of ekranoplan has been given in advance, it is unnecessary to only depend upon the aerodynamic design for solving the questions about the motion stability and maneuverability of ekranoplan, and through the optimization calculation of automatic control parameters the prescriptive performance index always may be achieved using the space motion equation of "Ekranoplan + Automatic Control System" and the optimum procedure.

Reference

[1] Chang-Hua Yuan, Yong-Lin Ye, 'Study on the space motion of Wing-In-Ground Effect Craft', FAST2003

[2] Jiang-Hua Huang, 'Practical computer simulation of chemical processes – MATLAB's application in chemical engineering (in Chinese)', Chemical industry press, China, 2004

The Doppler effect of load applied to SWATH and the analysis method of hull strength

YUAN Jianhong[1] CHEN Zhijian[1] Hu Yong[2]

(1.Dept. of Naval Architecture and Ocean Eng., Naval Univ. of Engineering, Wuhan, 430033,China;

2. The 701 research institute ot China boat industry)

Abstract: The main hull of Swath has special structural configuration, its load induced by wave is distinguished with conventional ship. Using wave theory, the load induced by wave applied to swath is discussed, it shows that the load induced by wave applied to swath has characteristic of apparent Doppler effect, that is, the transverse bending moment induced by side pressure is dynamic load, its variation frequency vary with ship speed but its amplitude is not to approach to zero. Be based on the analysis of speed and wave frequency and state of sailing, it is pointed that If the dynamic stiffness of SWATH is not enough, Doppler effect can destroy the structural configuration of SWATH seriously.

Key words: SWATH the load induced by wave Doppler effect total strength

1. Introduction,

Structural configuration and hydrodynamic response of wave and stress of SWATH is more complicated than that of mono-hull ship and conventional catamaran. The design method of hull strength is key technique to develop this kind of ship successfully. Design and process of SWATH in our country is still in elementary phase, at present, we have not possessed a series of reliable structural design and construction criterion. In order to solve the problem on strength design in actual engineering, we usually chose partly foreign correlative criterion. At present all criterion of foreign classification society evaluate the load induced by wave in angle of static analysis and have not considered dynamic characteristic of load induced by wave. In fact, using a large design load value try to control many aspects of hull strength in these criterions, Hull strength have two aspects, one aspect is static strength, the other is dynamic strength, whether static strength or dynamic strength is not enough, hull structure will be damaged n appearance. By author's study, it shows that the load induced by wave applied to swath has characteristic of apparent Doppler effect, if design of hull structure is not reasonable, Doppler effect can destroy structural configuration seriously, so strength design of SWATH must be considered dynamic characteristic of load induced by wave.

2. Doppler effect of load applied to SWATH

Spectrum density of wave can be denoted by nature frequency, when ship is sailing in wave at a certain speed, its load frequency applied to ship is different with nature frequency, the frequency of disturb force is often named as encounter frequency. As variety of speed, the load frequency will increase. This effect of load is similar with Doppler effect in Physics; it is named as Doppler effect of load in this paper.

If the SWATH march at the speed of V and in the direction that the included angle between the ship and the wave propagation direction is β as shown in Fig.1, the encounter frequency which the ship encounter with the wave is:

$$\omega_e = \omega - \frac{\omega^2 V \cos\beta}{g} \tag{1}$$

in equation (1), ω is wave frequency, V is ship speed, g is gravity acceleration.

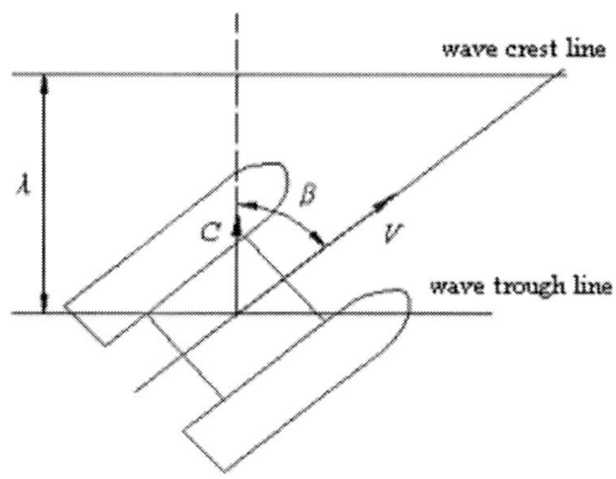

Fig.1 the relationship of encounter frequency and ship speed and course

3.Characteristic of load induced by wave of Swath

3.1 Conventional mono-hull

In overall, conventional mono-hull can be liked as a slightness beam floating on water with variable cross-section, its maximum load will appear when ship sail on trochoidal wave at speed of wave speed and is in state of sagging and hogging as it is shown in Fig.2. When ship speed is lager than wave speed, the relative place between ship and wave will change as it is shown Fig. 3, there are one more wave crest or trough applied to ship at same time, the higher ship speed, the more wave crest or trough applied to ship there are. Wave bending moment of this state will less than that of computational state, if sailing speed is enough larger so as to make the amount of wave crest or trough reach a relative value, the additional wave bending moment will approach to zero.

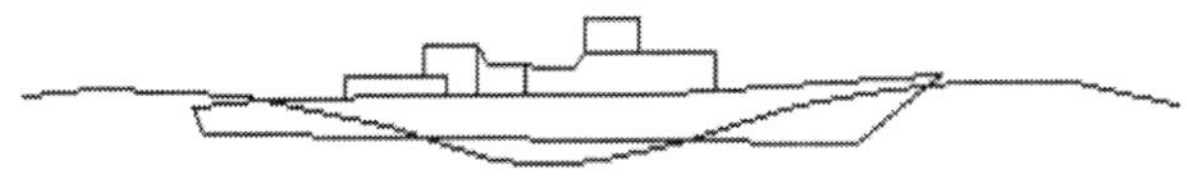

a sagging

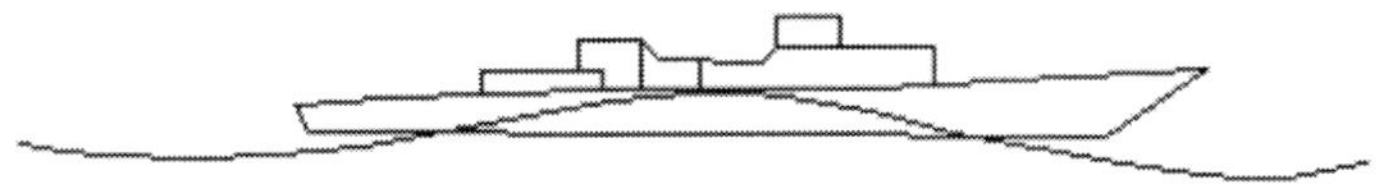

b hogging

Fig.2 relationship of conventional mono-hull and wave when ship speed equal to wave speed

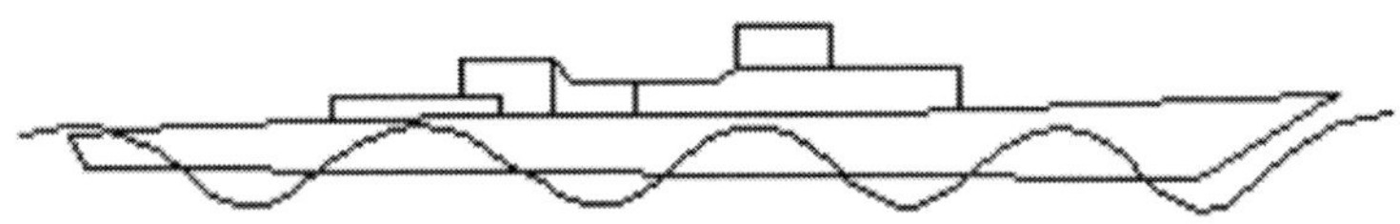

Fig.3 relationship of conventional mono-hull and wave when ship speed is larger than wave speed

The vertical bending vibration frequency of conventional mono-hull is usually larger than wave frequency, when ship is in the state of Fig.2, its change frequency of load induced by wave from sagging to hogging is always less than its overall vibration frequency, so though the wave load reach its apex value in state of sagging and hogging, it can't lead ship to resonate. Equation (1) shows that when ship speed is lager than wave speed and reach to a certain value, its encounter frequency will probably approach to its inherent frequency, but at this time, the additional wave bending moment will approach to zero, so ship wouldn't resonate too. Be based on above analysis, Doppler effect of load applied to conventional mono-hull is not apparent.

3.2 SWATH

SWATH structure consists of underwater body(submerged body)、abovewater body(cabinet)、backup post、torsion box（crossstructure）,superstructure. The main hull of SWATH is 'Π' framework, When it sail in the state of oblique sea as it is shown in Fig5 and Fig6, one bow of two sidehull lies on wave crest, the other lies on trough, one stern of two sidehull lies on trough, the other lies on wave crest, the maximum torsion moment will induced by wave. When it sail in the state of oblique sea and ship speed is increasing, the relative place between hull and wave couldn't translate from Fig2 to Fig3 as conventional mono-hull. Fig4 shows the additional wave load applied to SWATH. When it sail in the state of oblique sea, the additional wave buoyancy can be divided into two load components: buoyancy in vertical direction and side pressure on backup-post.

The buoyancy in vertical direction is denoted as $F_{\text{bow-left}}$、$F_{\text{stern-right}}$、$F_{\text{bow-leftt}}$、$F_{\text{stern--right}}$,when SWATH and wave lie to the position as Fig.4, the direction of $F_{\text{bow-leftt}}$、$F_{\text{stern--right}}$ is upwards and positive buoyancy, the direction of $F_{\text{bow-left}}$、$F_{\text{stern-right}}$ is downwards and minus buoyancy. The four components can form the non synchronous pitching torsion moment applied to SWATH When wave crest or trough pass amidships, it can form sagging or hogging moment applied to sidehull. It is similar with conventional mono-hull that the additional wave bending moment will approach to zero if speed of SWATH is enough large.

The amplitude of transverse moment induced by side pressure applied to torsion box can't approach to zero with increasing speed. The application point of resultant force of side pressure will move, the instantaneous value of moment will vary with frequency ω_e as shown in equation (2). It is named as load Doppler effect of SWATH that change frequency of transverse moment induced by side pressure will vary with speed but its aptitude can't approach to zero. The dynamic bending moment described in equation (2) is named as Doppler effect load. Be based on above analysis, the Doppler effect of load applied to Swath is very apparent.

$$M_{bt} = M_{b0}\sin(\omega_e t + \varepsilon) \qquad (2)$$

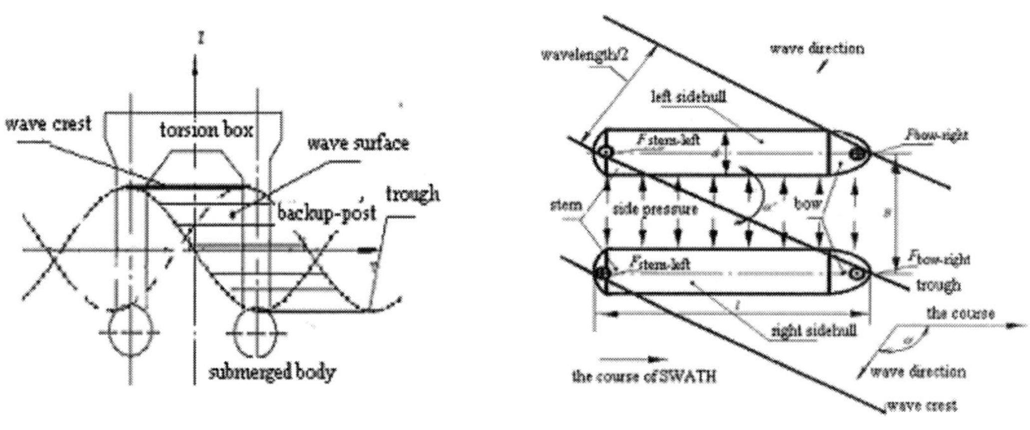

a b

Fig.4 the load when SWATH sail in the state of oblique sea in regular wave

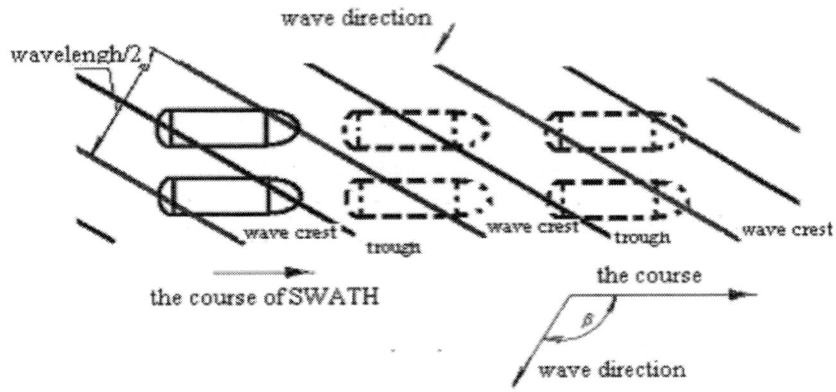

Fig.5 SWATH sail in the state of oblique sea in regular wave (low speed)

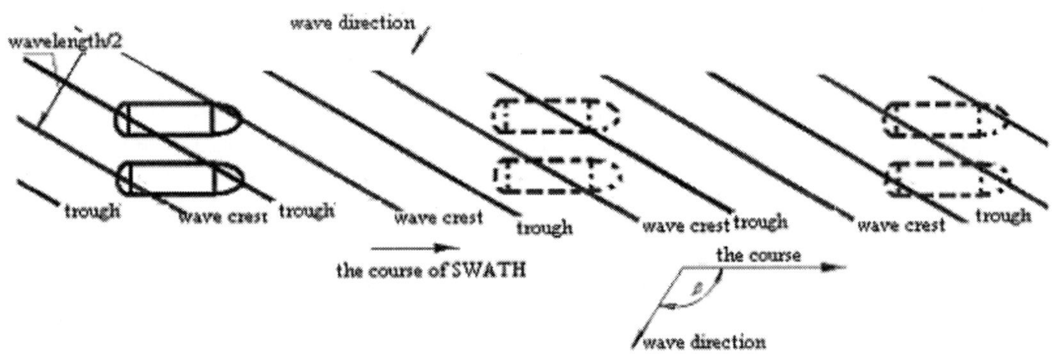

Fig.6 SWATH sail in the state of oblique sea in regular wave (high speed)

4.The relation of Doppler effect load and total strength

When the SWATH march along the direction that is equal to minus or positive 135 degrees, the relationship between encounter frequency and wave frequency is shown in table 2:

The probability in table 2 is chosen from the data of Western Pacific and East China Sea listed in the literature [4].

Table 2 the relationship between encounter frequency and frequency of waves

Wave			ship speed (knot)	encounter frequency (Hz)
frequency of waves (Hz)	wave period (s)	probability of occurrence		
0.1	10.0	4.3%	18	0.14
0.2	4.00	when <7s		0.37
0.3	2.33	total 68.9%		0.68
0.4	2.50			1.07
0.5	2.00			1.55
0.6	1.67			2.11
0.7	1.43			2.76
0.8	1.25			2.49
0.9	1.11			3.30
1.0	1.00			4.20

1.0	1.00			6.83

If the overall vibration nature frequency of SWATH is small, and the encounter frequency will approach to its nature frequency, at this time the hull structure will resonate.

The author ever performed modal analysis of SWATH, it is proved that first order nature frequency of the boat supervised and managed by the customs is smaller and is about 1.155Hz.If SWATH sail in the state of oblique sea at the speed of 18 knot in the regular wave whose frequency is 0.4Hz, the encounter frequency is 1.07Hz, the stagger ratio with the first order frequency of natural overall vibration of hull is 7.9 percent. If SWATH sail in the state of oblique sea at the speed of 18 knot in the regular wave whose frequency is 0.5Hz, the encounter frequency is 1.33Hz, the first order frequency of natural overall vibration of hull less than encounter frequency, it means that SWATH will appear wave-induced vibration, Table 1 indicated that the probability of occurrence of regular wave whose frequency is 0.4Hz is much greater, so the probability of wave-induced vibration is much greater.

When the hull structure resonates, its dynamic inner stress is $1/2\xi$ times larger than static inner stress subjected to external aptitude load. The value ξ of mental material is 0.01～0.02, the working stress will be enlarged 25～50 times than static stress subjected to external aptitude load When the hull structure resonate. Once this phenomenon appear, stress of the hull structure will go beyond the limit value so as to make structure be damaged though there are larger design load margin.

5.Conclusions

(1).The load subjected to SWATH have very apparent Doppler effect.

(2).If design of hull structure is not reasonable, the first order global frequency of SWATH is probably lower than or approach to its encounter frequency in regular wave, SWATH would appear wave-induced vibration, at this case, the hull structure have safety hidden trouble.

(3).It is advised that the load induced by wave subjected to SMATH should be accessed properly not only in anger of static analysis but also in anger of dynamic analysis. Its dynamic stiffness should be regarded as an important index which control its hull structure design

6.References

[1]CHEN Zhijian YUAN Jianhong The total inherent vibration and the strength of SWATH China International Boat Show & High Performance Marine Vehicles (HPMV) conference, 2004.

[2]Zhen ming Li baiqi. The high performance ship of 21th century ocean. Beijing: Publishing House of Defense Industry, 2001

[3]Tao yaosen. The seaworthiness of ship, Shanghai, Publishing House of Shanghai JiaoTong University, 1983.

[4]Huang dingliang, The performance theory of SWATH, Publishing House of Defense Industry, 1993

Calculations of Resistance, Sea-keeping Characteristics and Optimization of High Speed Trimaran

Huang De-bo, Duan Wen Yang, Zhou Guang Li,

Li Yun Bo, Han Kai Jia, Li jiang Hua

(Harbin Engineering University, China)

Abstract: There is an urgent demand for higher speed in marine transportation. Trimaran has become a heated research topic and turned out to be a promising displacement hull form due to its various merits in hydrodynamic performances and other aspects over conventional hull form, such as speed and seakeeping, as well as general arrangement and other attractive characteristics. Presented in this paper are some of our research works on theoretical evaluation methods of trimaran's ship resistance, seakeeping, hull-form and configuration optimizations. Related model tests showed that the methods presented are effective. The conclusions confirmed that trimaran is an advantageous displacement type of ship if designed properly.

Key words: trimaran, resistance, seakeeping, optimization

1. Introduction

Peoples in all times have been attempting for higher speed and better sea-keeping performance vessels. The development of internationalizing world trades and water transportation require faster vessels, to compete with the air shipping and trucking services. For example, containers are to be speeded up to 40 to 60 kn, as opposed to today's usual 20 to 30 kn. Some naval vessels have the same trend. Many countries have, since the end of last century, worked out projects of fast ship development. Through hundreds of years of effort and with the great progress in construction and ship hydrodynamics etc. and the advent of more advanced and powerful computer, and some related numerical techniques, regarding hull form, the well-developed shipbuilding has reached to such an extent that conventional hull forms has left little room for improvement— a very few percentage of speed increase at the same power will be considered a remarkable success. And that is due to the limitation by the natural rules for the conventional type of form: it must be wide and full enough for adequate stability, spaces and displacement among other requirements, which demand unacceptably great power for vessels at high speeds. Un-conventionalizing of hull forms is a solution in this regard. Various types of novelties appeared in recent decades, New techniques on planing or semi-planing, air cushion, hydrofoil assisted, catamaran and multihull ships burgeoned in these days, as well as hybrid versions of above new concepts, but some of them are not suitable to high seas. The ocean

Author: Huang De Bo, (1943-) professor, Harbin Eng. Uni. Main interests: Ship Hydrodynamics, ship resistance and optimization. Transportation requires more fast displacement vessels.

Among the displacement type of ships, trimaran or multi-hull form, have been paid increasing attention. Though the concept of trimaran can be traced back to the outrigger-assisted canoe of ancient Pacific islanders, and some researches could be found in 1960's, more serious development appeared only a few years ago, and the great advance has made it a heated topic. Many advantages of trimaran over the mono-hull and catamaran are attracting the eyes of both commercial and navy ships owners, which include excellent speed characteristics, high performance in sea-keeping, larger deck area and many others.

The trimarans like QinetiQ's RV Triton and Austel's high speed ferry Benchijigua Express have set a new horizon of displacement type of ships. More trimarans will come into being in a few years. Multi-hulls are even considered to be the "future type of ships", and some sort of "breakthrough in water transportation". The United States, U. K. France, Russia, Italy, Australia and some other countries have made some efforts in the development of this form.

We have studied trimaran since 1996, by developing methods of resistance and seakeeping performance, optimization of hull form and configuration, and carried out some model tests, obtaining encouraging results.

2. Theoretical Methods

The method of wave resistance developed by Huang De Bo and Li Yun Bo [11], based on Noblesse's New Slender Ship Theory[10] , and the method of wave steepness restriction was generalized for trimaran.

A ship of length L is assumed to move at velocity U in calm water. In the far field $\xi \to \infty$, the Green function

$$G = \frac{1}{\pi F_n^2}\int_{-\infty}^{\infty} \mathrm{Im}\, e^{kL(\xi+z)+ikL(\xi-x)\cos\theta+(\eta-y)\sin\theta}\, dt \tag{1}$$

From Green Theorem, the potential can be derived as

$$\phi(\vec{\xi}) = \int_h (Gn_x - \phi G_n)da + F_n^2\int_c \left[G\left(n_x^2 + t_x\phi_l - n_z t_y \phi_d\right) - \phi G_x\right]_y dl \tag{2}$$

where h and c are respectively the average wetted surface and its intersection line with the mean free surface, n and **t** are respectively the hull's normal and tangential vectors, and **d**=n×t.

After some manipulation, we have wave resistance

$$C_W = R/(0.5\rho U^2 L^2) = 1/\pi\int_{-\infty}^{\infty}|K|^2\sqrt{(1+t^2)}dt \tag{3}$$

where R is the dimensional wave resistance, K is the Kochin function

$$K(t) = \frac{1}{F_n^2}\int_h (En_x - \phi E_n)da + \int_c\left[E\left(n_x^2 + t_x\phi_l - n_z t_y\phi_d\right) - \phi E_x\right]_y dl \tag{4}$$

and

$$\begin{aligned} E(x, y, z) &\equiv e^{kLz - ikL(x\cos\theta + y\sin\theta)} \\ &= e^{kLz - ik_0L\left(\sqrt{1+t^2}x + t\sqrt{1+t^2}y\right)} \end{aligned} \tag{5}$$

$\left(k = k_0 \sec^2\theta = k_0\left(1+t^2\right),\ k_0 = g/U^2\right)$

On the other hand, C_w can be written as

$$C_W = 4\pi^3 F^4 \int_{-\frac{\pi}{2}}^{\frac{\pi}{2}} \left|A(\theta)/\lambda(\theta)\right|^2 \cos^7\theta d\theta \tag{6}$$

where $A(\theta)$ is the wave amplitude, l is wave length of the elementary wave. As stated in [11], in the framework of potential theory, the discrepancy of numerical and test results could be considered stemming from the over–estimation, for some θ, of the elementary wave steepness $A(\theta)/\lambda(\theta)$, which is due to the neglecting of nonlinearities and viscosity. To compensate these, A/λ could be restricted somehow. A technique was introduced, and shown to be effective. From the relationship between A and K, the restriction can be applied to K instead of A/λ, and the expression of the integrand in (3) was expanded in special functions for more effective calculation, resulting in an effective and fast prediction method. This method was then extended to trimaran case.

3. Optimization of Hull Form and Configuration of Trimaran

By use of Thin Ship Theory, the relation of wave resistance C_w with half-breadth function $f(x, z)$ could be obtained. And the latter could be expressed in Tent Functions, and C_w became a quadratic form of the changing-permissible offsets $y_{ij}=f(x_i, z_j)$. Then a nonlinear programming was used to minimize C_w, subject to some prescribed constraints, such as block coefficient, midship section and water surface coefficients etc. In the optimization process, to prevent the hull from distortion, as occurred frequently in some hull optimization techniques, some penalty methods were introduced and they showed satisfactory effects. In our method, a "mixed resistance coefficient" C'_w involving as well the frictional resistance effect was used as the objective function instead of C_w itself. And a revision of this optimization method was created for the catamaran case. These two versions were applied respectively to the centre hull and the outriggers, and the optimized components were combined and with the method of configuration optimization to find the optimal trimaran. The prediction of C_w was completed by the better theoretical method as stated in 1 rather than the Thin Ship Theory.

As to the configuration optimization of the trimaran, since the forms, sizes, locations of the three components have significant impact on the resistance (for some range of speed). Some researchers used a serious, a great number of tests with various combinations, to find the optimal configuration, and produced practical chart for design use. We used both the Enumeration and Genetic algorithm to search the optimal combination and configuration of the three hulls, which were made possible by our effective and fast resistance prediction method. The results showed that, in low range of Froude number F_n, the location of the outriggers affected sensitively on C'_w, but not significant in magnitude; whereas in higher F_n range, resistance was found smaller when the outriggers were located far off the midship, towards the two ends of the centre hull. And it was found that the length of the outrigger should better not excess 1/3 of that of the mail hull, to avoid worsen total resistance due to greater frictional resistance.

4. Sea-keeping Performance Prediction

A 2-and-a-half-Dimension method was developed to predict the sea-keeping characteristics of trimaran.[12] Assuming that the speed is high and the conditions required for the Strip Theory are met, $\phi_j (j=1\sim 7)$ satisfy

$$\begin{cases} \dfrac{\partial^2\phi_j}{\partial y^2}+\dfrac{\partial^2\phi_j}{\partial z^2}=0 \quad (in D) \\ \left[(i\omega-U\partial/\partial x)^2+g\partial/\partial z\right]\phi_j=0 \quad (z=0) \\ \dfrac{\partial\phi_j}{\partial n}=\begin{cases} i\omega n_j+Um_j & (j=2\sim 6) \\ -\partial\phi_0/\partial n & (j=7) \end{cases} \quad (on S) \\ \phi_j=\partial\phi_j/\partial x=0 \quad (x>x_0) \end{cases} \tag{7}$$

X_0 being the bow position.

Introducing the transforms $t(x)=(-x+x_0)/U$ and $\psi_j(t)=e^{i\omega t}\phi_j(t)$ one has

$$\begin{cases} \partial^2\psi_j/\partial y^2+\partial^2\psi_j/\partial z^2=0 \\ \partial^2\psi_j/\partial t^2+g\partial\psi_j/\partial z=0 \quad (z=0) \\ \dfrac{\partial\psi_j}{\partial n}=\begin{cases} (i\omega n_j+Um_j)e^{i\omega t} & (j=1\sim 6) \\ -\partial\psi_0/\partial n e^{i\omega t} & (j=7) \end{cases} \quad (on S(t)) \\ \psi_j=\partial\psi_j/\partial t=0 \quad (t=0) \end{cases} \tag{8}$$

The integral equation of source distribution function can be derived from Green Theorem by satisfying the boundary condition.

$$2\pi\psi_j(p,t)=-\int_{s(t)}\sigma(q,t)\ln(r/r')ds_q+\int_0^t d\tau\{\int_{s(\tau)}\sigma(q,\tau)\tilde{G}ds_q - -\frac{1}{g}\sigma\tilde{G}n_y\Big|_{\eta=A,\zeta=0}\left[\dot{A}(\tau)\right]^2+\frac{1}{g}\sigma\tilde{G}n_y\Big|_{\eta=B,\zeta=0}\left[\dot{B}(\tau)\right]^2\} \tag{9}$$

And

$$-\pi\sigma(p,t)+\int_{s(t)}\sigma(q,t)\partial/\partial n_p\ln(r/r')ds_q=-2\pi\partial\psi_j/\partial n_p+ +\int_0^t d\tau\{\int_{s(\tau)}\sigma(q,\tau)\frac{\partial\tilde{G}}{\partial n_p}ds_q-\frac{1}{g}\sigma\frac{\partial\tilde{G}}{\partial n_p}n_y\left[\dot{A}(\tau)\right]^2+\frac{1}{g}\sigma\frac{\partial\tilde{G}}{\partial n_p}n_y\left[\dot{B}(\tau)\right]^2\} \tag{10}$$

Where A and B are respectively the intersection coordinates of the hull with the free surfaces port and starboard. Solving these for σ and using Bernoulli equation

$p_j(x,y,z)=-\rho(i\omega\phi_j-U\partial\phi_j/\partial x)=-\rho e^{-i\omega t}\partial\psi_j/\partial t$, the hydrodynamic forces upon the hull will be

$$T_{ij}=\iint_S p_j n_i ds=-\rho\int_L e^{-i\omega t(x)}dx\left[\int_{S[x(t)]}\partial\psi_j/\partial t dl\right] \tag{11}$$

The diffraction force can be obtained likewise. And hence

$$\begin{cases} [(M + A_{33})(-\omega^2) + C_{33} + i\omega B_{33}]\xi_3 + [(A_{35})(-\omega^2) + C_{35} + i\omega B_{35}]\xi_5 = F_3 \\ [(A_{53})(-\omega^2) + C_{53} + i\omega B_{53}]\xi_3 + [(I_{22} + A_{55})(-\omega^2) + C_{55} + i\omega B_{55}]\xi_5 = F_5 \end{cases} \quad (12)$$

Where $\eta_3(t) = \mathrm{Re}[\xi_3 e^{i\omega t}]$ and $\eta_5(t) = \mathrm{Re}[\xi_5 e^{i\omega t}]$ are respectively heave and pitch displacements, and

$$T_{ij} = \omega^2 A_{ij} - i\omega B_{ij} \quad (13)$$

The STF and 2 1/2 methods were used for added masses, damping coefficients and diffraction forces F_j^D. Having these, (12) can be solved for the motion response.

5. Numerical Calculations, Optimization and Model Tests

The numerical methods above-mentioned were validated by some calculations, the confirmations showed that they are reliable. And optimization method was used to obtain some optimum trimarans, e.g. FA1 and FA2. Relevant models were made, and some different configurations were arranged for both resistance and seakeeping tests, like trimaran models FA1-A, FA1-B, FA2-A, FA2-B etc, with -A indicating the outriggers near the midships, and -B, near the sterns.

A serious of comparisons between prediction and test results showed satisfactory confirmation, cf. Fig.2 and Fig.6, indicating that the hull optimization and performance prediction methods we developed are effective, reliable, with very little CPU time in calculation. The confirmation in total resistance between predicted and tests in Fig.2 is remarkable, (in which, the last higher test point bodies, it should be discarded).

Fig. 1 Model test of optimized trimaran FA2-A and FA2-B

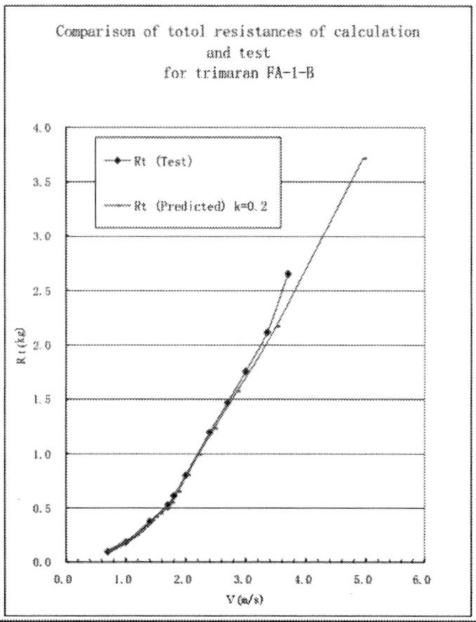

Fig. 2. Comparison between predicted and measured (shorter curve) total resistances of trimaran model FA1-B

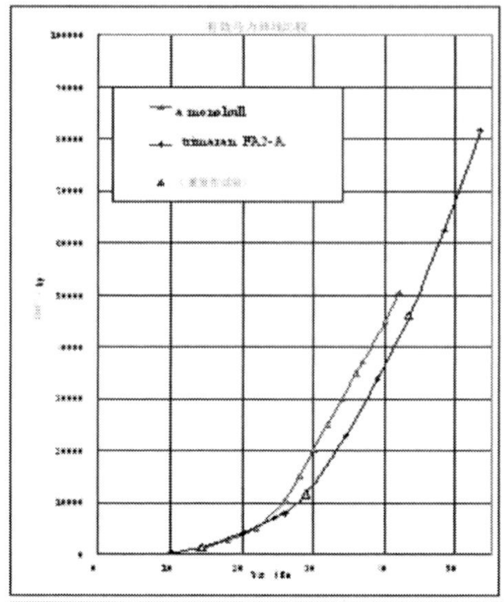

Fig. 3. A typical comparison between EHP of a monohull (shorter curve) and trimaran FA2-A, having the same displacement

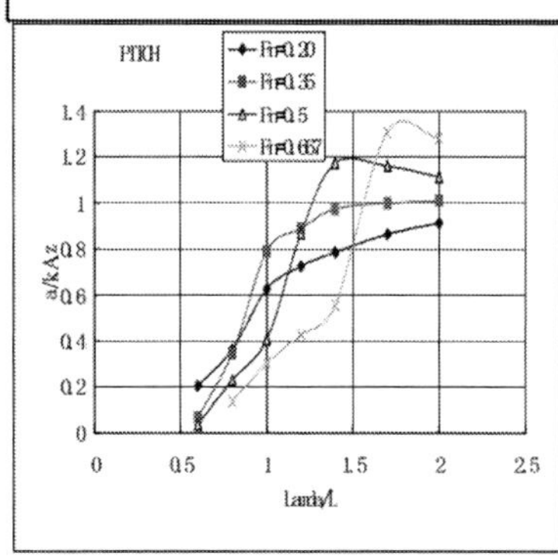

Fig. 4 ROA of pitch of Trimaran model FAI-B

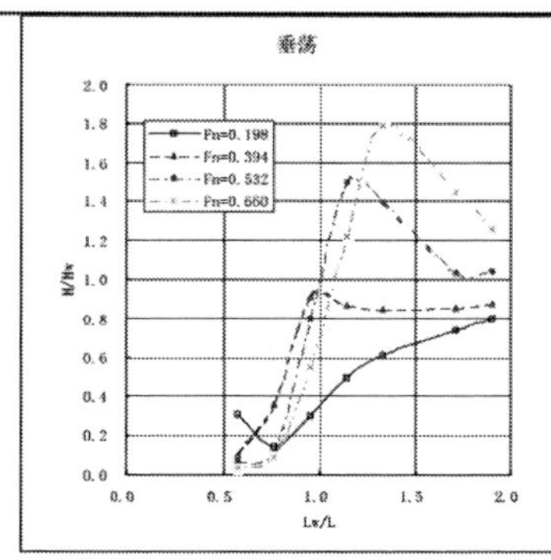

Fig. 5 ROA of heave of Trimaran model FA2-B

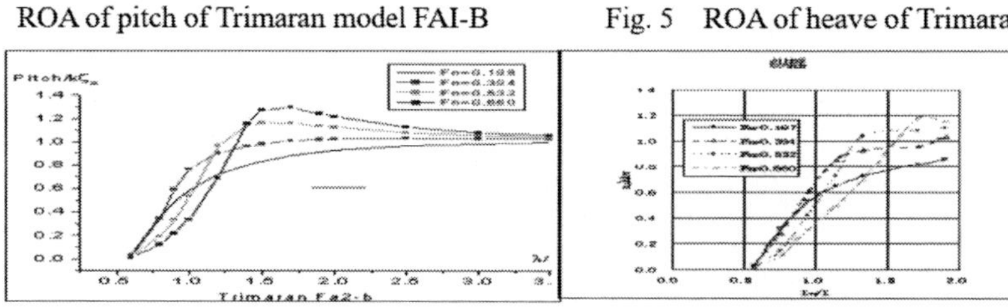

(a) (b)

Fig. 6 Comparison between calculation (a) and test (b) results of pitching of Trimaran model FA2-B in regular waves.

The resistance tests for the optimized trimaran models showed that they have very low resistance in high speed range, with 15-20% less effective horse power than the mono-hull form having the same displacement. But in lower speed range, they appear no better than the later, mainly due to their larger wetted surface area. As to the sea-keeping, e.g. pitch, the trimaran showed attractive performance, especially at higher speeds, cf. Fig.4.

6. Conclusion

The authors' theoretical and experimental studies on calculations of resistance, seakeeping, optimization of hull form and configuration of trimaran showed, that the methods we developed are effective and reliable, and that trimaran, if designed properly, will have very good hydrodynamic performances. In addition to these performances, trimaran has, as shown by other researchers, many other attractive features, which makes it a highly promising type of high speed, high seakeeping performance displacement vessel.

Except for these researches, our group has completed some investigations of flow calculations with some viscous hydrodynamic solvers, and more are still going on. They will be reported later.

We believe that in the near future, trimaran will become one of the most popular ship forms; and, in regard to oceangoing fast displacement vessels, it will be "A dominant ship of the new generation".

Reference

1. "An Investigation of the Resistance Characteristics of Powered Trimaran Side-Hull Configurations", B. Ackers et al. SNAME Trans. Vol. 105, 1997 pp349.
2. " Some Research on the Wave Resistance of a Trimaran". S. Narita, International Seminar on Wave Resistance. 1976, Japan, p381.
3. "Trimaran Ships, The Configuration for the Frigate of the Future". D. Andrews & J. Zhang, Naval Engineering Journal, May 1995.
4. "R & D of a Displacement Type Ship (Part2)", Takarada, N. et al. 1993, FAST 1993.
5. "Super-slender Mono-hull with Outriggers", J. Linstrom et al. FAST 1995. p295.
6. "Fast Displacement Ships -An Economical Option for High Speed Transportation". H. Langenberg. FAST, 1995.
7. "R & D of a displacement-type high speed ship (part 1: A conceptual design study)" N. Takarada et al .日本造船学会论文集 170 号.
8. "Studies on Minimization of Wave Making Resistance for High Speed Trimaran with Small Outriggers", K. Suzuki, M. Ikehata. 日本造船学会论文集 177 号.
9. "A Study of Trimaran Ovoid Type Ship". Huang De-bo, Han Kai Jia, Zhao Wei. The 3 rd International Conference for High Performance Marine Vehicles. 2000, Shanghai.
10. "A Slender-ship Theory of Wave Resistance ". F. Noblesse. J. S. R. Vol. No. 1, March 1983.
11. " Ship Resistance based on Noblesse's Slender Ship Theory and Wave Steepness Restriction". Huang De-bo, Li Yun Bo. S.T.R. Vol. 44. 1997.
12. Comparison of Two Sea-keeping Prediction Methods for High Speed Multi-hull Vehicles", Wen Yang Duan, De-bo Huang, D. Hudson & G. Price. FAST 2001, 4-6 th. Sept. 2001, Southampton, UK.
13. "Optimal Ship Forms for Minimum Resistance". C. Hsiung, J. S. R. Vol. 25, No2, 1981.
14. "Study of the Relation Between Configuration of Trimaran and its Resistance". Li Jiang Hua, Huang De-bo, J. of Harbin Engineering University, No. 4, vol. 23, 2002. (In Chinese)
15. "A Study of Trimaran Ovoid Type Ship", Huang De-bo, Han Kai Jia, Zhao Wei, The 3-rd INTERNATIONAL CONFERENCE FOR PERFORMANCE VEHICLE. Nov. 2000, Shanghai.
16. "Recent Development of High Speed Performance Displacement Vessels" (in Chinese). Huang De-bo. The 7 th National Conference on Hydrodynamics, Sept. 2005. Harbin.

An Investigation on Small Water-line Area Single Hull Wave Making Resistance and Ship Form Optimization

Zheng Yi Lu Xiaoping Dong Zushun

(Dept. of Naval Architecture & Ocean Eng., Naval Univ. of Engineering, Wuhan 430033, China)

Abstract: Present paper launched an investigation on the ship form optimization for the SWASH (Small Water-line Area Single Hull), which is a new kind of high performance ship. With a preliminary concept design ship form used as the parent ship, the wave making resistance theory and quadratic programming method are combined to establish a new ship form optimization method for SWASHs, in which the source-sink intensity distribution is used as the optimization variable and wave making resistance is used as the object function. As the prospect of engineering application, the wave resistance of the optimized ship form developed out by this method could reduce 20%, and the total resistance 8%, compared to the one of the parent ship form.

Keywords: SWASH, wave-making resistance, ship form optimization, quadratic optimization

1. Introduction

The concept of the SWASH, which is commonly known as Small Water-line Single Hull, was brought forward in early 1980's. The ship form of the SWASH is approximately the half of the SWATH's twin hull[1]. Then some exploratory researches on the hydrodynamic performance of SWASH were introduced as follow. Krylov Shipbuilding Research Institute began their works in 1980's, and did some hydrodynamic performance tests on both SWASH and SWATH[2][3]. There were also some research results published in recent years, which promoted the research of the SWASH.

There are many advantages on the sea performance of SWASH, and the Submerge-Emerge Small Water-line Area Single Hull(S-E SWASH)[4][5] can overcome some of the disadvantages of conventional SWASH to a certain extent and brought out a wide application potential. But the technical information of SWASHs is deficient, and much research work should start from blank, so that it would be rather difficult. Nevertheless, on the other hand, the difference of the research level in this field is not obvious, so that the innovative results could be made out by some efforts. The research work on ship form and resistance performance would be of great importance both in practical engineering and theoretical domain.

The research on the ship form and resistance performance of SWASHs was conducted in present paper to establish the foundation for the development of this new type of High Performance Marine Vehicles. In the earlier work [6], according to the analysis of the ship form and design calculation, two plans of the SWASH ship form were put forward and the moulded lines of one plan was also presented. Present paper using the concept form in [6] as the parent form, the wave making resistance theory and quadratic programming method are combined to establish a new ship form optimization method for SWASHs. The source-sink intensity distribution is used as the optimization variable and wave making resistance is used as the object function. The wave making resistance of the optimized ship form developed out by this method could reduce 20%, and the total resistance 8%, compared to one of the parent ship form.

2. Wave Making Resistance Calculation of SWASH

2.1 Form Parameters of the Concept Boat

The SWASH in present paper mainly consists of an underwater main hull and a strut together with some stabilizing fins on the main hull. By all appearances, these are also the main influence factors to the resistance performance. As a initial research stage on ship design, feasibility and the main hull form and strut shape optimization should be first conducted. Then, according to the results from the bare hull form in the first step, scale, shape and positions of the stabilization fins could be determined, which in turn offer the interim ship form for the next step. So it is the iteration procedure to develop a satisfactory ship form.

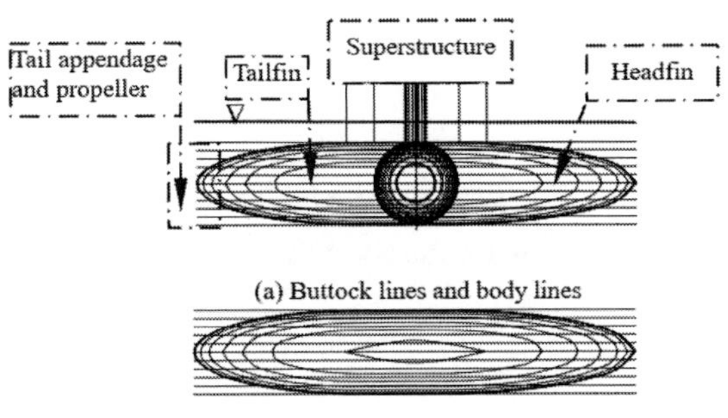

Fig.1 Simple moulded lines of SWASH

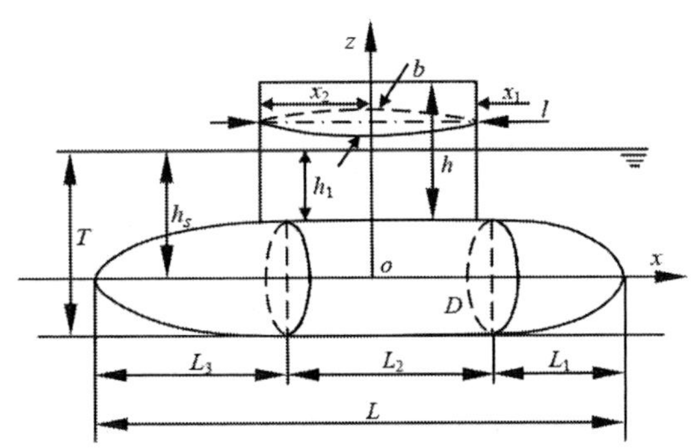

Fig.2 Principal dimensions of SWASH

Table1 Principal dimensions of the main hull

The dimension nomenclature	Symbols
Length overall	L
Length of entrance	L_1
Length of parallel middle	L_2
Length of run	L_3
Maximum diameter	D
Longitudinal coordinate of buoyancy center	x_c
Displacement of float	Δ_0
Displacement of full immersion	Δ
Draft (or immersion depth of main hull axes)	$T(hs)$

As the main hull and the strut have great influence on the resistance performance, so an optimization aimed at the resistance and a balance on some navigation performance should be taken into account before a practical ship form is achieved. In present paper, a preliminarily designed ship form in literature[6] was used as a parent form. A revolution form hull was used as main hull and vertical cylinder as the strut. The main hull is composed of three sections, the entrance, the parallel middle and the run section. The parallel middle section is a cylinder, and the entrance and run sections could be spheroids or hyperbola or parabola revolution form. Then the main hull form is among the torpedo, the submarine and the one piece hull of the SWATH. The moulded lines were shown in Fig.1, and the main dimensions nomenclature and the set of coordinate system was shown in Fig.2.

Table.2 Principal dimensions of the strut

Nomenclature	Symbols
Length	l
Maximum breadth	b
Immersion depth	h_1
Height	h
Longitudinal coordinate of the leading edge and the trailing edge	x_1, x_2

2.2 Calculation Method of the Wave Making Resistance

Up to the present, there is not a well accepted method to calculate wave making resistance of the SWASH. Considering that the linear theory can get satisfactory results for the SWATH, and [6] also use this method, here in present paper, the linear wave making resistance theory was used once again.

The wave making resistance Rw of the SWASH in present paper consists of three parts: the Rw of main hull, the Rw of strut and the Rw of their interfere. Here in present paper Discrete Source Point Method[6] was introduced to carry out the numerical calculation. The main hull and the strut can be seen as a whole, and the interference effect is spontaneously embodied in the numerical calculation process. According to literature[6], wave making resistance of the SWASH can be written as:

$$R_w = \frac{8\rho g^2}{\pi V^2}\int_0^\infty \left[I^2(u)+J^2(u)\right]ch^2u\,du \tag{1}$$

$$\begin{matrix} I(u) \\ J(u) \end{matrix} = \iint_S \frac{\partial f(x,z)}{\partial x} \begin{matrix} \cos \\ \sin \end{matrix} (K_0 x ch u)\exp(K_0 z ch^2 u)\,dx\,dz \tag{2}$$

According to the main dimension nomenclature of the SWASH in Table.1 and Table.2, and the coordinate system in Fig.2, numerical calculation grids can be made on the central longitudinal plane of the ship hull, which is shown in Fig3. By using this way to plot out the grids, the geometry character of the ship hull surface can be shown more exactly, thus a more exact result of Rw can be expected.

The cell grid on the central longitudinal plane and the cell grid on the hull surface have a relationship of projection. Discrete sources are distributed in the geometry center of the central longitudinal plane cell grids, but the source intensity data are calculated from the surface cell grids parameter. Then from (1) and (2), a numerical calculation expression can be achieved as below:

$$R_w = \frac{8\rho g^2}{\pi V^2}\int_0^\infty \left[I^2(u)+J^2(u)\right]ch^2u\,du \tag{3}$$

$$\begin{matrix} I(u) \\ J(u) \end{matrix} = \sum_{s=1}^{N} m_s \begin{matrix} \cos \\ \sin \end{matrix} (K_0 x_s ch u)\exp(K_0 z_s ch^2 u) \tag{4}$$

m_s in (4) represent the volume source intensity of each grid:

$$m_s = \frac{V}{2\pi}\iint_{S_0} \frac{\partial f(x,z)}{\partial x} dxdz \tag{5}$$

Based on the surface integral theory, m_s in each cell grid can be written as:

$$m_s = \iint_{S_0} \sigma(x,y)dS = \frac{V}{2\pi}\iint_{S_0} \frac{\partial f(x,z)}{\partial x} dS = \frac{V}{2\pi} n_{x0s} A_s \tag{6}$$

n_{x0s} represents the x-component of the unit normal vector of each hull surface grid, and A_s represents the area of each hull surface grid. Then with (3), (4) and (5), an expression of the wave resistance of SWASH can be obtained :

$$R_w = \frac{16\pi\rho g^2}{V^4}\int_0^\infty \left[I'^2(u) + J'^2(u)\right] ch^2 u du \tag{7}$$

$$\begin{matrix} I'(u) \\ J'(u) \end{matrix} = \sum_{s=1}^{N} \frac{V}{2\pi} n_{x0s} A_s \begin{matrix} \cos \\ \sin \end{matrix} (K_0 x_s chu) \exp(K_0 z_s ch^2 u) \tag{8}$$

2.3 Numerical Calculation and Validation

In order to confirm the accuracy of the program based on (7) and (8) a ship form in literature[4] is calculated by the programs, the results is shown in Fig.4. Compared with the model test data, the numerical calculation results correctly reflect the regularities of the ordinary curve of C_W vs. Fn. And for the ship form in Fig.1, numerical calculation results is shown in Fig.5, .

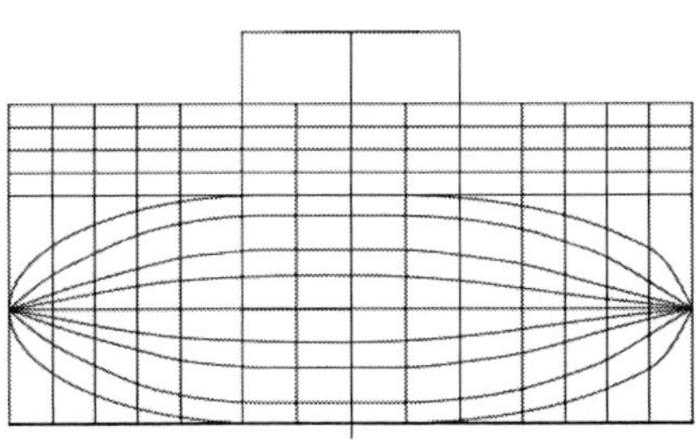

Figure 3 SWASH wave resistance calculation panel

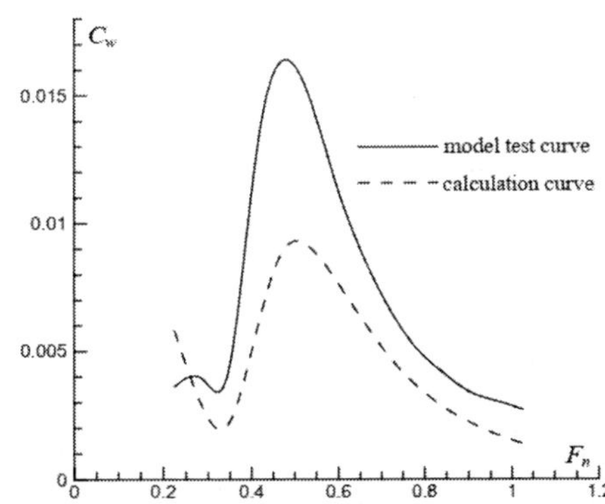

Figure 4 The comparison between the calculation wave resistance curve and model test wave resistance curve

3. Ship form Optimization Based on the Wave Making Resistance Theory Using the Quadratic Programming Method

3.1 Quadratic Programming and Ship Form Optimization

Nonlinear optimization is widely used in the engineering design nowadays, it's mainly a problem about extremes with constrains—both equality and inequality constrains. Among the ways to solve the nonlinear optimization problems, there are some well developed methods, such as Most Feasible Direction Method, Penalty Function Method and General Multiplication Method etc. The nonlinear optimization problems are commonly expressed as:

$$\min_{x \in R^n} f(x) \tag{9}$$

$$\text{s. t.}\quad C_i(x) = 0 \qquad (i = 1, 2, \ldots, m_s) \tag{10}$$

$$C_i(x) > 0 \qquad (i = m_s + 1, \ldots, m) \tag{11}$$

In the nonlinear optimization theory, the methods to solve the quadratic programming problems are well developed and had been used in ship form optimization. The matrix form of the quadratic programming problem can be expressed as below:

$$\min_{x \in R^n} \frac{1}{2} x^T H x + f^T x \tag{12}$$

$$\text{s. t.}\quad Ax \le B \tag{13}$$

$$Cx = D \tag{14}$$

In 1980s, Prof. Hiong used the tent function methods to express the hull surface function, and the quadratic programming was used in the ship form optimization. The wave making resistance was calculated using the Michell integral, and the friction resistance was calculated using ITTC recommended formula. By using the tent function, the wave making resistance and friction resistance were reduced to a quadratic form expressed by the ship hull offsets. The y data of the hull offsets was used as the design variable, together with several constrains, a quadratic programming was established to achieve ship form with minimum resistance. After that, several researchers also used this method in their research, such as the minimum resistance ship form research for the transom ships[10], and the ship form optimization of the wave piercing catamarans[11]. But as for the irregular underwater ship form, such as protrude in the bow or in the stern, chine line on the hull etc, this tent function methods has its limitation. Further more, this method use offsets y as the optimization variable, so when the optimal solution is achieved, the offsets are also fixed. The results lack some relaxing factor so that can not be adjusted, and some results even have big deviation from the conventional ship forms and can not be used in actual engineering. On the other hand, the ship form achieved form this methods only has a linear relationship with the source-sink distribution, so it is not convenient for the further research using the nonlinear source-sink distribution methods.

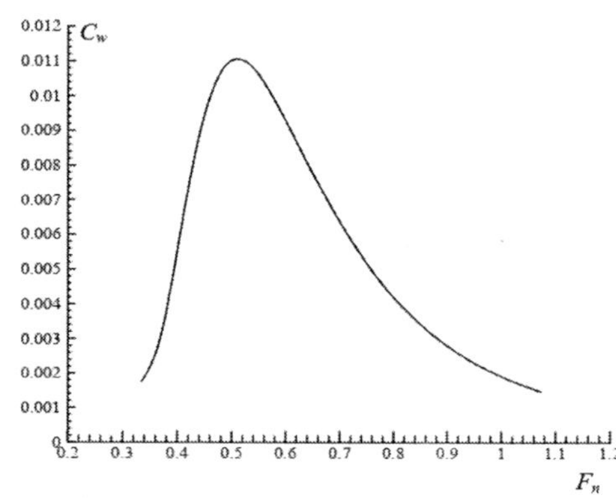

Figure 5 wave resistance curve of SWASH initial design conception in present paper

3.2 The Object Function for Optimization Problem

Present paper use expression (3) as the object function for optimization problem, then a quadratic programming problem is established. The quadratic programming model based on the Michell theory has an advantage of higher degree resolution and lower computational complexity, and the result is also acceptable. In this method, the cell grid source-sink intensity m_s are used as the optimization variable. As shown in Fig.3, if the amount of the cell grids is N, the amount of optimization variable, the source-sink intensity of the cell grids in another words, also is N. According to (6) and (7):

$$\frac{I(u)}{J(u)} = \sum_{s=1}^{N} m_s \frac{C(s,u)}{S(s,u)} \tag{15}$$

Where:

$$\frac{C(s,u)}{S(s,u)} = \frac{\cos}{\sin}(K_0 x_s chu)\exp(K_0 z_s ch^2 u) \tag{16}$$

Then:

$$R_w = \frac{8\rho g^2}{\pi V^2}\int_0^{\infty}\{\sum_{p=1}^{N}\sum_{q=1}^{N} m_p m_q \left[C(p,u)C(q,u) + S(p,u)S(q,u)\right] ch^2 u\} du \tag{17}$$

Through further simplification, then:

$$R_w = \frac{8\rho g^2}{\pi V^2}\sum_{p=1}^{N}\sum_{q=1}^{N} m_p m_q \int_0^{\infty}\{\left[C(p,u)C(q,u) + S(p,u)S(q,u)\right] ch^2 u\} du \tag{18}$$

Set:

$$D = \int_0^{\infty}\{\left[C(p,u)C(q,u) + S(p,u)S(q,u)\right] ch^2 u\} du$$

Then the wave making resistance can be expressed in a matrix form:

$$R_w = \frac{8\rho g^2}{\pi V^2} M^T H M \tag{19}$$

The H in the expression is a N-dimension column vector consists of m_s. After some constrains are set, a quadratic programming problem which form is shown above in (12), (13) and (14) can be set up:

$$\min_{x\in R^n} R_W = \frac{8\rho g^2}{\pi V^2} M^T H M \tag{20}$$

$$\text{s. t.} \quad AM \le B \tag{21}$$

$$CM = D \tag{22}$$

After some regulation, a more typical quadratic programming problem form can be achieved. The quadratic programming problem can be a convex programming problem if the character of matrix H is positive definite. And a convex programming problem can be easily solved. This method use m_s as the optimization variable, then the results of the optimization is a distribution of m_s, which satisfies definite constrains and has the minimal wave making resistance. According to the m_s distribution, the offsets of the ship can also be solved. As the offsets are not solved directly, the present method has relaxing factors compared with former methods. Thus, more fair moulded lines can be achieved. Besides, there are many ways to calculate the offsets form m_s, and which to be adopted can be determined according to the demands. The nonlinear wave making resistance theory can also be adopted to establish a relationship between the hull form and the source-sink distribution if needed.

3.3 From Source-sink Distribution to Offsets

Use the source-sink distribution to solve the offsets is an important step of the optimization method in present paper. According to the potential flow theory of hydrodynamics, m_s distribution on the central longitudinal plane and the ship form has a definite relation. The most convenient corresponding rule between m_s and offsets is the linear wave making resistance theory of thin ship, by which a linear ship form can be achieved. The transform relation between the source-sink distribution and the offsets can be expressed as:

$$\sigma_s = 2\frac{\partial y}{\partial x} \tag{23}$$

σ_s is the area source intensity of the central longitudinal plane panel grids. After integrating formula (23), the hull form function, the half width of the ship hull in another word, can be solved:

$$y = \frac{1}{2}\int_{FP}^{x_s} \sigma_s dx \tag{24}$$

The ship form solved from (23) and (24) is so called "linear thin ship". According to the former investigations, this ship form construction method has enough accuracy for ship form optimization, and the ship forms for the practical engineering use can be achieved by this method. Therefore this simple method is adopted in present paper to solve the ship form offsets. As the m_s used in the wave making resistance and the quadratic programming expressions are volume source-sink intensity, the area source-sink intensity σ_s can be expressed approximately as:

$$\sigma_s = \frac{m_s}{\Delta S_s} \tag{25}$$

Put(22)in(21):

$$y = \int_{FP}^{x_s} \sigma_s dx \approx \int_{FP}^{x_s} \frac{m_s}{\Delta S_s} dx \tag{26}$$

Here ΔS_s is area of the m_s panel grid.

3.4 Interpolation of m_s

According to expression (27), the surface offsets can be solved. As the m_s don't have the same z height, so in actual calculation, an interpolation process is needed to find out the m_s on the same height. The waterlines calculated through the interpolation are shown in Fig. 6.

3.5 Constrains

Constrains are important for the optimization problem. According to the practice of ship design, it is flexible to choose constrains. Also, it is an innovatory work for this new ship type. The initial result of the optimization can be studied to establish more feasible constrains. As the cell grids on the central longitudinal plane were plot out, so the ship length L is fixed automatically.

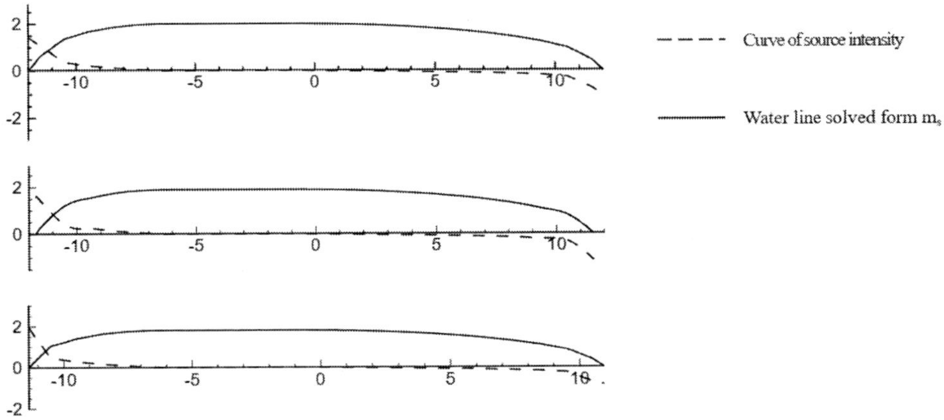

Fig. 6 Source intensity distribution and Y waterline solved

Maximum breadth is also an important constrain of the optimization, and the surface offsets y should be less than the maximum breadth of the parent ship. According to (26), when on the waterline of same height:

$$y_s = \int_{FP}^{x_s} \frac{m_s}{\Delta S_s} dx \approx \sum_s \frac{1}{2} \Delta x_s \left(\frac{m_s}{\Delta S_s} + \frac{m_{s+1}}{\Delta S_{s+1}} \right) \le D/2 \tag{27}$$

And D in the expression is the maximum breadth of the ship, Δx_s is the segment length of the cell grid of m_s. As m_s corresponds to the offsets points of the ship hull, they all satisfy (27). After simplification, a matrix form can be achieved:

$$A_1 M \le B_1 \tag{28}$$

A_1 is the maximum breadth constrain matrix which could be expressed by Δx、m、ΔS. For m_s, it's a linear inequality constrain.

Displacement constrain is also an important constrain in the optimization process. The displacement can be express approximately as[13]:

$$V = 2\int_{-\frac{L}{2}}^{\frac{L}{2}} \int_0^T ydzdx \tag{29}$$

In numerical calculation, the expression can be numerically expressed as below:

$$V = 2\int_{-\frac{L}{2}}^{\frac{L}{2}} \int_0^T ydzdx \approx \sum_j \sum_i \frac{1}{2} (y_{0,i} + y_{1,i}) \Delta z_i \Delta x_j \tag{30}$$

where i is the index of the waterlines and j the index of the cross sections. Δx_j、Δz_i is the x、z segment length of the panel grids. This expression can be converted to the matrix form:

$$C_1 M = D_1 \tag{31}$$

where C_1 is the displacement constrain matrix in relation of Δx、Δz、m、ΔS, and C_1 a row vector; $D_1 = V$. This is a linear equation constrain about m_s.

Beside these, the entrance angle constrain should be subjected. As m_s is used as the optimization variable, the entrance angle constrain appears as the numerical constrain on m_s at the head point of the waterline. The matrix form is:

$$A_2 M \leq B_2 \tag{32}$$

According to the former ship form optimization experience, these constrains are not enough, the result may be a waveform that can not be used in actual design. A common way to solve this problem is to give direct numerical constrains to m_s:

$$A_3 M \leq B_3 \tag{33}$$

$$C_2 M = D_2 \tag{34}$$

Put (29)、(32)、(33)、(34)and (35)together, a quadratic programming problem like (21)、(22)and(23) is established.

3.6 Prospect of the Optimization Result

With the quadratic programming problem consists of (21)、(22) and (23) solved, m_s distribution for the minimum wave making resistance can be obtained, then the ship form can be conducted from the source intensity distribution. In present paper, the Visual FORTRAN IMSL function (based on the improved Powell method) is used to carry out trial calculation, combined with former optimization experience, a preliminary result is obtained: According to the method established in present paper, an optimization ship form developed from parent with 20% reduction of wave making resistance and 8% reduction of total resistance is obtained. As the mathematical method the Visual FORTRAN IMSL optimization function used is a general purpose method, which is not efficient for the optimization problem in present paper. More suitable and efficient methods will be introduced in subsequent work, such as Lemke methods [14] etc.

4. Conclusions

Based on the research work of present paper, it could be concluded that:

①In the optimization process using the wave making resistance theory and nonlinear programming, the indirect method using m_s as the optimization variable can obtain satisfying ship form. It's acceptable for SWASH's form optimization, and applicable in actual engineering practice.

②The ship form obtained by present method would achieve 20% wave making resistance reduction compared with the parent form, correspondingly, 8 % total resistance reduction.

The SWASH is a new kind of high performance ship which brings out a wide application potential. But there will be a long way to go before the actual application of SWAHS. As for the ship form optimization aspect, more suitable and more efficient mathematical and engineering methods should be undertaken in the subsequent research, and model test should be carry out to validate the performance of the optimized ship form.

Reference:

[1] Yi Shaolin. Ship Resistance [M]. Beijin: National Defense Industry Press. 1985.12.

[2] Victor A. Dubrovsky, Konstantin I. Matveev. Small waterplane area ship models： Re-analysis of test results based on scale effect and form drag [J]. Ocean Engineering. 2005, 1(1).

[3] Schenzie P. The HSVA systematic SWATH model series. Third International Conference on Fast Sea Transportation [C], Germany 1995.

[4] Dong Zushun. Report of the SWASH Model Test(1)[P]. Wuhan: Dept. of Naval Architecture & Ocean Eng., Naval Univ. of Engineering. 2005.6..

[5] Lu Yao. Research on the Key Technique of S-E SWAHS[D]. Wuhan: Master's degree thesis of NEU. 2005.12.

[6] Dong Zushun. Lu Xiaoping. An investigation of SWASH ship forms and resistance. Journal of Ship Hydrodynamics [J](Adopted)

[7] LU Xiaoping, LI yun, DONG Zushun. The Comparison of the Resistance for Several Displacement High Performance Vehicles[J]. Journal of Hydrodynamics, Ser. B 2005,17 (3):379-389.

[8] Yuan Yashang. Computational Methods for Nonlinear Optimization[M]. Shanghai: Science and Technology Press. 1992.3.

[9] Hsiung C C. Optiamal ship form for minimum wave resistance, JSR, 1981, 25: 95-116

[10] Huang Debo. Optimization Design for Transom Ship Form[J]. Journal of the Harbin Ship Engineering College [J]. 1985.No.2.

[11] Zhao Lianen. Hydrodynamic Principle and Design for High Performance Marine Vehicle [M]. Harbin: Harbin Industry University Press. 2000.1.

[12] Cheng Tianzhu. Wave Making Resistance and Application in Ship Form Design[M]. Wuhan: Huazhong Institute of Science and Technology Press. 1985.12.

[13] Zhu Jun. The Statics of Ship [M]. Changsha: National Defense Technology University Press. 2001.12.

[14] Zhang Guangcheng. Computational Methods for Nonlinear Optimization [M]. Beijing: Higher Education Press. 2005.1.

Long term prediction of wave load and structural strength analysis of SWATH on Entire ship FE model

Liu Jian-hua Ren Hong Wang Hui-cai Huang Shou-cheng

Marine Design & Research Institute of China, Shanghai, 200011

Abstract:

As an example, calculation of a large SWATH ship is introduced in this paper. By a 3-D radiation-diffraction hydrodynamic program, the long-term prediction of wave induced motion and load is performed, based on which a group of design waves are defined. The stress in each part of main structural members of ship hull on design waves is calculated by an entire ship FE model. The assessment of structural strength is preformed by an FE model of entire ship. The properties of wave load of SWATH and the way of defining design waves are described.

Key Words: Ship structural strength, Wave load, FE calculation, Design waves

1 Introduction

The SWATH ship is the new hull form which developed in 20th century later periods. Because of the spacious deck area and the fine movement performance, the SWATH ship receives shipowner's favor. But the promotion of this hull form faces many questions continuously. The fine performance of the SWATH ship has much difficulty in realizing because of the complex design variables, the navigation posture not easy to control, the design method and the test condition not easy to satisfy, as well as the difficulty on the correct evaluation of wave-induced load and ship structure optimization design.

The SWATH ship mainly includes five major parts, namely lower hull, vertical struts, upper hull, wet deck, and cross deck. The structural form, the wave hydrodynamic response and the load conditions of the SWATH ship are much complex than that of the mono-hull ship. The lower hull means round or oval torpedo-like structure positioned at the base of vertical struts submerged in water. The pressure characteristic of the lower hull is much like that of the pressure tube body. The cross deck is very weaker while the SWATH ship navigates in quartering sea. The rigidity of the vertical struts which is long and narrow is the weakest part of the entire ship, but the pressure that

Biography: Liu Jian-hua（1975.4-）,male, engineer of Maric.

the vertical struts withstand is the biggest. At the same time, the buoyancy provision of the SWATH ship only depends on lower hull and the partial vertical struts. Moreover, the water plane area of the SWATH ship is very narrow. So it is very sensitive to weight for the SWATH which brings about higher requirement in design and construction process of the SWATH.

Generally speaking, the SWATH ship has unusually large beam ratio, double vertical struts and double decks. So the longitudinal strength and rigidity are enough. But the SWATH ship has wide cross contacted area and special cross section shape, the wave induced load (transverse bending moment and vertical shearing force) in beam sea and torsional moment in quartering sea may achieve a suitable magnitude, thus high stress will occurs in main structural members of ship hull. Moreover, according to "Guide for building and classing SWATH vessels" of American bureau of shipping, large amplitude of pitch and heave movement of the SWATH ship in head seas and following seas or zero speed usually occurs, which causes damage of ship structural members[1].

In view of the fact that the present guidelines of the various classification societies related the SWATH ship still are not mature, it is necessary to apply the direct analysis method to ship structure design.

This paper takes a large-scale SWATH as an example. With the aid of the DNV SESAM software and entire ship FE model, long term prediction of wave load and ship motion is carried on by spectrum analytic method. The wave-induced hydrodynamic pressure and inertia force are computed on a group of design wave. By structural strength analysis of SWATH on Entire ship FE model, the accurate stress distribution of the SWATH ship is obtained in detail. The assessment of structural strength is preformed by an FE model of entire ship. So it will provide the basis for the future structure optimization design for SWATH.

2 Long term prediction of wave load

The wave load is the most important load which ship structural design must consider. Its correct computation is the important premise to reasonably evaluate of ship structural strength and optimize structural design. In this paper, wave induced load is calculated by a 3-D radiation-diffraction hydrodynamic program, which a hull wet surface hydrodynamics model and an entire ship mass model need to set up. The mass model is used to accurately describe the mass distribution of ship structure mass, the mass of machine and ship loading. The set up of mass model is used to calculate the hydrodynamic pressure and inertial force accurately. The hull wet surface hydrodynamics model can directly use the hull shell grid of finite element model, See Figure 2.

2.1 Wave induced motion and frequency response functions of dominant load

In general, the interaction system of ship and ocean waves usually is taken as a linear system to carry on statistics prediction using spectrum analytic method. It is assumed that the hydrodynamic response of ship to wave action is linear, namely wave is a steady normal distribution stochastic process and also the wave induced motion and loading on ship is a steady normal distribution stochastic process.

The SWATH ship motion and hydrodynamic pressure on elements of wet surface under regular waves are calculated. The input data are determined as follows: (1) The number of wave frequency (or wave length) is to be taken as 25 cases, range of wave frequency taken as 0.1-2.5, and the increments taken as 0.1. (2) Seven wave heading angles are to be taken, including 0^0(head sea), $30^0, 60^0, 90^0, 120^0, 150^0, 180^0$ (following sea). A total of 175 regular waves of unit wave amplitude are selected. A series of dominant load are selected as frequency response function variables, which also are selected as response function variables of short- and long-term prediction. In addition, these dominant loads are used as control function to determine the design wave parameter. The heading of frequency response functions of dominant load under different heading angles of regular waves is calculated using a 3-D radiation-diffraction hydrodynamic program. The wave frequencies where dominant load are at the maximum amplitude are determined according to frequency response functions. Figure 1 shows frequency response functions of transverse bending moment of centerline plane. It is demonstrated that the wave frequency is equal to 1.0 rad/s and the heading angle is equal to 90^0 where transverse bending moment of centerline plane is at the maximum amplitude, see Figure 1.

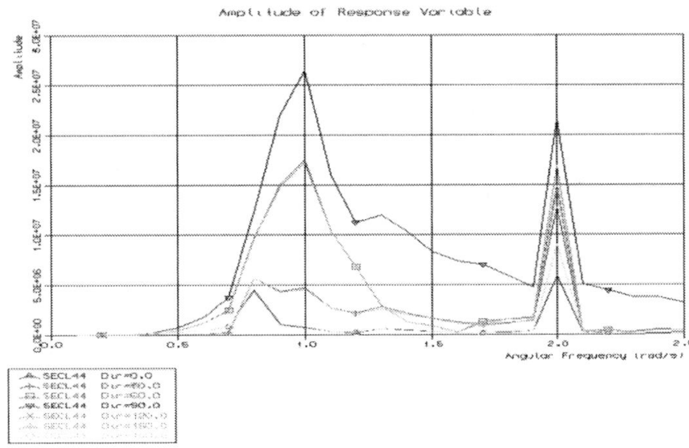

Fig.1. Frequency Response functions

2.2 Long term prediction of wave induced motion and dominant load

The prediction of wave induced motion and dominant load mainly includes the following two processes:

(1) Based on mathematical statistic theory, the short-term prediction of ship motion and wave load is performed using frequency response function and ocean wave spectrum.

(2) The long-term prediction of ship motion and wave load corresponding to prescribed probability level is performed using frequency response function, ocean wave spectrum and wave scatter diagram.

The aim for the short- and long-term prediction of wave load is similarly to determine possibility or design extreme responses of wave load for prescribed sea state. The short-term prediction of wave load is referred to predict statistics value of wave load in the relative short time. So it is supposed that ship loading, speed, heading and sea state are constant. The global wave data, which is divided into 104 groups according to sea state, is expressed as function of significant wave height and mean zero-crossing wave period[2]. Only appropriate wave spectrum is selected, the short-term prediction of load in prescribed sea area can be performed using frequency response function.

The long-term prediction of wave load is referred to predict statistics value of wave load in the relative long time, when ship loading, speed, heading and sea state are variable. Summation of probabilities over all anticipated ship loading, speed, headings and sea states is used to evaluate long-term statistics. No matter what a different combination of ship loading, speed, heading and sea state, all may be considered as one kind of short-term condition.

For ships designed for unrestricted service, the prediction of the maximum wave load likely to be experienced during their service life covers mainly sea conditions to be encountered during navigation in sea areas around the world, expressed in the probability of waves occurring in various periods and heights. For the SWATH ship in this paper, Pierson-Moskowitz wave spectrum and global wave scatter diagram are adopted for calculation of values for long-term prediction of ship's motion and wave-induced load.

2.3 Design waves group

In principle, the assessment of structural strength can be performed by the long-term prediction of stress of hull specific position using stress response function. But the computation workload and the data quantity are too huge to practice in present. A reasonable feasible method is that the ship hull structural stress is calculated on a determined regular wave, on which the computation result is equal to the extreme load of long-term prediction. This kind of determined regular wave is called design wave.

Because the high stress region of the ship structure in different wave is different. It is needed that the assessment of structural strength of main structural members of ship hull must be on a group of design wave (wave frequency, wave amplitude as well as heading angle), which are determined according to different dominant load parameter that attain maximum value.

Design wave method[3] includes the following two steps, namely the selection of dominant load parameters and the determination of design waves.

For assessment of the SWATH ship structure, the dominant load parameters are to be taken as follows:

- vertical bending moment, vertical shearing force and rolling torsional moment of middle cross section;
- rolling torsional moment of aft $L/4$ cross section and fore $L/4$ cross section;
- transverse bending moment, pitching torsional moment, transverse split force and

vertical shearing force of centerline plane;

- Pitching torsional moment and vertical shearing force of demi-hull centerline plane.

Determination of design waves includes the following parts, namely wave frequency, wave amplitude and wave direction. For prescribed condition (including heading), the frequency response functions are obtained by a 3-D radiation-diffraction hydrodynamic program. The wave frequency and heading angle where load response attains maximum value is namely the frequency and direction of design wave. The design wave amplitude is calculated according to long-term extreme responses of prescribed probability level and frequency response function. Table 1 shows 22 design wave parameters of the SWATH ship of this paper, see Table 1.

Table 1 Design wave parameter

Condition	Dominant control Load parameter	Wave amplitude (m)	Wave frequency (rad/s)	Wave length (m)	Wave direction $(^0)$	Phase$(^0)$
1	vertical bending moment of middle cross section	2.14	0.8	96.2	0	-158.924
2	vertical bending moment of middle cross section	2.14	0.8	96.2	0	21.076
3	vertical shearing force of middle cross section	3.58	0.5	246.2	180	-101.182
4	vertical shearing force of middle cross section	3.58	0.5	246.2	180	78.818
5	rolling torsional moment of middle cross section	3.60	0.5	246.2	180	78.832
6	rolling torsional moment of middle cross section	3.60	0.5	246.2	180	-101.168
7	rolling torsional moment of aft $L/4$ cross	2.78	1.6	24.04	120	-57.986
8	rolling torsional moment of aft $L/4$ cross	2.78	1.6	24.04	120	122.014

9	rolling torsional moment of fore $L/4$ cross	2.95	1.6	24.04	60	114.8
10	rolling torsional moment of fore $L/4$ cross	2.95	1.6	24.04	60	-65.2
11	transverse bending moment of centerline	2.07	1.0	61.5	90	96.147
12	transverse bending moment of centerline	2.07	1.0	61.5	90	-83.85
13	pitching torsional moment of centerline	2.45	1.7	21.3	90	76.153
14	pitching torsional moment of centerline	2.45	1.7	21.3	90	-103.847
15	pitching torsional moment of demi-hull	3.67	0.5	246.2	0	-79.060
16	pitching torsional moment of demi-hull	3.67	0.5	246.2	0	100.94
17	transverse split force of centerline plane	1.95	1.0	61.5	90	96.25
18	transverse split force of centerline plane	1.95	1.0	61.5	90	-83.75
19	vertical shearing force of demi-hull centerline	2.30	0.8	96.2	0	-159.571
20	vertical shearing force of demi-hull centerline	2.30	0.8	96.2	0	20.429
21	vertical shearing force of centerline plane	2.65	1.7	21.3	90	-102.102
22	vertical shearing force of centerline plane	2.65	1.7	21.3	90	77.898

3 Structural strength analysis of SWATH on Entire ship FE model

3.1 FE model of entire ship

The SWATH ship's global three-dimensional finite element model covers hull structure within ship's complete length and full breadth, including all decks, side shell and bottom, inner bottom, girder, longitudinal and transverse bulkheads, floor, beam, pillar and watertight bulkhead stiffener, etc. All primary plating such as deck, shell, inner skin, girders, transverse bulkhead structure including its vertical and horizontal stringer webs are simulated plate elements. Secondary members such as frames on plating are simulated beam elements, with the eccentric section properties for connection of plating.

The finite element model, which has 22238 nodes and 44792 elements, includes all of the principal structural members, which contribute to the hull structural stiffness, see Figure 2.

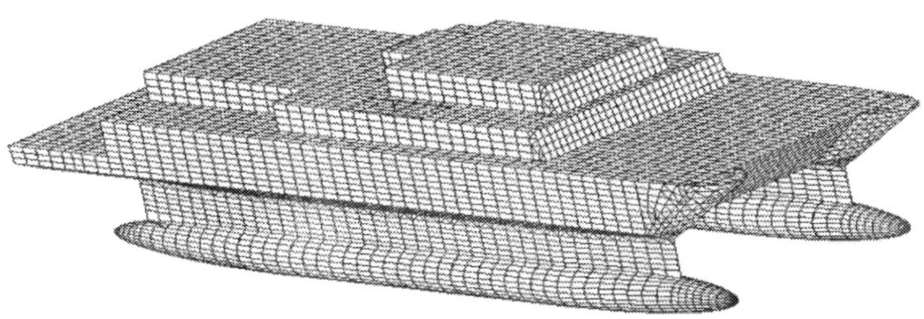

Fig.2. Structural Finite Element Model of the Swath Ship

3.2 Load on the SWATH ship

The load on the SWATH ship mainly includes the following four kinds of loads:

(1) Weight. The SWATH ship's finite element model may be divided into a series of areas along distribution curve of light weight in longitudinal direction and use the different material density to simulate the light weight by controlling of material density. The weight of machines in their respective areas acts as joint force on corresponding model nodes. Other secondary equipment and machinery are also simulated by material density in their areas. According to load distribution areas, the weight of ship loading acts on corresponding nodes.

(2) External hydrostatic pressure, which acts, according to draft under the calculation, on external wet surface of hull;

(3) Wave induced hydrodynamic pressure. The hydrodynamic pressure on elements of

wet surface is to be obtained by calculation program based on 3-dimensional Radiation and diffraction theory and applied on shell elements of hull.

(4) Inertial force, to be obtained by multiplying mass model consisting of structural mass and loading mass with acceleration of nodal motion.

3.3 Boundary conditions

For the global finite element analysis, the hull structural model is to be in equilibrium state (in balance) under any load combination for all forces. So the boundary supports is to satisfy the following two requirements: namely the reaction forces are to be as zero as possible and the whole model is without any rigid motion. At the same time special attention is to be paid to reduce the effect of calculation result for additional stresses and deformation caused by boundary conditions.

Boundary conditions are to be determined based on the corresponding structure response. To reduce the bad effect on cross deck and corresponding structure from calculation as much as possible, it is determined that boundary supports and constriction condition for global model analysis are to be as follows:

Node A: longitudinal, transverse and vertical constraint:

$$X = Y = Z = 0$$

Node B: transverse and vertical constraint:

$$Y = Z = 0$$

Node C: vertical constraint:

$$Z = 0$$

Figure 3 shows the diagram of boundary conditions.

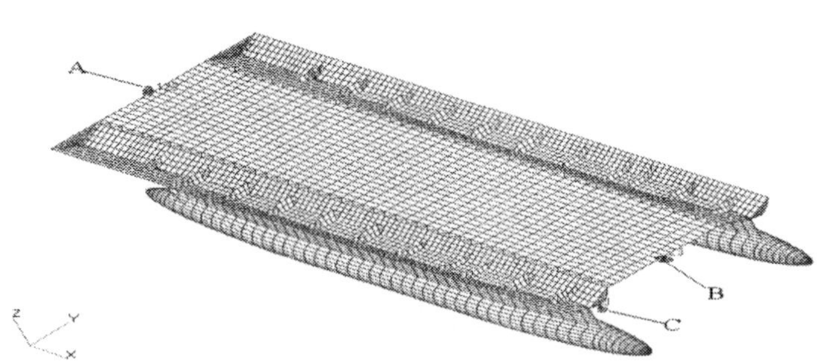

Fig.3. Diagram of Boundary Conditions

3.4 Assessment of structural strength

Wave induced hydrodynamic pressure and inertial force derived from design wave calculation is applied to the entire ship finite element model. The stress in each part of main structural members of ship hull on design waves is calculated by a finite element model, thus the stress distribution and quantity of structural members of ship hull are obtained in detail. The stress of structural members consists of total stress and local stress. After all the computed results on a group of design wave are obtained, it can derive envelope of the maximum stress of structural members of ship hull. The assessment of structural strength of main structural members is preformed according to the related allowable stress criteria. Figure 4 shows the distribution diagram of Von Mises stress of typical transverse bulkhead in beam sea. Figure 5 shows the Von Mises stress distribution along ship length direction in Beam Sea.

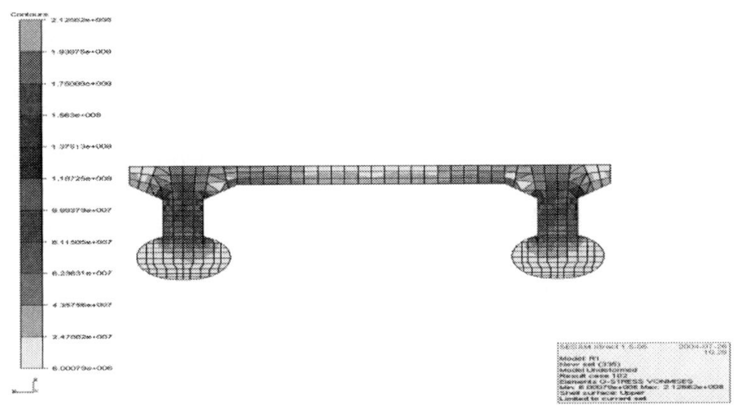

Fig.4. Distribution Diagram of Von Mises Stress of Typical Transverse Bulkhead in Beam Sea

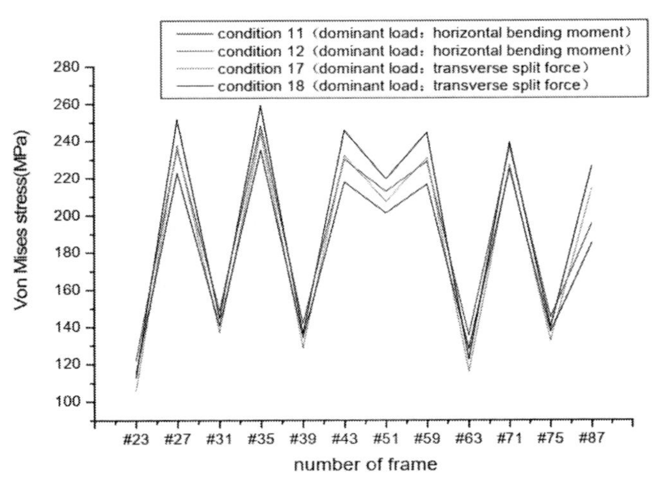

Fig.5. The Von Mises Stress Distribution along Ship Length Direction in Beam Sea

4 Conclusions

This paper takes a large-scale SWATH as an example. By a 3-D radiation-diffraction hydrodynamic program, the long-term prediction of wave induced motion and loading is performed, based on which a group of design waves are defined. The structural strength of SWATH on Entire ship finite element model is analyzed. It can be concluded that:

(1) The most dangerous load condition of the SWATH ship occurs in beam sea, next in head or following seas. The stress of main structural members of ship hull is relative small in quartering sea. The load condition of beam sea should be the key load condition for ship structural design and optimization.

(2) Von Mises Stress and shear stress are relative small in amidship and relative big in fore or aft $L/4$ position in head or following seas. For beam sea, Von Mises Stress is relative big in transverse bulkhead and relative small in transverse strength frame. Shear stress is slightly big in fore or aft $L/4$ position in beam sea. For quartering sea, Von Mises Stress and shear stress are relative small in other regions except in amidship and fore or aft $L/4$ position. Von Mises Stress and shear stress of amidship is smaller than that of fore or aft $L/4$ position in quartering sea.

(3) For beam sea, high stress mainly occurs in the joint of upper hull and vertical struts of transverse bulkhead and transverse strength frame. For head or following sea, high stress mainly occurs in the joint of lower hull and vertical struts.

(4) By the structural strength analysis, Von Mises Stress and shear stress of main structural members of ship hull all not exceed the allowable stress. It suggests that structural design of the SWATH ship satisfy the related strength criteria.

(5) Stress concentration of the SWATH ship is extremely prominent, which mainly occurs in transverse bulkhead, transverse strength frame, the joint of upper or lower tip of vertical struts and ship hull shell structure. The strengthening of these structural members should be specially considered in ship structural optimization design.

Reference

[1] American bureau of shipping, "Guide for building and classing SWATH vessels"(2002).

[2] Bitner-Gregersen E M, Cramer E H, Loseth R. Uncertainties of load characteristics and fatigue damage of ship structure, Marine Structures,1995,8;97-117.

[3] China classification society, "Guidelines for Direct Calculations of Strength of Container Ship"(2003).

[4] China classification society, "Guidelines for Construction of Small Waterplane Area Twin Hull Craft "(2004).

The online overall optimization of rapidity and manoeuvrability of high-speed monohull ship based on chaos algorithm

DONG Liang YANG Song-lin

(School of Naval Architecture and Ocean Eng. , Jiangsu University of Science and Technology, Zhenjiang Jiangsu 212003,China)

Abstract: Through overall demonstration and analysis, the optimization mathematic model on rapidity and manoeuvrability of high-speed monohull ship is established in this paper. Then based on the MATLAB6.5, program the mutative scale chaos algorithm to on-line optimize the simulation model of rapidity and manoeuvrability of high-speed monohull ship. The results shows: mutative scale chaos algorithm is effective and feasible to the overall optimization of rapidity and manoeuvrability of high-speed monohull ship.

Key words: high-speed monohull ship; rapidity; manoeuvrability; online optimization; mutative-scale chaos optimization algorithm

The determination of wave induced loads for 2500t grade SWATH vessel

Lin Ji-ru You Guo-hong Tang Tian-cheng Shi Li-guo

(CHINA SHIP SCIENTIFIC REASERCH CENTER)

ABSTRACT

The 2500t grade SWATH vessel is a Oceangraph Survey Vessel for China Academy of Sciences. The most significant primary wave induced load for a SWATH vessel is encountered in the beam waves, which is absolute different from the conventional monohull ship. There are two ways to determine the beam wave induced load, one is based on theoretical research and the formula regressed from a lot of model test results and full-scale test results, another is carrying out the scaled model test to get the exact wave loads. This paper presents the result of the wave induced loads calculation for the Oceangraph Survey Vessel.

Keywords: SWATH vessel, wave induced loading

1 the approaches to determine the wave induced loads

The 2500t grade SWATH vessel is a new integrated scientific review ship designed by CSSRC for China Academy of Sciences. This ship must serve over the globe and there is a need to ensure structure integrity under 11 class wind according to owner's demand. In this case, the reliability of hull strength is the key aspect for the structure design.

The research results from lots of researcher and practical sailing shows: the most significant primary wave induced load acting on a SWATH vessel is the beam wave induced load; by contrast, the longitudinal wave load is the primary load on a conventional monohull ship. Therefore predicting rational seaway loads is the principal task for the structure design.

There are three ways to predict seaway loadings on SWATH vessels. One is based on strip theory and source-sink distribution method[1], accounting for hydrodynamic interaction between two demi-hulls. But because of the nonlinear relationship between lateral force and wave height, the prediction results will be uncertain. This method is available for the coastal navigation SWATH not beyond high sea condition. The second method is undertaking model tank test under irregular wave to obtain the wave induced load. The third method is using mathematic formula regressed from lots of model test data and full scale test data, which is available for severe North Atlantic sea conditions. This method is adopted by ABS and LR. In term of form of the formula provided by ABS, CCS also gives a regression formula from model test dada including 12 America single strut SWATH vessels、4 tandem strut SWATH vessels、and 5 China SWATH vessels. For the SWATH vessel which Δ>2000t , the difference between lateral load obtained by CCS formula and ABS(1999) formula is not greater than 3%. For the SWATH vessel which Δ<2000t , the prediction result obtained by CCS formula is between the ABS(1990) and ABS(1999), see table 1. From the point of view of safe, the formula for design always gives conservative results.

Table 1 comparing side forces obtained from formula given by ABS（1999）and CCS

draught (t)	50	100	200	400	800	1500	3000	5000	10000
$\frac{F_{CCS}}{F_{ABS}}$	0.7872	0.8063	0.8299	0.8596	0.8982	0.9449	0.9809	0.9760	1.0249

2 Calculating side force in accordance with CCS *Guidelines for Construction of Non-high Speed SWATH Craft,* which is referred to herein as the "*CCS*"

2.1 calculating according to *CCS* section 2.2.2.3

2.1.1 Original parameter

displacement: $\Delta=2569$t

struts length: ls=50.78m

draught: d=6.5m

2.1.2 side force

$D=3.24-0.55\log 2569=1.3646$

$T=1.754\times 6.5\div(2569)^{1/3}=0.8324$

$L=0.75+0.35\text{th}(1.65\times 50.78\div(2569)^{1/3}-6.0)=0.7912$

$F_y=9.81DTL\Delta=9.81\times 1.3646\times 0.8324\times 0.7912\times 2569=22650\text{kN}$

2.2 comparing side forces according to different formulae provided by several ship classification societies

table 2 comparison of side forces according to different formulae

Ship classification society	Ship type	Side force（kN）	$\frac{F_y(\text{other})}{F_y(\text{CCS})}$
CCS	SWATH	22650	1
ABS	SWATH（1999）	22913	1.0166
LR	Catamaran（2004）	22913	1.0166
DnV	Catamaran	27362	1.2080
Model test	science review SWATH vessel	19410	0.8570

3 side force obtained from model tank test

3.1 similarity rule

the following parameters must be similar: geometry , Froude number, gyration radius and position of neutral axis.

3.2 model

According to the wave-making capability of CSSRC tank, the scale is taken as 1:25 to make irregular waves up to 9m height.

The model was made up of wood-frames with GRP shell。

Primary dimensions of model and ship are listed in table 3.

3.3 wave

ISSC wave spectra was used with significant wave height $H_{1/3}$ taken as 1.25m、2.50m、4.0m、5.0m、6.0m、7.0m、8.0m、9.0m and the corresponding period T_{01} taken as $3.86\sqrt{H_{1/3}}$.

In order to verify the safety margin of the results from irregular wave test, the maximum lateral force to the regular wave was also measured, with the ratio of wave height / ship breadth between 2.5 and 4.5.

3.4 test cases

Self-propulsion model tests for 772 ton and 1600 ton SWATH vessel have been undertaken previously to study the influence of the speeds and headings on beam wave induced forces and torsional moments, and come to the following conclusions:

a. the maximum lateral force is encountered at 0 speed in beam sea condition, and acts at the mid-draught of the vessel。

b. lateral forces varies with headings. It come up to maximum in beam sea condition (90°), and drop to half in oblique sea， then equal 0 in head sea.

c. lateral forces at 0 speed are greater than the value at a specified speed in same headings.

d. torsional moments from test are consistent with the value according to ABS（1999）。

These conclusions are consistence with those driven by other researchers. Hence the tank test was undertaken only at 0 speed in beam seas to study the influence of wave height on lateral force, then to find the critical sea state to get exactly lateral force for ship design.

Table 3

item	ship	model
scale	1	1:25
Overall length (m)	62.100	2.484
Waterline length (m)	53.320	2.133
Moulded breadth (m)	26.000	1.040
Moulded depth (m)	10.500	0.420
draught(m)	6.50	0.260
displacement	2501(t)	freshwater156(kg)
Length of lower hull(m)	53.54	2.142
Distance between centerline of lower hulls(m)	19.250	0.770
Breadth of lower hull(m)	6.681	0.267
Height of lower hull(m)	5.500	0.220
wetdeck height at mid craft(m)	9.100	0.364
Deck height at mid craft(m)	13.500	0.540
Maximum thickness of struts(m)	3.000	0.120
Longitudinal location of cg(m)	0.2017 before midcraft	0.0081 before midcraft

Height of cg(m)	9.210	0.3684
Longitudinal gyration radius	$0.26L_{WL}$	$0.26L_{WL}$
transverse gyration radius	0.38B	0.38B
Mounting angle for the fore fin	8.5°	8.5°
Mounting angle for the aft fin	−4.5°	−4.5°
Natural period for Rolling in still water (s)	21.0	4.20
Natural period for pitching in still water (s)	15.6	3.12
Natural period for heaving in still water (s)	9.5	1.90

3.5 conclusions

Model test results scaled to ship of 2501 ton were listed in table 3 and fig.1, and come to the following conclusions.

a. wave reduced lateral force varied with wave height: In the lower sea state and medium state , lateral force increased with the wave height, and up to maximum between wave height 5m ~5.5 m, then decreased with the wave height because the portions of long waves were in the majority[8, 9].

b. the maximum of significant lateral force was 5900kN, which was encountered at wave height 5m.

Model test results in regular wave were linear extrapolated to the extreme state $H/\lambda=0.1$,see table 4 and fig.2. The maximum of lateral force was 23842kN, equivalent to four times maximum of significant value in irregular wave, when the ratio of wave length / ship breadth was about 3.0. This value was 8.63% greater than the value in irregular waves 21948kN, because linear extrapolation produce conservation results.

Table 3 significant lateral force results in irregular wave at 0 speed

Significant wave height $H_{1/3}$ (m)	1.295	2.550	3.850	4.800	4.950	5.000	6.050	6.900	7.475	9.000
Mean period of wave T_{01} (s)	6.75	8.30	8.95	9.70	9.70	8.15	10.10	11.15	12.40	12.90
Lateral force (kN)	3094	5391	5719	5844	5188	6328	5688	4469	3859	4266

Transverse bending moment at central longitudinal section (kN.m)	17227	30977	32148	32891	28711	35469	31016	24219	20977	23125
a_{cg} (m/s^2)	0.14	0.51	0.82	0.91	0.96	1.03	1.07	1.09	1.05	1.17
Rolling angle (°)	1.88	3.22	6.09	6.64	7.32	7.68	9.45	9.46	7.51	11.5
Pitching angle (°)	0.57	0.48	0.75	0.77	0.91	1.13	1.09	1.09	1.17	1.42
heaving (m)	0.320	1.135	1.982	2.490	2.575	2.600	3.375	3.575	4.025	4.725

Table 4 Lateral force in regular wave with $H/\lambda = 0.1$ at 0 speed

Wave height (m)	6.75	7.725	9.225	10.475	11.600
Wave length (m)	67.50	77.25	92.25	104.75	116.00
Wave height/ship breadth	2.60	2.97	3.55	4.03	4.46
Lateral froce (kN)	23219	23842	19563	16844	12219
Transverse bending moment at central longitudinal section (kN.m)	115266	130583	108116	93109	65332
a_{cg} (m/s^2)	8.72	5.80	6.26	7.00	7.29

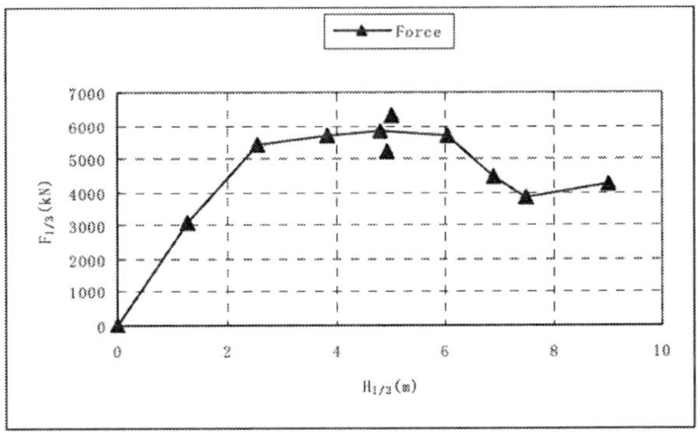

fig.1 Lateral force vs. significant wave height

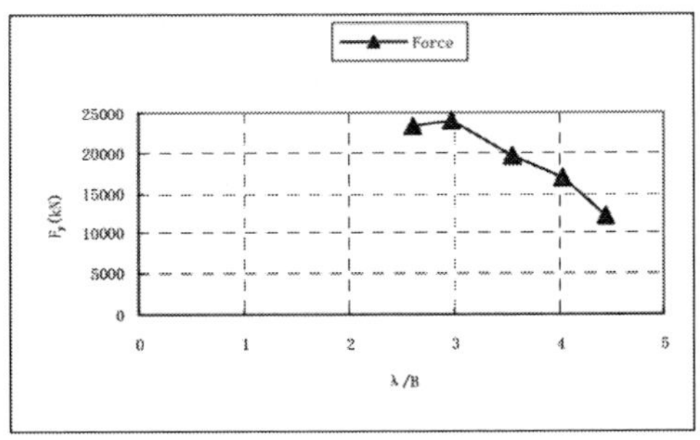

fig 2 Lateral force vs. ratio of regular wave length / ship breadth at (H/ λ =0.10)

4 Extrapolate model test results for ship design

4.1 Correction because of the adjust of primary dimensions

The primary dimensions of ship adjusted from Δ=2501t and ls=53.32 m at preliminary level design stage to Δ=2569t and ls=50.78 at detail level design stage.

Lateral force was corrected according to formula in *CCS*.

$$\frac{F_y(2569t)}{F_y(2501t)}$$

$$=\frac{9.81\times(3.24-0.55\log 2569)\times 1.754\times 6.5\div 2569^{1/3}\times(0.75+0.35th(1.65\times 50.78\div 2569^{1/3}-6.0))\times 2569}{9.81\times(3.24-0.55\log 2501)\times 1.754\times 6.5\div 2501^{1/3}\times(0.75+0.35th(1.65\times 53.32\div 2501^{1/3}-6.0))\times 2501}$$

$$=0.8844$$

4.2 lateral force for ship design

Fy（2569t）=2×0.8844×（1.86×5900）=19410kN

Where :2 is the safety factor，accounting for the following factors：

a. the spectra shape and characteristic period of the practical sea condition will differ from those in tank.

b. the change of stowages will affect the lateral force.

c. emergency turn will enlarge the wave loads.

d. different data processing methods will affect the lateral force.

The lateral force in different headings should be modeled as distributed load along ship length and draught in accordance with *CCS* Section 2.2.2 and Section 2.2.3

a. in beam sea (90°)

in beam sea (90°), the longitudinal distribution of side force is assumed to be trapezoidal. Lateral force and horizontal torsional moment can be written as:

$$F_Y = \frac{W_{11}+W_{12}}{2} \times d \times l_s$$

$$M = \frac{1}{2} \times \left| \frac{W_{11}-W_{12}}{2} \right| \times d \times \frac{l_s}{2} \times \frac{2}{3} l_s = 0.02 F_y \times l_s$$

Then distributed pressures were obtained as:

$$W_{11} = 1.12 \frac{F_y}{d \cdot l_s} = 1.12 \times \frac{19410}{6.5 \times 50.78} = 65.862 kN/m^2$$

$$W_{12} = 0.88 \frac{F_y}{d \cdot l_s} = 0.88 \times \frac{19410}{6.5 \times 50.78} = 51.749 kN/m^2$$

b. in quartering seas (45° /135°)

In quartering seas (45° /135°), lateral force and torsional moment can be written as :

$$0.6 F_Y = \frac{W_{21}+W_{22}}{2} \times d \times l_s$$

$$M = \frac{1}{2} \times \left| \frac{W_{21}-W_{22}}{2} \right| \times d \times \frac{l_s}{2} \times \frac{2}{3} l_s = 0.10 F_y \times l_s$$

Then distributed pressures were obtained as:

$$W_{21} = 1.20 \frac{F_y}{d \cdot l_s} = 1.20 \times \frac{19410}{6.5 \times 50.78} = 70.567 kN/m^2$$

$$W_{22} = 0$$

4.3 design value of horizontal torsional moment

In oblique seas, the lateral force acting on the hull non-uniformly produces horizontal torsional moment. According to *CCS* Section 2.2.2 and Section 2.2.3, the horizontal torsional moment will be produced as:

a. torsional moment in 90° beam seas

$$M = \frac{1}{2} \times \left| \frac{W_{11}-W_{12}}{2} \right| \times d \times \frac{l_s}{2} \times \frac{2}{3} l_s = \frac{1}{12} \times (65.862 - 51.749) \times 6.5 \times 50.78^2$$

$$= 19713 kN \cdot m$$

b. torsional moment in 45° /135° quartering seas

$$M = \frac{1}{2} \times \left| \frac{W_{21}-W_{22}}{2} \right| \times d \times \frac{l_s}{2} \times \frac{2}{3} l_s = \frac{1}{12} \times 70.567 \times 6.5 \times 50.78^2$$

$$= 98564 kN \cdot m$$

4.4 pitching connection moment

In oblique seas, the pitching connection moment and rolling moment also will be produced because of pitching connection between two demi-hull.

Pitching connection moment can be estimated according to *CCS* Section 2.2.4, with the vertical acceleration at the gravity center of the ship taken as the maximum test result times a safety factor 2.

$a_{cg} = 2 \times 1.86 \times 0.717g = 6.24 m/s^2$

$M_{p1} = 0.125 \Delta a_{cg} L_w = 0.125 \times 2569 \times 6.24 \times 50.78 = 1.0175 \times 10^5$ kN.m

$M_{p2} = 0.25 \Delta a_{cg} b = 0.25 \times 2569 \times 6.24 \times 19.25 = 7.715 \times 10^4$ kN.m

M_p=the greater of M_{p1} and $M_{p2}=1.0175\times10^5$ kN.m

Distributing M_p according to *CCS* Section 2.2.4 will produce rolling torsional moment besides the pitching torsional moment。 Uniform pressure P will be:

$$P=\frac{0.125\Delta a_{cg}L}{0.25L^2}=\frac{\Delta a_{cg}}{2L}=\frac{2569\times6.24}{2\times50.78}=157.84kN/m$$

The M_p mentioned above will be encountered in heading 20° . In quartering seas 45° /135° , only 80%of M_p will act on the hull, and in beam seas no torsional moment will be needed to be taken into account..

4.5 vertical share force

4.5.1 at central longitudinal section

$Q_1=0.25F_y=0.25\times19410=4.8525\times10^3$kN

4.5.2 at box/strut intersection

$Q_2=0.25F_y+1.35(\Delta/4)=0.25\times19410+1.35\times9.81\times2569/4=1.3358\times10^4$kN

5 longitudinal wave induced loads

Because the length of waterline of the ship is only 50.78 m and the ratio of length / breadth equals 1.95, the longitudinal wave load will not be too large. The longitudinal strength of the hull can be verified using approximate methods instead of a rigorous way. Longitudinal wave load can be estimated by the formula in *CCS* Section 2.2.6

6 Slamming pressure

The slamming event number、 value and region on SWATH vessels will vary with the sea states、 speeds、 headings、 wet deck clearances and the motion attitude of the craft. It is a complex random process. On the study of the trial records of five SWATH vessels (57t Halcyon、 73t Frederick Creed、 190t Kaimalino、 3500tT-AGOS19 and 5400t T-AGOS23), Jerome P. Sikora presented a formula[3] to predict the design slamming pressure of a SWATH in lifetime of 5400days (30 years). This formula was adopted by ABS. But the precondition to use this formula is that the slamming rate must be less than 10 slams per hour. In other words, the wet deck clearance must be appropriate to ensure lower slamming rates.

The number of slamming events in this model test was not too great and the maximum slamming pressure measured in this test was only 113kPa (occurring at 0 speed , $H_{1/3}=6$m, $T_{01}=10$s, on outboard haunch)[10]。Because our study and test records on slamming are insufficient, the slamming design pressure for this craft was estimated by the formula in *ABS*（1999）, with correction n=6.78 in accordance with the age of vessel of 20 years. The design pressure was 1.77% less than the one from *ABS*(1999).

7 Load combinations in strength analysis

The combined load conditions in table 5 must be taken into consideration in structural analysis. The most sever heading for horizontal torsion is 18 degree, and the maximum of pitching connection moment will be encountered in oblique sea 60° . Because the lateral force is the primary seaway load for the SWATH vessel, torsional strength will be verified in quartering seas 45° /135° . In this case the horizontal torsional moment and the pitching connection moment will be 80% of the peak values, and the lateral force will be 60% of the maximum.

Table 5 combined load conditions

Load case	heading	Case no.	Load kind								
			buoyancy	Lateral force(inward)	Lateral force(outward)	Torsional moment	Pitching /rolling moment M_P	Longitudinal bending moment $M_L(+)$	Longitudinal bending moment $M_L(-)$	gravity	Vertical shear force
Transverse strength verify	90°	1	0.65	√	/	Fore 1.0 W_{11} Aft 1.0 W_{12}	/	/	/	0.65	/
		2	1.35	/	√	Fore 1.0 W_{11} Aft 1.0 W_{12}	/	/	/	1.35	/
		3	0.65	√	/	Aft 1.0 W_{11} Fore 1.0 W_{12}	/	/	/	0.65	/
		4	1.35	/	√	Aft 1.0 W_{11} Fore 1.0 W_{12}	/	/	/	1.35	/
		5	/	/	/	/	/	/	/	/	1.0
Torsional strength verify	135° / 45°	6	/	√	/	Fore 1.0 W_{21} Aft 1.0 W_{22}	0.8	/	/	/	/
		7	/	√	/	Aft 1.0 W_{21} Fore 1.0 W_{22}	0.8	/	/	/	/
		8	/	/	√	Fore 1.0 W_{21} Aft 1.0 W_{22}	0.8	/	/	/	/
		9	/	/	√	Aft 1.0 W_{21} Fore 1.0 W_{22}	0.8	/	/	/	/
Longitudinal strength	180° /	10	/	/	/	/	/	1.0	/	/	/

verify	0°	11	/	/	/	/	/	/	1.0	/	/

Referances

1 E.T.Reilly, Y.S.Shin, E.H.Kotte,. "A Prediction of Structural Load and Response of a SWATH vessel in Waves." Naval Enginneers Journal , May 1998. P251~264.

2 Sikora J.P., Dinsenbacher A.L. "SWATH Structure: Naval Research and Development Applications ." Marine Technology , Vol.27 , July 1990.

3 Sikora J.P. "Design Algorithms for Primary and secondary Loads on SWATH vessels." Naval Engineers Journal，May 1995. P121~127.

4 American Bureau of Shipping , "Guide for Building and Classing Small Waterplane Area Twin Hull (SWATH) Vessels.," 1999.

5 Lloyd's Register, "Rules and Regulations for The Classification of Special Service Craft", July 2004.

6 中国船级社，"非高速小水线面双体船建造指南"，2005.2.

7 钱家玉，林吉如，李琪华，"SWATH 船在波浪中的波浪载荷及运动分析"，第二届全国海事技术研讨会文集，1996.

8 林吉如，尤国红，周德才，"1600t 小水线面双体船波浪载荷模型试验和分析、应用"，CSSRC 科技报告，2004.6.

9 林吉如，尤国红，周德才，"中科院新型综合科学考察船波浪载荷模型试验"，CSSRC 科技报告，2004.6.

10 邱强，"中科新型综合科学考察船砰击压力测试及初步分析"，CSSRC 科技报告，2004.9.

Material selection and structure design in High-Speed Aluminum alloy ships

Wang chen quan Liao yinbo

Wuhan nanhua highspeed ship engineering co.,LTD Wuhan 430064

ABSTRACT At the sight of anticorrosion, this article give a discussion about selection of aluminum alloy used in ships. Also have a Instruction about types and efficient of sheet with ribs, analyses the feature and structure design of sheet with ribs in aluminum alloy ships.

Key words: aluminum alloy property of anticorrosion sheet with ribs structure design

Instructions

As a high speed Aluminum alloy ship nowadays, it has a elegant and noble gesture, florid color, outstanding speed than others, which attracted us when it passed through our sight。 Which lead many of naval designers probed and soul-searching, even admirable to jealous?

Through the gloss, it's a real item for all of naval designers how to design this kind of high speed Aluminum alloy ship.

Comparing to normal steel ships, there isn't substantial different than modernize high-speed aluminum alloy ships. But it will be more complex than steel ships at the sides of selection of material and structure design. Today, high-speed ships developed very quickly worldwide, meanwhile the types of aluminum alloy material and new pattern of structure have been up copping successively. Near two years, as the quality of Aluminum alloy ships increased quickly in domestic. It has made the shortage of aluminum, other side, much of question in design of aluminum alloy ships which naval design have never meted, this consistent the development of High speed aluminum alloy ships in the state.

As a point, the selection of Aluminum alloy used in ships has been discussed in the article, and analyzed sheet with rib's advantage feature and discussed some questions in structure design.

1. Aluminum alloy selections in ships

1.1 Principle of selections

Much more than 450 aluminum alloy ships have been surveyed since 1970, and all of them had been designed as the rules of high speed of craft in DNV. Nowadays, there is two kinds of aluminum alloy , 5000 series and 6000 series, have been certificated by DNV, and allowed to use in ocean environment. See TAB 1.

Approved Aluminum alloy used in ship table 1

Strain hardened Aluminum alloy	Strength by heat treatment Aluminum alloy
NV—5052	NV—6060
NV—5154A	NV—6061
NV—5754	NV—6063

NV—5454	NV—6005A
NV—5086	NV—6082
NV—5083	
NV—5383	

As mainly factors, the location and corrosion environment must be considered in selection of structure design. For 5000series, seven-grade aluminum alloy supplied, they are different in mechanical property as the different heat treatments condition that offered a Varity choice to selection for navel designer. Same as 5000series aluminum alloy, the 6000 series have been sampled in table 1, and they also owned different stress property each other.

1.2 Property of anti-corrosion

The standard of aluminum alloy anticorrosion have been ruled in each states, compared them, they show a different in each other.

- DNV (pt.2, ch.2, sec.9)

  203 The use of 6000 series aluminum alloy in direct contact with seawater may be restricted depending on application, and corrosion protection system. Applying these alloys should be agreed with the society.

- CCS rules of material and welding Ch.8, sec .2

  8.2.1.2 Aluminum alloys should have a good welding ability and anticorrosion, expected permit, AlSiMg alloy is not allowed to contract with seawater directly. (Writer note: here AlSiMg mean 6082)

- BV (pt.D, Ch.3, sec.2)

2.2.4 The aluminum alloy of grades 6005A and 6061 in 6000 series don't contract with sea water directly unless offered a protection system, such as coating and/or catholic protection.

Upon the items, you will be of enchanting. In rules of DNV, 6000 series aluminum alloy have been restricted over all. There give a instruction about 6082 in CCS, but other grades in 6000 series have not been referred. Opposite, it's clear in rules of BV. Grades 6000A and 6061 are restricted, and 6082 is permit to use selectively. Standard of rules is different, so designer also is not easy to judge which type of material should be to select.

But the items followed could be sure:

1) Contract directly to seawater means permit surface of aluminum alloy contract to seawater.
2) If you carried out a protection system e.g. Coating and/or catholic protection means don't contracted directly to seawater.
3) Restricted aluminum alloy don't offered 5000 series.

Could it contract directly to seawater for 5000 series aluminum alloy? Pure aluminum will not be corroded in bounder of crystal .For aluminum alloys, including Mg larger than 3%, when you heat them, they will form a c material $Mg_5Al_2(Mg_2Al_3)$, composed β phase .along the bounder of crystal . β phase is easy to be corrosive. Meanwhile β phase zone will increase if the content of Mg included enlarge. So Lenar co., in USA suggested: just 5086 (including Mg 3.5%~4.5%) could be used for shell in ship structure. and 5456 aluminum alloy (including Mg 4.7%~5.5%)have been applied as a good tension property. but it need to be protected with coating or catholic.

Alloys H116 and H117 both temper, treated

for extra resistance to exfoliation corrosion in AlMg alloys. In the marine atmosphere, the alloys ENAW 5000 and 6000 can have pitting depths up to 200-300 um after 20 years and overage weight loss corresponding to 100 um. ENAW 7000 alloys have a somewhat higher weight loss. In the same way as for pitting corrosion, the corrosion speed will decline over time. In a marine atmosphere, corrosion will mainly be regarded from an aesthetic viewpoint and it will not normally be necessary to carry not surface treatment or other methods of corrosion protection

Experience has shown that ENAW5000 alloys will not corrode more than 5 um per year when they are submerged in seawater, calculated, from weight loss. In general this will be pitting corrosion which will decline in speed during the first few years so that an average speed estimated over 10 years will be around 1 um per year.

For ENAW 6000 alloys the corrosion speeds will be somewhat higher, and will primarily be seen in the form of more and somewhat deeper pits. A pitting depth of 0.2 to 1mm is normal after 10 years exposure both for ENAW 5000 and 6000.

NEAW 7000 alloys will corrode somewhat more and will normally not submerged unprotected in seawater

It has become a model thought to adopt 5000 series aluminum alloy as shells and 6000series as profiles used in inner ship structures. Lately, some ships in DNV adopting 6082 aluminum alloy T_6 to extrude into sheet with ribs owned double T-sections and wider sheet with ribs 2275mm X 1150 mm by FSW welding. This procedure have obtained certificate by DNV society. Now it has used in superstructures. The fact offered a change for our idea that 6000 series aluminum alloy couldn't use to shells.

All kind of aluminum alloy have been adopted in vary part of ships. Show in Fig 1. DNV offered.

Fig 1 DNV-approved alloy and locations for a ship structure

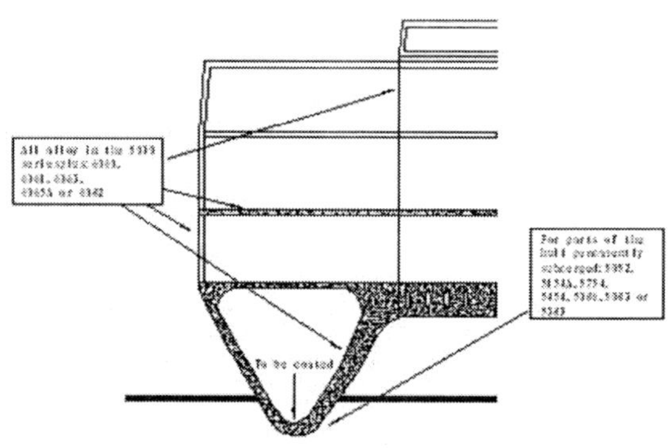

Figure 1

Figure 1 shows the alloys that can be used in the various sections of the hull. Basically, only the 5000 alloys can be used for the parts of the hull that are permanent exposed to seawater. The use of ENAW 6000 alloys for these parts mast be approved by DNV in each case For the remaining parts of the ship hull, a combination of ENAW 5000 and 6000 alloys is common, The ENAW 6000 alloys are mainly used for the hull structure above the waterline or for interior parts of the hull.

1.3 Property of welding

Aluminum alloy of 5000 series and 6000 series own a good property of welding each other .at he condition of correct WPS, suitable welding parameters and skill technician, it's easy to obtain a good quality seam. Meanwhile the quality of seams is satisfied and will meet the requirement of rules. The welding property between 5000 series and 6000 series or each other is the same.

2. Sheet with ribs

Compare with steel hull, Aluminum alloy ship's structure base on sheet with rib could be extrude into sections by module, so the plate with single extern section could be manufactured. hiss plate is sheet with ribs or rib .

2.1 Efficient of sheet with rib

Comparing the welding tech of MIG or TIG, it will be much more efficient and economy to weld the plate and sections together adopting the sheet with rib.

(1)It's no need to weld sections on plate in longitudinal, meanwhile decrease the workforce of structure welding and short the cycle of manufacture.

(2)adopting the sheet with ribs, the surface of plates outside after welding will be much more smooth and tidy than usual, and offered a pretty surface quality additional. Meanwhile it's no need to rectify the distortion of plate after welding. Usually it's not easy to rectifies because lack of theories and abundant experience.

(3)The sheet with ribs apply into ships will decline the consumption of weld material, aluminum alloy welding wire, argon gas and current contact nozzle. Other side, gas nozzle used in welding, welding time and welding arc is no existent. So it's much more economy. Accountant the cost of all, it is no fairly different to general welding method, such as TIG or MIG.

2.2 Types of sheet with ribs

(1). Wider sheet with ribs including mult-profiles, different in thickness, spanning and profiles are offered in table use for design guidance. This sheet of ribs made in Russia, and it's much expensive recently.

(2). Generally, single or couple sheet with rib could be welded by MIG to compose into wider. This process has been applied in many shipyards in the word.

(3).A new welding method, Friction Stir welding — FSW , could weld much more profiles in to a wider sheet with ribs. To answer the need of different spanning and vacancy of sheet with ribs, FSW is qualified to the process and the price of sheet with ribs is lower than the usual. This kind of process -FSW offered a variety selection and convenient for the application of sheet with rib used in aluminum alloy ship structures. It's a forward in future.

3. Structure design of sheet with ribs

3.1 Arrangement of applying

It could be used in parts of flat in ships .

(1)Every deck structure in superstructure (including flat or camber structures) and main decks

(2)Wall panels inner or flat outside panels

3.2 Profile direction of sheet with ribs.

In order to reveal the advantage property of sheet with ribs, it's needed to abide fixing method as usual. Traditionally, stiffeners and girders are assembly in vertical in bulkhead and sidewall panels. Adopting sheet with ribs, it's suitable to arrange the profiles in longitudinal. The girders and profiles will structure a complex system, which could increase stability and benefit for fixing and maintain of doors and windows.

3.3 Discussions about direction of profiles.

To select a suit type sheet with ribs, it's necessary to considerer enough about spanning, type of profiles and the structure arranged in horizontal. In order to strength the stability of anti buckling, the scantling of plate in section should avoid the rate of integral multiple, such as 1, 2, 3.....

3.4 Applications in translation structure

Some complex module, such as double curve structure, still should adopt a usual process to fix the structures. Between sheet with ribs structures restricted using in flat part, there must be a transition structure, the transition part also should be arranged as usual process, and how to design the transition structure suit and briefly, designers must soul to your heart.

3.5 Curving of sheet with ribs

(1) It's no problem to curve the sheet with ribs direction to the profiles in vertical. Just only the cuts of beam are correct toward the profiles and the camber also is easy to form, when sheet with ribs fixed.

(2) For larger curve rate structure in profile direction, it's available to obtain a curve model accord to the structures formed by heat or roll, there must be true, any double curve rate structure, applying a sheet with ribs, is unsuitable.

4. Mechanical behavior

There is an obvious advantage for sheet with ribs welded by FSW, well-suitable, lower cost, it direct a development in future. The quality of FSW process is unsuspicious, lots of articles have reported. Just there is less tech-reports or literatures about the load capability for structures welded by FSW. In order to supply a science foundation for sheet with ribs welded by FSW instead of usual welding process, we complete a bend test to the beam weld by MIG and FSW each other.

4.1 Test ample

Scantling and material selected for test shown in fig .2

Beam welded by FSW called FSW beam, and welded by MIG called MIG beam.

The WPS of MIG beam is same a usual in structure welding process, adopting double side discontinue weld. FSW beam is continuing welded along the seam.

Scantling and material selected for test fig .2

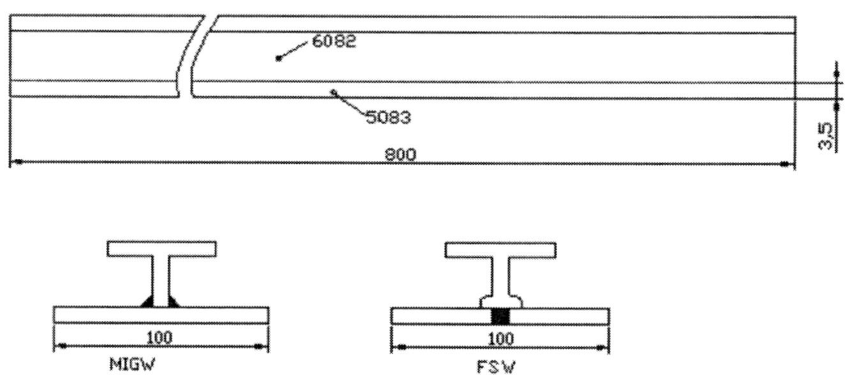

4.2 Bending test

Laid the test beam on the set of testing platform, between the spanning, exert a concentrated force on the beam; it will be curve down in direction of the force. Meanwhile the deflection will be recorded as the variety of force. When it load off, and reverse to initial, record the data and compare the record of deflection to beginnings, fix and monitor position of beam show in fig 3.

Fig 3 style of fixes and monitor position

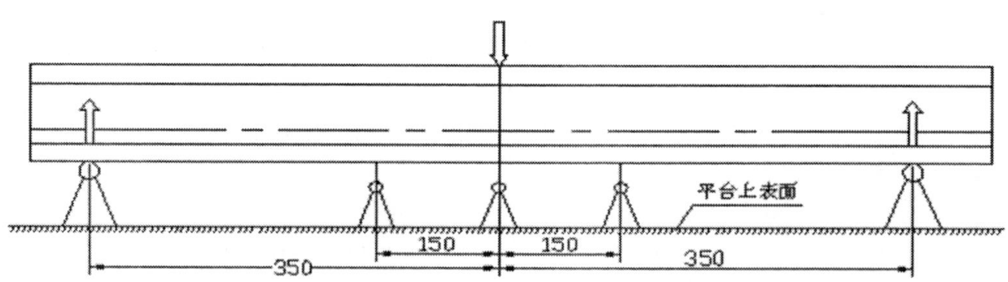

Spanning of beam in test is700mm (general almighty test machine) and 300mm (Auto control almighty test machine made in USA)

4.3 Experiment data

(1) Spanning 700mm

① Load increased from 1200N to 4200N gradually, the deflection recorded in tab2

Load and deflection record in tab 2

Test sample	load (N)					
	1200	1800	2400	3000	3600	4200
MIG beam	(0.8)	0.27	0. 3	0.83	1.17	1.52
FSW beam	(2.28)	0.24	0.47	0.72	0.95	1.19
D-value between MIG and FSW beam	——	0. 03	0.06	0.11	0.22	0.33

② Reverse to initials, the remainder deflection show in tab 3.

Remainder deflection tab 3

Test sample	Remainder deflection (mm)
MIG beam	0.90 – 0.80 = 0.1
FSW beam	2.31 - 2.28 = 0.03

Attention: the level of tensile must be lower than $R_{p0.2}$ and make sure the process of load off within the scope of flexibility

(2) Spanning 300mm

① The record of deflection and load in the process are completed automatically by test machine. We can have a learn that the stress loaded is larger than $R_{p0.2}$. the level of deflection recorded from 4KN to 20KN gradually show in tab4.

Load and deflection record tab 4

Test sample	load (KN)					
	0.0	4.0	8.0	12.0	16.0	20.0
MIG beam	0.00	0.80	1.30	1.80	2.68	-----
FSW beam	0.00	0.69	1.10	1.52	2.04	3.34
D-value between MIG and FSW beam	0	0.11	0.20	0.28	0.64	-----

Note: MIG beam shows a data when the loads exceed 20KN

② Trial curve of FSW beam and MIG beam show in fig 4.

Trial curve of FSW beam and MIG beam fig 4

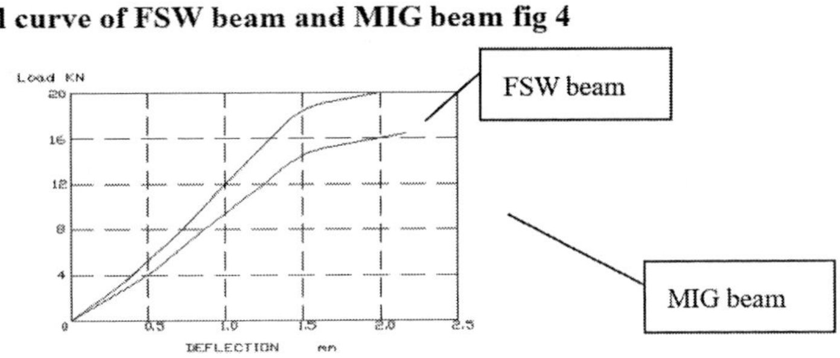

4 Conclusions

(1) At the same load, the deflection of Mig beam will e larger than FSW beam, which is the carrying capacity of FWS beam, is larger than MIG beam. Trial test show that there is no relative with the spanning.

(2) When the load is off to initial, the remainder of deflection MIG beam is 0.1mm, but FSW beam just is 0.03mm, that means the stability and maintain feature 'capacity of FSW beam is much more excellent than MIG beam. In a word, the FSW process used in sheet with ribs is quiet qualified and instead the MIG process. At any case of manufacture tech, efficient or mechanical property, the FSW process is feasible used in sheet with ribs.

Consulted literature

1. DNV society 《rules of shipbuilding 》1996
2. BV society Part D 《rules of material and welding》 1998
3. CCS 《rules of material and welding》1998
4. Wang chen quan ect.. 《Discussion on application of sheet with ribs in aluminum alloy ships》《ship engineering 》2003，4
5. Wang chen quan ect.. 《structure stability of high speed ship》 《ship engineering》2005，5
6. HYDRO 《aluminum in the marine environment》

Structure Fireproofing Design of Medical Cabin in Medical Ship

Yang ZhiQing[1] Ye Ming[2] Zhong ChenHua[1] Yimg MuDe[1]

(1:Navy University of Engineering

2: Warship Design Research Army Representative Office at ShangHai)

Abstract: There is no definitude prescript for fireproofing design of medical cabin of medical ship in warship specification. In this paper, it has been established that the mutual reference method of army specification and civil specification. Reference to cabin definition and fireproofing design criterion of SOLAS, it has been decided that the fireproofing design requirements of the medical cabin of medical ship. The conclusion of this paper is can be used for other special ship fireproofing design.

Keywords: medical ship, medical cabin, fireproofing design

When designing some special warship, we often contact with some type cabin that has not been definition by army specification but could been definition in civil specification. So, it's common things to refer to civil specification in warship design. In this paper, the reference to each other method has been established from the analyzing of the two specification similarities and differences. It is test feasible and reasonable from a case of some type medical ship structure fireproofing design.

The difference of two specification conclude: amount of cabin classified, fireproofing scale between cabins. The fireproofing scale could be expressed by the value of fireproofing separating. The value is list in table 1.

Table 1 each fireproofing scale value and calculating equivalent value

Fireproofing separating scale V	A-60	A-30	A-15	A-0	B-15	B-0	C
Fireproofing separating index value η	60	30	15	0	-15	-30	-60
calculating equivalent value η	120	90	75	60	45	30	0

The warship fire resisting division degree is established for analyzing the difference of army specification and civil specification. It reflects the importance of corresponding cabin in specification. Calculating the fire resisting division degree of two specification, it can be know that the difference of the fireproofing design in army and civil specification.

Consuming V_{ij} is the fire resisting division scale of type i cabin and type j cabin, η (Vij) is the value of V_{ij}, η max is the max value in all of η (Vij), $\sum_j \eta(v_{ij})$ is the sum of all the value of scale, n is the number of categories, then it can be calculated from underhood equation that the fire resisting division degree.

$$P_i = \frac{\sum_j \eta(V_{ij})}{n \cdot \eta_{max}} \quad i=1,2\cdots\cdots n;\ \ j=1,2\cdots\cdots n 。 \qquad (1)$$

2、 The method of warship design borrowing civil specification

In warship fireproofing design, we sometimes should refer to civil specification sfor some special cabin fireproofing design that has not been defined in army specification.

First of all, the similar type ship should be selected from civil specification, the value of the special type cabin fire resisting division degree can be calculated from the equation as follow.。

From interval proportion coincidence relation, the equation is as follow:

$$p = \frac{p_{max} - p_{min}}{p_{0\,max} - p_{0\,min}}(p_0 - p_{0\,min}) + p_{min} \qquad (2)$$

Where: p is the value of fire resisting division degree in army specifications;

p_{0max} is the max value of fire resisting division degree in civil specifications;

p_{0min} is the min value of fire resisting division degree in civil specifications;

p_0 is the value of fire resisting division degree of some type cabin in civil specifications;

p_{max} is the max value of fire resisting division degree in army specifications;

p_{min} is the min value of fire resisting division degree in armys specifications.

From this equation, the value of fire resisting division degree of the special type cabin in army specification can be calculated. Comparing this value and the value of other type cabin in army specification, we can conclude that this type cabin should be insert the some two types cabin between. Then from the two types cabin's fireproofing requirement, with the interpolation method, the fireproofing requirement of new type cabin can be reached out.

The fireproofing scale of new type cabin between the two types cabin that has been inserted is the value the two type cabin. When new type cabin border upon with itself, the value of fireproofing scale could be calculated from equation 3. When the new type cabin border upon with other type cabin, the value of fireproofing scale could be calculated from equation 4.

$$\eta = \frac{\eta(V_{1,1}) - \eta(V_{2,2})}{P_1 - P_2} \cdot (P - P_2) + \eta(V_{2,2}) \qquad (3)$$

$$\eta = \frac{\eta(V_{i,1}) - \eta(V_{i,2})}{P_1 - P_2} \cdot (P - P_2) + \eta(V_{i,2}) \qquad (4)$$

Calculating with these equation, we can calculate every fireproofing scale value of the new type cabin border upon other type cabin.

3、 The fire integrity requirement of medical cabin

For medical cabin in army medical ship is not defined completely sin army specification, we can class these medical cabin in solas specification as follow:

table 2 medical cabin class in solass

num.	medical cabin	**medical cabin class in solass**
1	serious injury salvage zone	great fire risky living room 8
2	operating room	middle fire risky living room 7
3	preoperative preparation room	middle fire risky living room 7
4	operation personnel bathhouse	middle fire risky living room 7
5	instruments ready room	middle fire risky living room 7
6	instruments scrubbing compartment	middle fire risky living room 7
7	antisepsis supply room	flammable liquid storeroom 14
8	X ray room	middle fire risky living room 7
9	CT rooms	middle fire risky living room 7
10	blood ready room/checkout room	great fire risky living room 8
11	cardiogram /special consulting room	middle fire risky living room 7
12	mouth diagnosis room	middle fire risky living room 7
13	eye ear nose throat diagnosis room	middle fire risky living room 7
14	helicopter rescue ready room	great fire risky living room 8
15	drugstore	flammable liquid storeroom 14
16	blood storeroom	great fire risky living room 8
17	sickroom	great fire risky living room 8
18	medical apparatus maintenance room	storeroom、workshop、pantry 13
19	medicine storeroom	flammable liquid storeroom 14
20	nurse office	great fire risky living room 8
21	doctor office	middle fire risky living room 7
22	central oxygen supply room	flammable liquid storeroom 14
23	central cacuum suction room compressed air supply room	flammable liquid storeroom 14
24	information center of hospital office	control position 1
25	ease room	great fire risky living room 8

From this table, we can conclude that: in solas specification, the medical cabin can be classed 5 type, namely as control position, flammable liquid storeroom, great fire risky living room, middle fire risky living room, storeroom、workshop、pantry.

From cabin identity, we think the control position in solas is corresponding with electronic equipment room in army specification, and the storeroom、workshop、pantry type cabin is corresponding with the other machine room in army specification. So, there are three types cabin has not definition in army specification, namely as flammable liquid storeroom, great fire risky living room, middle fire risky living room. We should borrow these three type cabin of solas to army specification. From above method, we can get their fireproofing design requirements.

(1) The fire integrity requirement of middle fire risky living room

cabin	middle fire risky living room

	bulkhead	bulkhead	deck			
	not fireproofing bound	fireproofing bound	not fireproofing bound	not fireproofing bound	fireproofing bound	fireproofing bound
			blow deck	above deck	blow deck	above deck
flammable liquid storeroom	A-30	A-60	A-60	A-60	A-60	A-60
missile room and gunner's room	A-30	A-60	A-60	A-60	A-60	A-60
type A machine room	A-0	A-60	A-30	A-60	A-60	A-60
control position and electronic equipment room	A-15	A-60	A-30	A-0	A-60	A-60
great fire risky service room	A-0	A-0	A-0	A-0	A-0	A-0
great fire risky living room	A-0	A-0	A-0	A-0	A-0	A-0
middle fire risky living room	A-0	A-15	※	A-0	A-0	A-0
other machine room	A-0	A-15	A-0	A-0	A-0	A-0
minor fire risky service room	A-0	A-0	A-0	A-0	A-0	A-0
living room	A-0	A-0	A-0	A-0	A-0	A-0
ladder and passage	A-0	A-0	A-0	A-0	A-0	A-0
open deck zone	※	A-0	※	※	A-0	A-0

（2）The fire integrity requirement of great fire risky living room

cabin	great fire risky living room					
	bulkhead	bulkhead	deck			
	not fireproofing bound	fireproofing bound	not fireproofing bound	not fireproofing bound	fireproofing bound	fireproofing bound
			blow deck	above deck	blow deck	above deck
flammable liquid storeroom	A-30	A-60	A-60	A-60	A-60	A-60
missile room and gunner's room	A-30	A-60	A-60	A-60	A-60	A-60
type A machine room	A-0	A-60	A-30	A-30	A-60	A-60
control position and electronic equipment room	A-15	A-60	A-30	A-60	A-60	A-60
great fire risky service room	A-0	A-15	A-0	A-30	A-0	A-60
great fire risky living room	A-0	A-15	A-0	A-0	A-0	A-60
other machine room	A-0	A-15	A-0	A-0	A-0	A-60

minor fire risky service room	A-0	A-0	A-0	A-0	A-0	A-30
living room	A-0	A-0	A-0	A-0	A-0	A-0
ladder and passage	A-0	A-0	A-0	A-0	A-0	A-0
open deck zone	※	A-0	※	※	A-0	A-0

（3）The fire integrity requirement of flammable liquid storeroom

cabin	flammable liquid storeroom					
	bulkhead	bulkhead	deck			
	not fireproofing bound	fireproofing bound	not fireproofing bound	not fireproofing bound	fireproofing bound	fireproofing bound
			blow deck	above deck	blow deck	above deck
flammable liquid storeroom	A-60	A-60	A-0	※	A-60	A-0
missile room and gunner's room	A-60	A-60	A-60	A-60	A-60	A-60
type A machine room	A-60	A-60	A-60	A-60	A-60	A-60
control position and electronic equipment room	A-60	A-60	A-60	A-60	A-60	A-60
great fire risky service room	A-60	A-60	A-60	A-0	A-60	A-0
other machine room	A-30	A-60	A-30	A-0	A-60	A-0
minor fire risky service room	A-15	A-60	A-30	A-0	A-60	A-0
living room	A-0	A-60	A-0	A-0	A-60	A-0
ladder and passage	A-0	A-60	A-0	A-0	A-60	A-0
open deck zone	※	A-0	※	※	A-0	A-0

4、 Conclusion

Medical ship is one type medical warship. It's a new job that it's medical cabin structure fireproofing design. Because the medical cabin has not completely defined in army specification, we start this research. From the medical cabin definition of solas, we borrow it to army specification with the method established in this paper. The results of this paper is helpful for other special warship cabin fireproofing design.

Reference:

1. 中国海军科技报告，《某型医院船医疗区域结构防火设计研究》，海军工程大学，2003 年 12 月
2. 仲晨华 冯明初 杨志青，《舰船防火设计总军民规范的差异及相互借用研究》，《中国造船》，2004.4
3. 仲晨华 孙洪江 杨志青 殷沐德，《基于人工神经网络模型的舰船防火设计方法》，《中国安全科学》，2005.6~7
4. GJB4000-2000，《舰船通用规范》
5. 《International Convention for the Safety of Life at Sea》，2004

Multi Relationship Research of Warship Cabin Fireproofing Characteristics

Zhong ChenHua[1] Wang DongTao[2] Yang ZhiQing[1] Ying MuDe[1]

(1: Navy University of Engineering

2: Army Representative Office at HuDong ZhongHua Shipyard)

Abstract: There are many types cabin in warship, and the relationship of cabin is very complex. In now fireproofing design, only direct and mutual relationship can be pay attention, and the subsidiary relationship is be ignored. So, there is difficult design according to now criterion in some special condition. In this paper, from system analysis of warship cabin, the multiple relationships is established. In one case by using this relationship, the application condition has been illustrated. The conclusion of this paper can be used for warship structure fireproofing design and constituting ship firefighting criterion.
Keywords: warship, structure fireproofing, fireproofing characteristic

Warship cabin structure fireproofing design is very important in warship design. In now design, the subsidiary cabin relation has been ignored, and the main attention is in border upon cabin fireproofing division, fireproofing field deciding, combustible material restricting, and firefighting equipment selecting. Along with the increase of ship complexity, the ignored relationship should be take in account, and there is no item in specification, then, there is difficulty in warship fireproofing design, for example the smoke influence. For this condition, in this paper, the multiple relationship map has been established with the system analysis method. Then, the importance and key condition has been analyzed, and do a case design with one relationship.

一. The multiple relationship of warship cabin structure fireproofing characteristic

From the cabin identity、boundary、position and dynamic transformation, it have been formed that cabin identity relationship、space relationship、division material relationship、flowing relationship.

1. Comparative relationship of cabin identity

Cabin identity scale concludes: space dimension、the quantity and performance characteristic of equipments、vibration、noises、lighting、electronmagnetism radiation、combustibility、combustion load、air composition、etc. This identity can be separated into direct relationship and indirect relationship. The direct relationship can be separated series relationship、parallel connection relationship、irrelevance relationship. Indirect relationship is the signification for engineering object and person produced from identity. It can be separated into importance relationship、fatalness relationship、fireproofing characteristic. The cabin importance is refer to cabin's different significance to warship task, example campaign command room、club、refrigeratory、inhabitation room、machine room、magazine、oil storeroom、assembly room etc. Fatalness relationship is the identity of cabin accident impairing other cabin. Fireproofing characteristic is the guideline of fire influence cabin, such as sanitation cabin、passage and inhabitation room has different fireproofing chatacteristic.

2. Space relationship of cabin

Cabin space relationship conclude border and apart relationship. There are eight combination relationships, with the cabin wall, deck, and in or not in a same main fireproofing zone. The space relationship is also the design step in cabin structure fireproofing design.

3. Border separating material relationship

The aim of space compartmentation concludes vusual angle、 personnel flowing、 air separating、 sound、 vibration、 thermal、 electromagnetism、 waterproofing comparting, and decorative comparting. Different separating aim has different separating form、 material and require, and one field could be composed by different separating cover. Different separating cover material is decided by the cabins both sides of the separating face and it's identity. Cabin compartmentation has many requirements, so the separating material maybe make of multiple types, such as combination of fireproofing layer and electronmagnetism shield layer.

4. Flowing relationship

Flowing relationship is make up of matter flowing between cabins. such as personal flow、 physical distribution、 air flow、 thermal、 sound wave、 electronmagnetic wave etc. The importance and complexity of flowing relationship increase with the increasing of warship complexity and dimensions.

二、 The application of warship cabin multiple relationship

In the warship cabin multiple relationships, some relationships is not only the cabin relationship, but also the design parameter of fireproofing design, such as space relationship、 border separating material relationship、 flowing relationship. They form the relationships as follow:

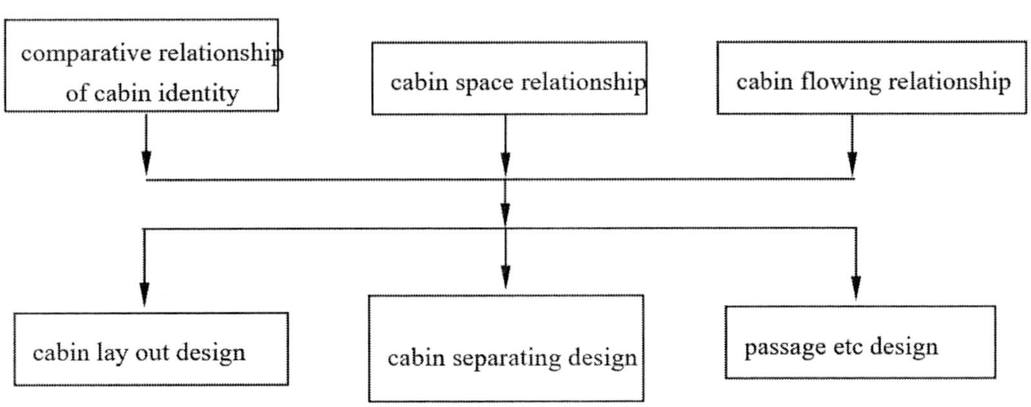

Graph 1 the sub relationship of cabin multiple relationship

One type relationship is corresponding with a design countermeasure, such as in table 1. For cabin flowing relationship, it is designed from person passage、 air pipe、 and water fire fighting system parameter etc. This relationship is very complex. In now warship structure fireproofing design, the fireproofing separating is mainly decided by neighbour cabin identity, and the not border upon relationship is thinked over in fireproofing field. Along with the big and special warship appear, the present design method is already fit practice requirement. Many research organizations has done some job about this.

Table 1

the kind of cabin identity comparative relationship	cabin lay out	border separating
series relationship	in same field, and border upon	lower fireproofing scale
parallel relationship	in the farthest field in same system	upper fireproofing scale
irrelevance relationship	no requirement	design by other relationship
importance relationship	lowest all losing principle, arrange in groups of the different importance cabin	fireproofing scale increasing with the importance
fatalness relationship	least dangerous separating border, concentrate lay out、airproofing lay out、open air lay out	fireproofing scale increasing with the fatalness
fireproofing identity	the requirement lower with the upper fireproofing identity	fireproofing scale decreasing with good fireproofing identity

三、Smoke influence analyzing inside the main hull of some type ship 2 deck

Smoke influence is one of cabin flowing relationship. In this paper, the case is about the main hull of 2 deck in some type ship. Because of no communication with outside, smoke will produce serious influence in fire procedure. The analyzing and treatment is as follow.

1. The danger of smoke

The danger of smoke is main concluded by smoke toxicity、thermal、block off light. Normal harm is list in table 2 and table 3. Base on statistic data, it take on 50~70% of death toll because of smoke.

Table 2 the venomousness gas in combustion

combustible substance	main venomousness gas	combustible substance	main venomousness gas
wood	CO_2、CO	nylon	aldehude、CO_2、CO
wool	CO_2、CO、HCN	hydroxybenzene colophony	ammonia、cyanide、CO
cotton、rayon	CO_2、CO	three assemble cyanogens ammonia-aldehyde colophony	ammonia、cyanide、CO
assemble four fluorin ethene	CO_2、CO	annulus oxygen colophony	acetone、CO_2、CO
polystyrene	benzene、toluene		

Table 3 venomousness gas in combustible substance combustion (speciman:1g, unit: g)

	free combustion							fume substance						
	CO_2	CO	aldehyde	phosgene	HCN、cyanide	ammonia	chlorine	CO_2	CO	aldehyde	phosgene	HCN、cyanide	ammonia	chlorine
polystyrene	0.192	0.142						1.698	0.504	0.003				
fibrin	2.294	0.440						0.202	0.172	0.012				
colophony	1.047	0.022					0.62	0.416	0.221					0.774
chloroethylene	0.433	0.229		0.0001			0.496	0.743	0.086		0.0008			0.479
nylon	1.226	0.304	0.0064		0.0076	0.032		0.907	0.355			0.0093	0.210	
rayon	1.836	0.116						1.130	0.225					
wool	1.641	0.446			0.007			0.650	0.138			0.003	0.035	
silk	1.352	0.634	0.0024		0.036	0.053		1.033	0.141	0.0012		0.009	0.308	
wood	1.626	0.270	mark					0.934	0.366	mark				
paper	1.202	0.135						1.001	0.273	mark				

2. Smoke analyzing of the 2 deck in some type ship

Deck smoke influence is one of cabin flowing relationship. In ship, combustible substance conclude: gasoline、alcohol、cabin furniture、medicine and person article for use、cabin decorate material etc. These combustible substance can produce a great deal CO_2、CO、soot and other macromolecule polymer smoke etc. Because of too much combustible substance, it cannot be calculated everything, so the calculation is main about smoke quantity, and the smoke quantity involve fire source and mix of smoke and air. In this paper, the case is gasoline combustion.

2.1 Fire source

In MSC/Circ-913, the examination fire source power is 1~6MW, so it's the condition of this paper case. The basis component of gasoline is C_8H_{18}, it's combustion chemistry equation is as follow:

$$C_8H_{18}+12\ {}^1/_2\ O_2=8CO_2+9H_2O$$

The combustion heat of gasoline is 44300KJ/Kg（10588Kcal/kg）. In this case, the combustion gasoline is 0.1354Kg/s, then the heat release from the combustion is as follow:

$$0.1354Kg/s\times44300KJ/Kg=5998KJ/s=5998KW$$

2.2 Mix smoke quantity produced from fire source

Base on isooctane combustion equation, burning 1Kg gasoline, there are 10.68 m^3 fresh air to be consumed. Base on combustion energy enactment, heat source is 6MW, gasoline consume is 0.1354Kg/s, so the smoke admixture is 1.446m^3. If it's complete combustion, combustion energy is all absorbed by smoke admixture, no radiation to outside, then the temperature of smoke is 2400℃. Specify the mix scale of combustion production and fresh air is 1：0～1：8, smoke exhaust temperature is 120℃, the cubage flux of combustion mixture is list in table 4~9.

Table 4 smoke evacuation volume flux at every mixing ratio（6MW fire resource） flux unit：m^3

mixing ratio	1:0	1:1	1:2	1:3	1:4	1:5	1:6	1:7	1:8
smoke evacuation volume flux	2.08	4.16	6.24	8.32	10.4	12.48	14.56	16.64	18.72

Table 5 smoke evacuation volume flux at every mixing ratio（5MW fire resource） flux unit：m^3

mixing ratio	1:0	1:1	1:2	1:3	1:4	1:5	1:6	1:7	1:8
smoke evacuation volume flux	1.73	3.47	5.2	6.93	8.67	10.4	12.13	13.87	15.6

Table 6 smoke evacuation volume flux at every mixing ratio（4MW fire resource） flux unit：m^3

mixing ratio	1:0	1:1	1:2	1:3	1:4	1:5	1:6	1:7	1:8
smoke evacuation volume flux	1.39	2.77	4.16	5.55	6.93	8.32	9.71	11.09	12.48

Table 7 smoke evacuation volume flux at every mixing ratio（3MW fire resource） flux unit：m^3

mixing ratio	1:0	1:1	1:2	1:3	1:4	1:5	1:6	1:7	1:8
smoke evacuation volume flux	1.04	2.08	3.12	4.16	5.2	6.24	7.28	8.32	9.36

Table 8 smoke evacuation volume flux at every mixing ratio（2MW fire resource） flux unit：m^3

mixing ratio	1:0	1:1	1:2	1:3	1:4	1:5	1:6	1:7	1:8
smoke evacuation volume flux	0.69	1.39	2.08	2.77	3.47	4.16	4.85	5.55	6.24

Table 9 smoke evacuation volume flux at every mixing ratio（5MW fire resource） flux unit：m^3

mixing ratio	1:0	1:1	1:2	1:3	1:4	1:5	1:6	1:7	1:8
smoke evacuation volume flux	0.35	0.69	1.05	1.40	1.75	2.10	2.45	2.80	3.15

2.3 Smoke layer thickness in passage

For main passage, the width is 1.9 m, length is 40 m, story height is 2.75~2.80 m, and the deck area of main passage is 76 m^2. The smoke layer average thickness is list on table 10 in different smoke quantity.

Table 10 smoke layer thickness in main passage

Smoke flux (m^3/s) / Smoke layer thickness (m) / time(S)	1	3	5	7	9	11	13	15	17	19
1	0.01	0.04	0.07	0.09	0.12	0.14	0.17	0.20	0.22	0.25
10	0.13	0.39	0.66	0.92	1.18	1.45	1.71	1.97	2.24	2.5
20	0.26	0.79	1.32	1.84	2.37	2.89	3.42	3.95	4.47	5.0
30	0.39	1.18	1.97	2.76	3.55					
40	0.53	1.58	2.63	3.68						
50	0.66	1.97	3.29							
60	0.79	2.37								
120	1.58	4.74								
180	2.37									
240	3.16									

2.4 The smoke layer thickness of main passage and cabin

In case of 1 cabin fire, smoke spreads into main passage. The smoke layer average thickness is list in table 11 and table 12.

Table 11 smoke layer thickness of main passage (main passage area is 76m^2, cabin area is 114 m^2)

Smoke flux (m^3/s) / Smoke layer thickness (m) / time(S)	1	3	5	7	9	11	13	15	17	19
1	0.01	0.02	0.04	0.05	0.07	0.08	0.10	0.11	0.13	0.14
10	0.08	0.23	0.38	0.53	0.68	0.83	0.98	1.14	1.29	1.44
20	0.15	0.45	0.76	1.06	1.36	1.67	1.97	2.27	2.58	2.88
30	0.23	0.68	1.14	1.59	2.05	2.5	2.95	3.41	3.86	
40	0.30	0.91	1.52	2.12	2.73	3.33				
50	0.38	1.14	1.89	2.56	3.41					
60	0.45	1.36	2.27	3.18						
120	0.91	2.72	4.55							
180	1.36	4.09								
240	1.82									
300	2.27									
360	2.73									

Table 12 smoke layer thickness of cabin (main passage area is 76m^2, cabin area is 114 m^2)

Smoke flux (m^3/s) / Smoke layer thickness (m) / time(S)	1	3	5	7	9	11	13	15	17	19
1	0.01	0.02	0.03	0.04	0.05	0.06	0.07	0.08	0.09	0.10
10	0.05	0.16	0.26	0.37	0.47	0.58	0.68	0.79	0.89	1.0
20	0.11	0.32	0.53	0.74	0.95	1.16	1.37	1.58	1.79	2.0
30	0.16	0.47	0.79	1.11	1.42	1.74	2.05	2.37	2.68	3.0
40	0.21	0.63	1.05	1.47	1.89	2.32	2.74	3.16	3.58	
50	0.26	0.79	1.32	1.84	2.37	2.89				
60	0.32	0.95	1.58	2.21	2.84					
120	0.63	1.89	3.16	4.42						
180	0.95	2.84								
240	1.26									
300	1.58									
360	1.89									
420	2.21									
480	2.53									
540	2.84									

3.2 Person and departure situation in this deck zone

In this zone, there are 148 persons, and 20 persons is medical care personnel. The departure must be from elevator. The lifting speed of medical elevator is 0.63~1.6 m/s. Assume the speed is 1 m/s, the lifting altitude is 20 m, then it needs 40 s to come and go. Add to two opening the door and person passing in and out, another 1 minute must be consumed. So it needs 3 minutes for two come and go. Base on table 4、11 and 12, the smoke in main passage is quite some after 3 minutes, so it must exhaust smoke for patient departure safety.

4. The advice about smoke vent

Smoke evacuation method could be smoke pipe and the old ventilating system of ship. The advice of smoke vent is as follow.

One imagination is straight tube which length is 20 m from 2 deck to outlet. Another imagination includes a 3 m length straight tube and a 10 m length horizontal tube. The flux of smoke vent is calculated underhood.

$$W_{浮} = \frac{1}{2} m v^2 + \Delta P_{途} \cdot \nabla + \Delta P_{局} \cdot \nabla$$

Where: $W_{浮}$ ——buoyancy work; $\Delta P_{途}$ ——immersed tube wall pressure loss; v ——velocity;

∇ ——tube innage; $P_{局}$ ——local pressure loss.

$$W_{浮} = (\gamma - \gamma_{热}) \cdot \nabla \cdot h_{直} \cdot g$$

Where: γ ——air denssity; $\gamma_{热}$ ——hot air density.

$$m = \gamma_{热} \cdot \nabla$$

$$\Delta P_{途} = \lambda \cdot l \cdot v^2 \cdot \gamma / 2g \cdot d$$

$$\lambda = l / (1.14 + 2\lg d/\delta)^2$$

Where: d—tube equivalent diameter; λ—firctional coefficient; l—tube length;
v—gas velocity; γ—gas specific gravity; g—acceleration of gravity;
δ—tube wall roughness.

$$\Delta P_{局} = \varsigma \cdot \omega^2 \cdot \gamma / 2 \cdot g$$

Where: ζ—coefficient of shock resistance;

coefficient of shock Sharp broaden resistance $\zeta_1=1.0$

coefficient of shock Sharp shrink resistance $\zeta_2=0.6$

coefficient of shock resistance of siphon $\varsigma_3 = 0.9$

Base on above formulation, velocity can be gain as below:

$$v_{直} = 10\sqrt{\frac{2d}{2.6d + \lambda h_{总}} \cdot \frac{9.8 \cdot h_{直}}{393}}$$

$$v_{弯} = 10\sqrt{\frac{2d}{3.5d + \lambda h_{总}} \cdot \frac{9.8 \cdot h_{直}}{393}}$$

The calculating result is list in table 13.s

Table 13

Flue section area		$1m^2$	$0.8m^2$	$0.6m^2$	$0.5m^2$	$0.4m^2$	$0.2m^2$	$0.1m^2$
Tube equivalent diameter d		1.128	1.009	0.874	0.798	0.714	0.505	0.356
Tube wall coefficient of		0.010	0.011	0.011	0.011	0.011	0.011	0.012
Straight tube	velocity	6.00	5.9	5.9	5.9	5.9	5.7	5.6
	flux	6.00	4.72	3.54	2.96	2.35	1.15	0.56
syphon	velocity	2.0	2.0	2.0	2.0	2.0	2.0	1.9
	flux	2.03	1.62	1.21	1.01	0.80	0.40	1.9

From table 4~12, it can be concluded that it needs 1 m^2 section area flue for 6MW fire source, and 0.5 m^2 section area flue for 1MW fire source. From table 13, the siphon is irrational.

四、Conclusion

The relationship of warship cabin is very complex. In this paper, only one feasible program has been presented from simple block diagram, the smoke influence relationship has not been analyzed. The method could be reference for warship structure fireproofing design.

Reference:

1．中国海军科技报告《××登陆舰防火分隔防火等级措施研究》，海军工程大学，2002年12月
2．仲晨华 冯明初 杨志青,《舰船防火设计总军民规范的差异及相互借用研究》,《中国造船》，2004.4
3．仲晨华 孙洪江 杨志青 殷沐德,《基于人工神经网络模型的舰船防火设计方法》,《中国安全科学》，2005.6~7
4．GJB4000-2000，《舰船通用规范》
5．《International Convention for the Safety of Life at Sea》，2004

A Summary on hydrodynamic research of high speed trimaran

Xiong, Xiaoqing Fan, Sheming Sun, Qun

(Marine Design and Research Institute of China, Shanghai 200011)

Abstract: The high speed trimaran is a new ship concept which has attracted increased attention from many countries lately. In this paper, a general introduction on hydrodynamic research of high speed trimaran in recent five years is provided. Then a comparison of current research situation between China and foreign countries is presented, together with outlook for future development on high speed trimaran hydrodynamics.

Key words: high speed trimaran; hydrodynamic performance

1. Introduction

Since the trimaran demonstrator RV Triton was built and deliverd by QinetiQ in 2000, multihull ships have become an important study direction for development of new ship concepts, and been considered to be a major solution for future surface vessels requiring high speed either for commercial purpose such as container carrier, Ro/Ro ship, or for military needs such as frigate, destroyer, small aircraft carrier, amphibious assault ship, elint ship, and so forth.

The high speed trimaran consists of a main hull and two smaller sidehulls, and all hull forms are very slender, which offers a number of potential advantages as follows:

1) Due to its slender subhulls, the trimaran has lower wave-making resistance, especially at higher speeds.

2) With support of the two small sidehulls, the trimaran shows superior stability and seakeeping performance than the equivalent monohull or catamaran, and keeps sailing with high speed even in severe sea states.

3) With the unique multihull configuration, it is easy to form a large deck area which provides much convenience for general arrangement. In case of naval ship, the sidehulls serve as effective shield of the main hull, and the commodious deck makes it possible to scatter weapon units optimally and gives much better helicopter availability.

Since hydrodynamic research of the high speed trimaran is fundamental for development of this new ship type, it has been an interesting subject focused by more and more scholars and institutes in marine design and research field.

2. Progress on Theoretical Research of High Speed Trimaran Hydrodynamics

2.1 Wave-making Resistance Calculations

Linear wave-making resistance theories, such as the improved thin-ship method of Michell and the new slender ship theory, have been widely employed in computational investigation of wave-making resistance of trimaran and wave interaction between subhulls, and in optimization of subhull forms as well as trimaran configuration.

By means of a linear Rankine sources panel method, Stefano, Dario Bruzzone, and Emilio[1] evaluated the wave making resistance of the main hull, the sidehull and the trimaran respectively for systematically varied configurations, and analyzed the effects of different hull form configurations on wave interactions between the main hull and sidehulls. Then, combined with a type of optimization algorithm, a numerical program was developed to find out the best trimaran configuration solution in terms of low wave-making resistance.

With a CFD technique SHIPFLOW, the method of which is of Rankine source type using higher order panels, SSPA[2] performed a theoretical study on how the shape and position of sidehulls affect the total resistance, wave making resistance and wave pattern of a trimaran at varied speeds. Meanwhile, the influence of speed variations on wave making interference between the hulls was analyzed.

The non-linear and viscous effects have also been properly accounted in calculation of wave making resistance for trimarans.

Prasanta, and Lawrence[3] developed a computational program Hydros for theoretical evaluation of trimaran resistance based on the thin-ship approach taking viscosity, surface tension, and surface elasticity into account. By applying both Hydros and SHIPFLOW for several configurations of a trimaran, a comparison of these two numerical approaches with model testing was given. It was shown the predictions of Hydros was more preferable than that of SHIPFLOW.

The PAFS code was developed by Kazuo, and Hisashi[4] to simulate the wave making resistance acting on the hull as well as the sinkage and trim of the hull for high speed monohulls, catamarans, and multihulls. Morino's panel method was extended to the free surface problems based on the 2nd order Kelvin type free surface condition with nonlinearity, sinkage and trim effects in this code. For a trimaran, the wave making resistance was computed by both classic thin-ship theory and PAFS code, measured by resistance model test and wave pattern probe. The four results above were compared to validate and improve the PAFS code.

2.2 Seakeeping calculations

Strip theory has been applied to theoretical predictions of trimaran seakeeping mostly.

Austal Ships[5] analyzed the vessel motions in varied sea states, speeds, and headings for an express trimaran ferry , using a 3-D time domain strip theory solver VERES developed by Marintek. The software gave very comparable results to the tank tests.

Numerical program TRIM was developed by Ermina, Cario, and Guido[6] for the trimaran seakeeping calculations adding to the potential flow 21/2D high speed theory the cross flow effects. Using this program, the seakeeping of varied hull form configurations was evaluated for a fast trimaran ferry.

By means of the 3D seakeeping code PRECAL, CETENA[7] calculated the vessel motions in head sea at design speed for different configurations of a trimaran frigate and analyzed the influence of sidehull position on trimaran seakeeping characteristics.

Using the seakeeping simulation program SEAMAN based on strip theory, SSPA[2] evaluated the seakeeping properties of two trimarans and a monohull concept with equivalent displacement and speed in all headings in certain sea state. The numerical results showed that the momohull experienced larger heave motions than the trimarans, and the difference was most prominent in bow and head seas. In addition, the effect of side hull shape and position on trimaran seakeeping ability was investigated.

2.3 Manoeuvrability Calculations

As far as it goes, studies on theoretical computation of the triamran manoeuvrability have been seldom. CETENA[7] specially adapted the simulation code SIMUP, which was developed for the prediction of the IMO manoeuvers for conventional monohull vessels, to predict the manoeuvrability of a trimaran. The turning circle manoeuver and ZigZag test were both simulated for varied configurations to investigate the influence of longitudinal position of sidehulls on the turning ability and course-keeping ability of the triamran.

3 Progress on Experimental research of High Speed Trimaran Hydrodynamics

3.1 Model Tests of Resistance

By means of model tests, the interaction between the hulls on a trimaran can be investigated by measuring the resistance component of subhulls and complete trimarans of varied configurations respectively. And through wave pattern measurements, the wave pattern resistance of triamrans can be determined for validation of theoretical predictions.

During the development of the RV Triton, QinetiQ[8] carried out considerable resistance tests with an 8 metre long model running in the large experimental tank at QinetiQ Haslar to determine key parameters, such as length-to-beam ratio of main hull, and length and position of sidehulls.

Lawrence, and Robert[9] performed systematical resistance model tests in the 270 meter long Towing Tank at QinetiQ Haslar for two trimarans. Not only the resistance and trim of the complete trimarans with varied configurations, but also the resistance of component subhulls with varied initial trim angles were tested at varied speeds. Finally, the influence of sidehull shape and position on trimaran resistance at each speed was analyzed according to the test results.

Diverse configurations of trimarans with two versions of sidehulls were tested at varied speeds in SSPA [2]. The resistance of main hull and side hulls were measured separately to investigate the effects of sidehull form and position on the wave interaction between the hulls of trimaran.

Antonio, Marco, Carlo, Ernina, Andrea, Dario and Sebastiano[10,11,12] cooperated in considerable model tests to study the influence of hull shape (hard chine or round bilge, symmetric or asymmetric) and relative position of subhulls (longitudinal and transversal) on the trimaran resistance at high speed. Meanwhile, the wave making resistances of the trimaran were identified by measuring the wave pattern in longitudinal cuts.

At present, the model resistance test results of high speed trimarans are usually extrapolated to full scale ship by two dimensional method (Froude method) or three dimensional method recommended by 15th I.T.T.C. The model-ship resistance correlation method should be improved further even for conventional vessels.

3.2 Model tests of seakeeping and manoeuvrability

Seakeeping model tests of high speed trimarans have been widely performed, however, few papers on the manoeuvrability tests for fast triamarans has been published.

Stefano, Marco, and Alberto[13] carried out systematic model tests for two design concepts of a fast trimaran ferry. The models were tested, in calm water and regular waves with different headings, to investigate the effects of hull shape and sidehull position to ship motions and dynamic stability.

To evaluate the seakeeping performance and the manoeuvrability of trimarans, Austal Shipyard[5] built a manned ship model Little Whisper in aluminium with the length of 10.67m and the design displacement of 1780Kg. Two side hulls could be moved relative to the main hull. The model was fitted with several accelerometers to record motions in six degrees of freedom, a GPS to show the speed, and a device to measure the thrust of propeller. The sea trial of this craft was conducted in waves corresponded to 10m prototype wave height. Ship motions in all sea headings were recorded, together with the wave height. The test results illustrated that the vessel could keep course in all sea headings, and showed satisfying seakeeping abilities.

Marintek[5] examined the seakeeping of a triamran at design speed in head and stern seas, and at zero forward-speed in beam and stern-quartering seas. A comparison between the trimaran and an equivalent catamaran demonstrated the superiority of the trimaran over a wide range of headings.

Marin[5] conducted seakeeping tests of three alternative designs for a fast trimaran with self-propelled models on which steering waterjets were installed. The effects of the variation of hull form and configurations to seakeeping performance of trimarans were investigated.

4 Hydrodynamic Research of High Speed Trimaran in China

4.1 Theoretical calculations

Based on the discrete source point method in linear wave resistance theory, a comparison research of the wave making resistance for high speed monohulls, catamarans and triamrans was deducted by Lu, Li, and Dong[14,15] to analyze the optimal speed range of each ship type. In combination with the orthogonal design method, the optimization of the principal dimensions, longitudinal position and transverse spacing of side hulls for a trimaran of Wiglley mathematic subhull form was investigated as well as the effect of midship section bottom slope and waterline entrance of side hulls on the wave resistance of high speed triamrans.

Three dimensional time domain Green function satisfying linear free surface and body surface boundary condition was employed by Li, Ye[16] to analyze the wave making resistance and wave profile of a trimaran of Wiglley mathematic subhull form.

4.2 Model tests

On the basis of rolling tests at zero forward-speed in beam seas, Shanghai Jiao Tong University[17] used a linear method to predict the rolling characteristics of a trimaran, of which the pitching and heaving motion were also investigated by seakeeping tests.

5. Conclusion Remarks

In general, the research of high speed trimaran hydrodynamics have been extensively performed. Definite object vessels are used for numerical calculation. Many numerical programs are specially developed for theoretical predictions of trimarans. Systematical model tests are usually carried out to investigate the high speed trimaran hydrodynamics, and often the test results are used to validate and improve the theoretical calculation methods.

However, as a new concept, following subjects about the high speed trimaran hydrodynamics should be studied deeply in future,

1) Numerical simulation of the transom hollow[9] created by the transom stern.

2) Application of the non-linear wave-making resistance theories in numerical simulation for high speed triamrans.

3) Application of different deterministic or stochastic optimization algorithms, such as enumeration method, orthogonal method, and genetic method in optimization of trimaran configurations.

4) Theoretical predictions of trimaran manoeuvrability.

5) Model-full scale correlation method for high speed trimaran.

6) Self-propulsion trial and steering experiment for high speed triamran propelled by waterjets or podded propulsors.

References:

[1] Stefano B., Dario B., Emilio T., Automatic Optimization of A Trimaran Hull Form Configuration, Proc. of the 8th International Conference on Fast Sea Transportation FAST 2005, St-Petersburg, Russia.

[2] Daniel Flodén, Keunjae Kim, Peter Ottosson, A Computational/Experimental Investigation on Resistance and Seakeeping Characteristics of Trimaran Configuration in Comparison with Monohull, The Ninth International Symposium on "Practical Design of Ships and Other Floating Structures- PRADS", Luebeck-Travemuende, Germany, September 12-17, 2004.

[3] Prasanta K. Sahoo, Lawrence J. Doctors, The Waves Generated by A Trimaran, Proc. of the 8th International Conference on Fast Sea Transportation FAST 2005, St-Petersburg, Russia.

[4] Kazuo Suzuki, Hisashi Kai, Numerical Estimation of Wave Making Characteristics of High Speed Ships Taking Sinkage and Trim Effects into Account, Proc. of the 8th International Conference on Fast Sea Transportation FAST 2005, St-Petersburg, Russia.

[5] Tony A., Kjell H.,A New Generation of Large Fast Ferry—From Concept to Contract Reality, Proc. of the 7th International Conference on Fast Sea Transportation FAST 2003 ,Naples, Italy.

[6] Ermina B., Carlo B., Guido B., Seakeeping Assessment of Trimaran Hulls, Proc. of the 7th International Conference on Fast Sea Transportation FAST 2003 ,Naples, Italy.

[7] Luca S., Roberta D., Michele V., Roberto D., Design Project of A Trimaran Multipurpose Frigate—Study of The Hydrodynamic Aspects, Proc. of the 7th International Conference on Fast Sea Transportation FAST 2003 ,Naples, Italy.

[8] Åse P. Thirud , Triton Treat at ONS Festival ,issue 7/8 , 2002, SOGM(SCANDINAVIAN OIL-GAS MAGAZINE)..

[9] Lawrence J., Robert J., The Optimization of Trimaran Sidehull Position for Minimum Resistance, Proc. of the 7th International Conference on Fast Sea Transportation FAST 2003 ,Naples, Italy.

[10] Antonio C., Marco F., Carlo P., Influence of Hull Shape on The Resistance of A Fast Trimaran Vessel, Proc. of the 7th International Conference on Fast Sea Transportation FAST 2003 ,Naples, Italy.

[11] Stefano B., Marco C., Marco F., Carlo P., Effect of Hull Form Variation on Hydrodynamic Performance of A Trimaran Ship For Fast Transportation, Proc. of the 8th International Conference on Fast Sea Transportation FAST 2005, St-Petersburg, Russia.

[12] Ermina B., Andrrea B., Dario B., Sebastiano C., Co-operative Investigation into Resistance of Different Trimaran Hull Forms and Configurations, Proc. of the 8th International Conference on Fast Sea Transportation FAST 2005, St-Petersburg, Russia.

[13] Stefano B., Marco C., Alberto F., Effect of Hulls Form Variations on The Hydrodynamic Performances in Waves of A Trimaran Ship, Proc. of the 7th International Conference on Fast Sea Transportation FAST 2003 ,Naples, Italy.

[14] Lu Xiao ping, Li Yun, Dong Zhu shun, The Comparison of The Hydrodynamic Features For Several Displacement High Performance Vehicle, Proc. of the 10th China International Boat Show & High

Performance Marine Vehicles (HPMV) Conference 2005, Shanghai, China.
[15] Li Yun, Lu Xiao ping, Dong Zhu shun, An Investigation of The Outer Hull Form Optimization of High Speed Trimarans, Proc. of the 10th China International Boat Show & High Performance Marine Vehicles (HPMV) Conference 2005, Shanghai, China.
[16] Li Guo an, Ye Heng kui, Wave Making Computation in Time Domain for Multi-hull Ships, Proc. of the 7th International Conference on Fast Sea Transportation FAST 2003 ,Naples, Italy.
[17] Li Yun, Lu Xiao ping, Dong Zhu shun, A Summary of Research on High Speed Trimarans, Ship Hydrodynamics Conference 2004, Wuhan, China. (In Chinese)

numerical simulation of the hydrofoil's hydrodynamic characteristics based on Fluent

Yang Pei-qing, Guang Yi-feng Song Peng-ke
(School of Naval Architecture and Ocean Engineering,
Jiangsu University of Science and Technology,
Zhenjiang , Jiangsu 212003)

Abstract: This paper presents numerical simulation of the hydrofoil's hydrodynamic characteristics, which includes resistant coefficient and lift coefficient by changing attack angle and different Reynolds. It can provide some advice for fixing the hydrofoil in the high-performance vessel such as yacht hydrofoil.
Keywords: Fluent Hydrofoil Hydrodynamic characteristics

1、MATHEMATICAL MODEL

When hydrofoil moves through the water, it belongs to the problem of incompressible liquid. So the Governing equation is list as following.

mass conservation equation:

$$\frac{\partial u_i}{\partial x_i} = 0 \tag{1.1}$$

momentum conservation equation

$$\frac{\partial u_i}{\partial t} + \frac{\partial}{\partial x_i}(u_i u_j) = -\frac{1}{\rho}\frac{\partial p}{\partial x_i} + \nu \frac{\partial}{\partial x_i}\left[\frac{\partial u_i}{\partial x_j} + \frac{\partial uj}{\partial x_i}\right] \tag{1.2}$$

$u_i (i = 1,2,3)$ —— velocity along the coordinate

$x_i (i = 1,2,3)$ —— the coordinate

p —— pressure of water

ρ —— density of water

ν —— viscosiy of water

t —— time

1.2 Turbulence model:

we use standard $k - \varepsilon$ model, which is based on Tur.Kinetic Energy k and Tur. Dissipation Rate ε.

Tur.Kinetic Energy:

$$\frac{\partial(\rho k)}{\partial t}+\frac{\partial(\rho k u_i)}{\partial x_i}=\frac{\partial}{\partial x_i}\left[\left(\mu+\frac{\mu_i}{\sigma_k}\right)\frac{\partial k}{\partial x_i}\right]+G_k+G_b-Y_M-\rho\varepsilon-Y_M+S_k \quad (1.3)$$

Tur. Dissipation Rate:

$$\frac{\partial(\rho\varepsilon)}{\partial t}+\frac{\partial(\rho\varepsilon u_i)}{\partial x_i}=\frac{\partial}{\partial x_i}\left[\left(\mu+\frac{\mu_i}{\sigma_k}\right)\frac{\partial\varepsilon}{\partial x_i}\right]+C_{1\varepsilon}\frac{\varepsilon}{k}(G_k+C_{3\varepsilon}G_b)-C_{2\varepsilon}\rho\frac{\varepsilon^2}{k}+S_\varepsilon \quad (1.4)$$

About how to choose the parameter, I refer to other people [1]

$C_{1\varepsilon}=1.44, C_{2\varepsilon}=1.92, C_\mu=0.99, \sigma_k=1.0, \sigma_\varepsilon=1.3$

二、PHYSICAL MODEL

This paper chooses a 2- Dimensional Hydrofoil. α is the angle of attack.

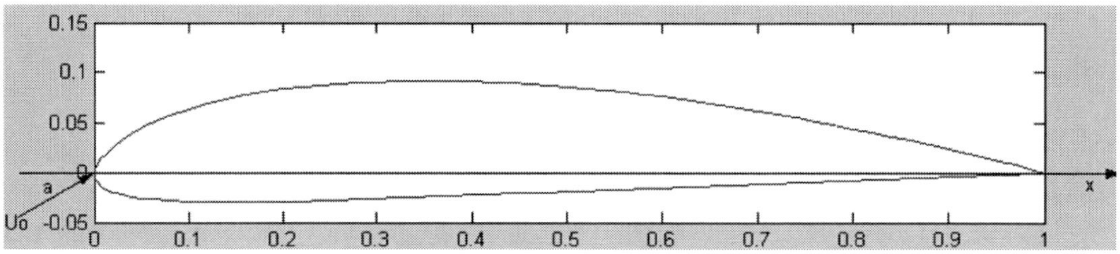

Figure 1 model

三、NUMERICAL MODEL

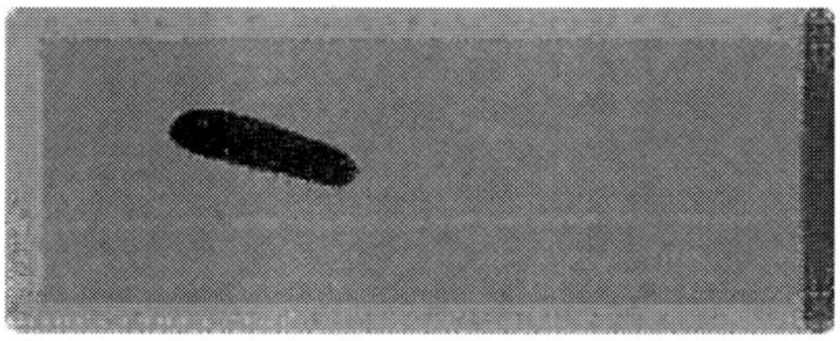

Figure 2 field and boundary conditon

Boundary condition:

The left, top, bottom is Velocity inlet. In this case, the inlet velocity is one string length ahead of hydrofoil.

The right is pressure outlet, which is three string length after hydrofoil.

The hydrofoil is a wall, the standard wall function is introduce.

四、THE RESULT OF CALCULATION

4.1 The change of the lift coefficient with the change of the angle o f attack

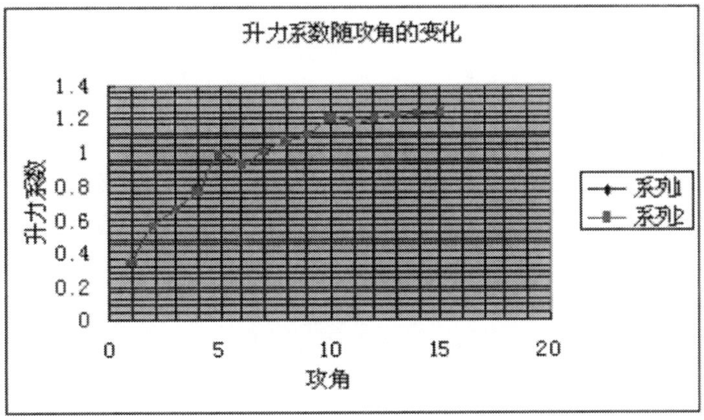

Figure 3

4.2 The change of resistant **coefficient with the change of the angle o f attack**

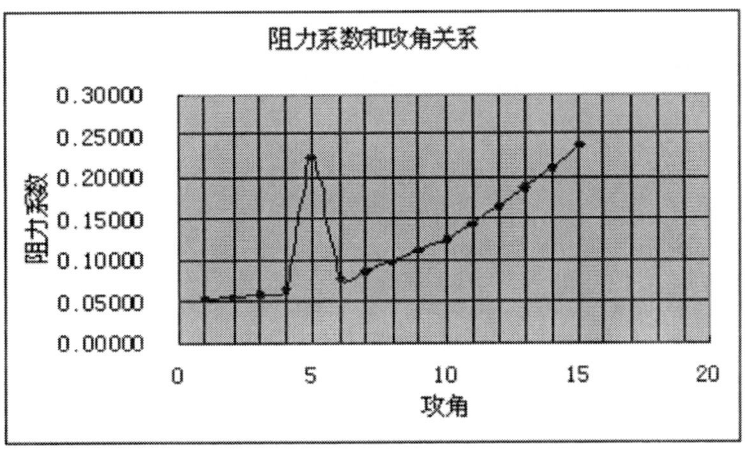

Figure 4

4.3 When the angle of attack is 15 degree, the distribution of pressure with the change of Re

(1) Re=2000

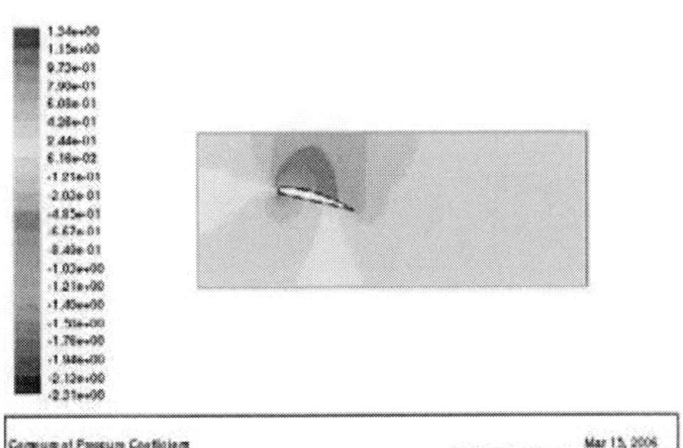

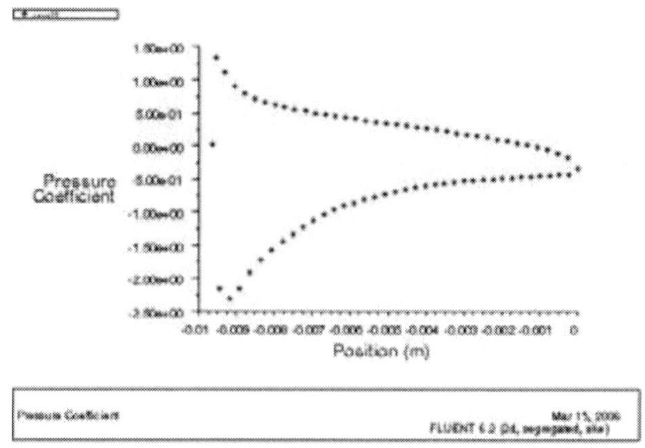

Figure 5

(2) Re=4000

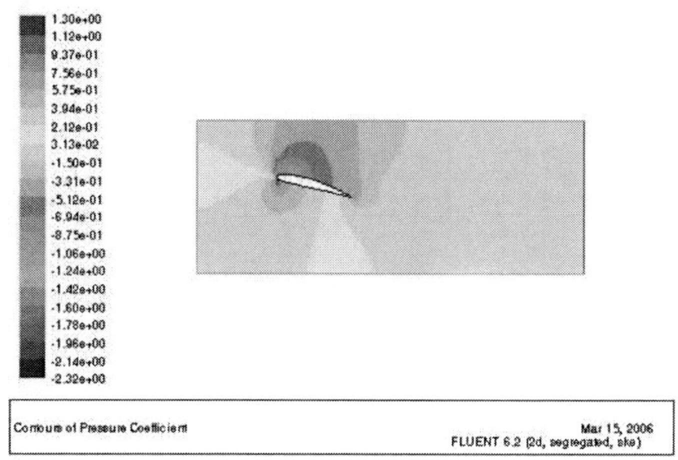

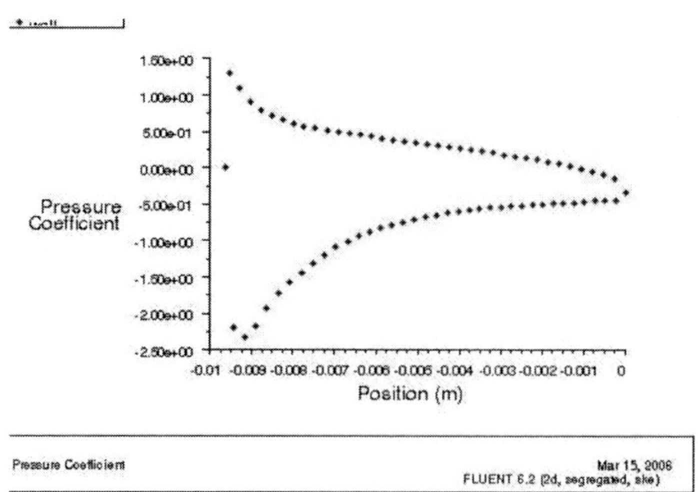

Figure 6

(3) Re=6000

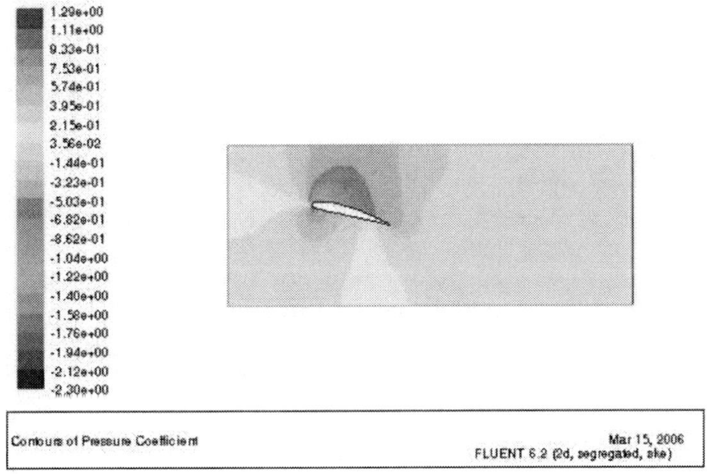

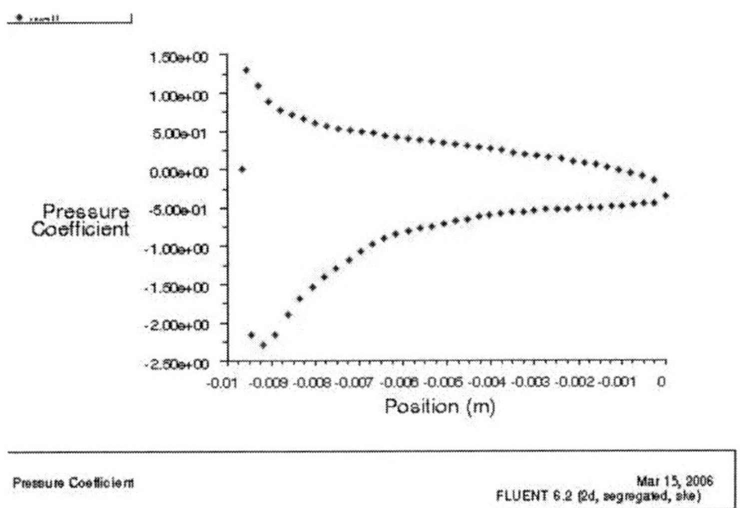

Figure 7

(4) Re=8000

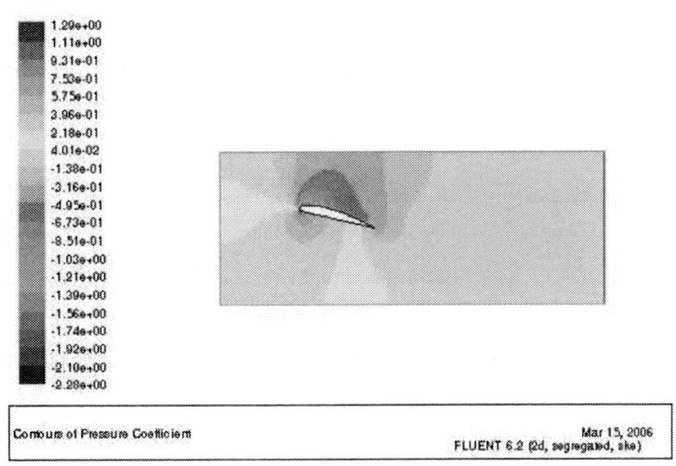

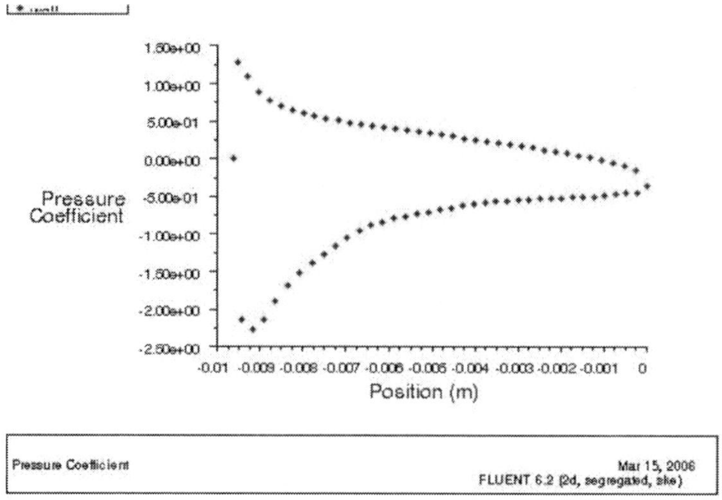

图8

五、CONCLUSION

(1) Through the result from the figure 3, we find there is a place where the angle of attack is 5 degree, the lift coefficient decreases. So ,we should notice it when we fix a hydrofoil on a vessel .

(2) We use CFD to simulate the motion of hydrofoil, which is more economical than test. But, we should refer to test, when we choose parameters in Fluent.

Authors' Brief Introduction:

Yang Pei-qing (1982-), Male, Postgraduate Student of Jiangsu University of Science and Technology, China.

Guang Yi-feng (1964-), Male, Associate Professor of Jiangsu University of Science and Technology, China.

Song Peng-ke (1982-), Male, Postgraduate Student of Jiangsu University of Science and Technology, China.

REFERENCES

[1] Wang Zhi-dong、Zhu Ren-qing. Hydrodynamic Characteristics of 2- Dimensional Hydrofoil Moving near to Free water Surface. Ship Engineering. 2004, 26 (3).

[2] Zhang Jun. Flow Visualization Experiments and CFD Simulation on Flow around Airfoil at Large Attack Angle . Shipbuilding Of China, 2000, 2(41)

[3] Wang Fu-jun. The dynamic Analysis of CFD .The press of Tsing-hua University.

[4] Fluent Inc.,FLUENT USER' S Guide. Fluent Inc.,2003

Study on the stability of WIG in the take off and landing process

SHI ya-jun, YUAN chang-hua, XU xin-fa

(China Ship Scientific Research Center, Wuxi 214082, China)

Abstract: Wing-In-Ground effect craft is a new type of ship, which can take off, land on and cruise steadily proximity to water surface. Its in-flight stability is just cared for a long time, but its stability in the take off and landing process is also important. In the take off and landing process of WIG, the motion is hydrodynamic and aerodynamics coupling one. We introduce the space motion equation of WIG for the analyses motion characteristic in take off and landing process. The motion have decoupled into longitudinal one and lateral one. The course stability and longitudinal stability are studied respectively by small perturbation method. The effect of wind and wave is also considered.

Key words: Wing-In-Ground effect craft; stability; take off and landing process

Regulatory Framework for high-Speed Craft

Du Aiping American Bureau of Shipping, Singapore
A. K Seah American Bureau of Shipping, Singapore

Synopsis

Since the early 1990s, there has been a steady development in commercial high-speed craft (HSC), especially those carrying passengers. Although operational history does not suggest HSC have significantly higher incident rate compared to conventional vessels, studies indicate that they do pose unique risk profile and in this sense greater risks. New challenges therefore present themselves in the safety and regulatory requirements of HSC.

This paper provides an overview of the regulatory framework now in place. It will discuss safety issues of HSC, and how international regulations intended for conventional vessels, such as SOLAS and Loadline Conventions, are being changed and new safety philosophies embodied in the HSC Code. No less important are national regulations, such as those of UK MCA and USCG, as many HSC are confined to operations in the territorial waters of a single state. Enforcement of the regulations and limitation of the applications will be discussed vis-à-vis feedback/concerns from shipyards and owners. Some topics of special interests, such as having to consider raking damage, in addition to stability, for survivability in assumed damaged conditions will be highlighted. Current safety initiatives such as considerations of human factors and risk assessment will also be discussed. The paper will conclude with some suggestions for effective regulatory implementation.

1. History of High Speed Craft Operations

Since their contraption in the 1950s, high-speed craft have gone through many technological innovations, in hull design, propulsion system and other aspects, aimed primarily at improving vessel speed. Various hull forms came into being: hydrofoils, hovercraft, and wing-in-ground effect craft, as well as novel forms of propulsion using marine gas turbines and waterjet propulsors.

The 1980s saw another period of significant innovations in high-speed craft. There were the adoption of catamaran hull form, and the use of aluminum alloy in hull construction. And the other significant development was the re-employment of diesel engines along with waterjet propulsion systems. Compared to the earlier dynamically supported craft, these innovative designs made high-speed craft relatively simpler to design and build and, more importantly, they became cost effective to operate and henceforth made high speed craft commercially viable.

Since the early 1990s, there has been a rapid expansion and development of commercial high-speed craft (HSC), especially for passenger service. Although operational history did not indicate greater risk for HSC than conventional vessels, some recent studies concluded otherwise. For example, a formal safety assessment conducted by the British Maritime and Coast Guard Agency (MCA) [4] indicates that the projected accident rate for HSC is nearly five times that of conventional craft, due primarily to the increased risk of collision in high-density traffic areas. The results of this formal safety assessment give credence to the growing concerns about the safety of high speed craft. These concerns point toward the safe operation of HSC, and associated human factors.

New challenges for the safety of HSC come with the growth of commercial HSC. In response to this challenge, IMO, in 1994, developed the International Code of Safety for High-Speed Craft (known as the 1994 HSC Code). This code is a progression from the Dynamically Supported Craft (DSC) Code, which was developed by IMO in 1977 as an optional safety standard for high-speed and small vessels. The HSC Code was made mandatory through its adoption in May 1994 by a new Chapter X of the SOLAS Convention. The new Chapter and the HSC Code came into force internationally on 1 January 1996 for high-speed craft (including hovercraft) on international voyages built on or after that date.

The adoption of the 1994 HSC Code did not dampen innovative ideas as more novel types and sizes of HSC have been developed. This led IMO to make a thorough revision to the HSC Code, particularly in the provisions for design, stability, construction, fire safety equipment and operation. The new International Code for the Safety of High-Speed Craft, 2000 (the 2000 HSC Code) came into force on 1 July 2002.

2. The International Code for the Safety of High-Speed Craft, 2000

To date, the HSC Code 2000 is the most developed set of international rules for building and operating HSC. The philosophy of the Code is based on "the management and reduction of risk as well as the traditional philosophy of passive protection in the event of an accident". The Code encourages mathematical analysis to assess risk and determine validity of safety measures imposed. Countries signatory to SOLAS Convention 1974 are obliged to enforce the requirements of the HSC Code by implementing new Chapter X of SOLAS.

The HSC Code 2000 shall be applied as a complete set of comprehensive requirements to the following high-speed craft engaged in international voyages with the keel laid date on or after 1 July 2002:

.1 passenger craft which do not proceed in the course of their voyage more than 4 hours at operational speed from a place of refuge; and

.2 cargo craft of 500 gross tonnage and upwards which do not proceed in the course of their voyage more than 8 hours at operational speed from a place of refuge when fully laden.

There are two different principles of protection and rescue embodied in both 1994 and 2000 HSC Code.

The first of the principles was originally foreseen at the time of development of the DSC Code. This principle recognizes that where rescue assistance is readily available and the total number of passengers is not more than 450, a reduction in passive and active protection may be permitted. Such craft are called "assisted craft" and form the basis for "category A passenger craft" of HSC Code.

The second principle takes into consideration the further development of HSC into larger craft. This principle recognizes that where rescue assistance is not readily available or the number of passengers is more than 450, additional passive and active safety precautions are required. These additional requirements provide for an area of safe refuge on board, redundancy of vital systems, increased watertight and structural integrity and full fire-extinguishing capability. Such crafts are called "unassisted craft" and form the basis for "cargo craft" and "category B passenger craft" of HSC Code.

In essence, the 2000 HSC Code has been heavily revised. The new requirements for hull construction are summarized in Table 1.

Table 1: New Requirements of HSC Code 2000 - Hull

-Mandatory on/after 1 July 2001

Sections	**Changes**
2.2	New requirements of watertight & weathertight integrity for doors, windows, hatchways and other openings
2.3	Revised intact stability criteria in displacement mode --- new definition of multihull, revised weather criteria for multi-hull in Anne 7 and new Annex 8 for monohull.
2.4	Intact stability in the non-displacement mode ---requirements for turning calculation/tests deleted.
2.6~2.13	Revised damage zone and damage criteria (Annex 7 & 8)
4.2.5	Requirement for emergency instruction deleted
4.3~4.4	Revised design acceleration & collision load for accommodation
7.2	SFP materials now refer to Fire Test Procedure Code.

7.3	New spaces added into major, moderate and minor fire hazard areas. Revised SFP separation notes.
7.4	Revised SFP requirements.
7.6	Added ducting requirements for galley. Added duct material requirements. Revised fire damper requirements.
7.8.1.3 & 7.9.3.5	New fire door requirement for special category spaces and ro/ro spaces.
7.11.1	New dimension requirement on alternative safe areas for Cat. B passenger craft.
8.7.7~8.7.9	Additional survival & rescue boat launching requirements.
8.9	Revised operational, maintenance and inspection requirements for LSA equipment.
8.10	Only craft <30m may be exempt for rescue boat. (20 m in 1997 Code)
8.11	New requirements for helicopter pick-up areas
Annex 6	Revised wind pressure for stability of hydrofoil craft
Annex 7	Revised wind heeling moment for stability of multihull craft
Annex 8	Revised damage stability criteria for monohull craft
Annex 10	Revised seat test requirements
Annex 11	Additional testing requirements for open reversible inflatable liferafts iaw MSC.81(70)

2.1 Intact Stability of High-Speed Craft

Intact stability of monohull and multihull in the displacement mode is to comply with requirements and criteria as specified in Chapter 2/2.3, Annex 7 and Annex 8. The concept and philosophy of these requirements and criteria are the same as those for conventional vessels, but with consideration of the effect of GMT, waterplane area and heeling due to high-speed turning. Planning effect is not considered here.

Intact stability requirements in the non-displacement mode are stipulated in Chapter 2/2.4 and Chapter 2/2.12, with specific heeling angle limitations but without details of consideration for stability reduction in following seas and unique motion characteristics such as chine walking, sudden large heel and porposing. For these unique characteristics, there is still a need in further development of stability theory for high speed state.

To take account of the unique characteristics of faster high speed craft, Chapter 18 and Annex 9 of the HSC Code require the roll and pitch stability be qualitatively assessed during operational safety trials. The results of such trials may indicate the need to impose operational limitations. Intact stability in the transitional mode is also to be noted to minimize the time to pass from the displacement mode to the non-displacement mode and vise versa unless it is demonstrated that no substantial reduction of stability occurs during this transition. Hydrofoil craft shall comply with the relevant provisions of Annex 6.

Other means of demonstrating compliance with the requirements of any part of Chapter 2 may be accepted, provided that the method chosen can be shown to provide an equivalent level of safety. Such methods may include:

1) mathematical simulation of dynamic behavior;
2) scale model testing; and
3) full-scale trials.

The new cross application of monohull and multihull intact stability criteria considers the effect of the waterplane area and GMT. Table 2 shows the alternative criteria:

Table 2. Alternative Intact Stability Criteria

GM_T	$\frac{B_{WL} A_{WP}}{\nabla}$	
	≤ 7	> 7
≤ 3	Annex 8	Annex 8 or Annex 7
>3	Annex 8 or Annex 7	Annex 7

Where:

BWL = maximum waterline beam at the design waterline (m); for multihull, this shall be taken to the outside of sidehulls

AWP = waterplane area at the design waterline (m2)

∇ = volume of displacement at the design waterline (m3)

GMT = transverse metacentric height in the loading condition corresponding to the design waterline, corrected for free surface effects (m)

2.2 Damage Stability of High-Speed Craft

The significant change in 2000 HSC Code on damage stability is the inclusion of bottom raking damage. This requirement has increased the bottom damage extent and basically is a requirement of double bottom.

The specific damage stability criteria stipulated in 2000 HSC Code are for craft in displacement mode. The Administration may permit the use of low-density foam or other media to provide buoyancy in void spaces. However, the extent of assumed damage zones will have substantial impact to small craft, which will be discussed later.

1) Side damages are assumed anywhere on the periphery of the craft, i.e. including damage to the transverse watertight bulkheads, i.e. two compartment damage. (Both 1994 and 2000 HSC Code)

2) Raking damage is to be considered in 2000 HSC Code. The damages are assumed at any part of the surface hull(s) to be vulnerable to raking damage stability, See Figure 1.

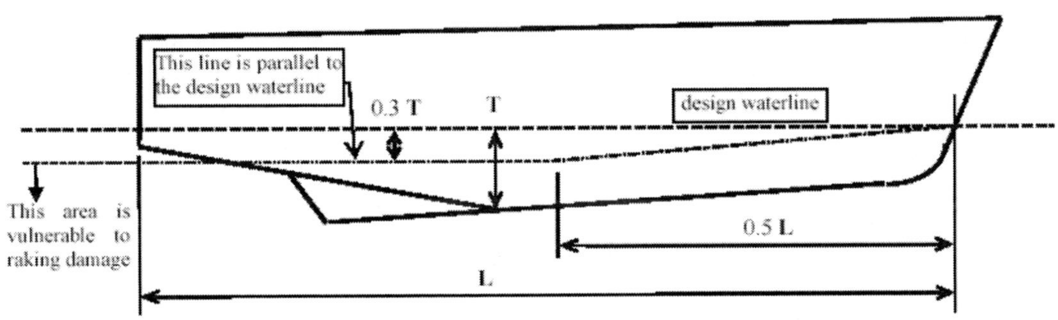

where: T = maximum draught of the hull (each hull considered individually in the case of multihulls) to the design waterline, excluding any non-buoyant structure.

Figure 1. Raking Damage

The above two requirements require closer spacing between watertight subdivision bulkheads and a double bottom of considerable height.

2.3 Accommodation and Escape Measures

Chapter 4 of HSC Code requires public spaces and crew accommodation be designed and arranged so as to protect the occupants from unfavorable environment conditions and to minimize the risk of injury to occupants during normal and emergency conditions.

Superimposed vertical accelerations above 1g at longitudinal centre of gravity should be avoided unless special precautions are taken with respect to passenger safety.

Mounting of large masses such as main engines, etc. shall be proved by calculation to be able to withstand, without fracture, the design accelerations. See Table 3.

Table 3 Design Acceleration as Multiples of g

Types of craft / Direction	All HSC except amphibious ACVs	Amphibious ACVs
Forward direction	g_{coll}	6
After direction	2 or g_{coll} if less	3
Transverse direction	2 or g_{coll} if less	3
Vertical direction	2 or g_{coll} if less	3

where:

g_{coll} = the collision design acceleration expressed as a multiple of the acceleration due to gravity (9.806 m/s^2)

The public spaces, control stations and crew accommodation of high speed craft are to be designed to protect occupants in the design collision condition. The plan projected area of craft energy-absorbing structure forward of the transverse plan is not to be less than required by 4.4.1 of 2000 HSC Code. 4.42 to 4.6 have detailed guidelines on seat, seatbelts etc.. Annex 10 provide criteria for testing and evaluation of revenue and crew seats.

Evacuation time is calculated based on structural fire protection time (SFP). An evacuation procedure, including an evacuation analysis are to be provided. The achievement of the required evacuation time shall be verified by a practical demonstration conducted under controlled conditions in the presence of the Administration.

2.4 Directional Control System

High speed craft shall be provided with means for directional control of adequate strength and suitable design. Directional control may be achieved by air or water rudders, foils, flaps, steerable propellers or jets, yaw control ports or side thrusters, differential propulsive thrust, variable geometry of the craft or its lift-system components or by a combination of these system.

The probability of total failure of all directional systems shall be extremely remote when the craft is in normal operating conditions. A failure mode and effect analysis shall include the directional control system.

The limits of safe use of any of the control system devices shall be based on demonstrations and a verification process in accordance with Annex 9.

2.5 Fire Safety

Chapter 7 of HSC Code stipulates requirements for structural fire protection, ventilation and fire fighting. The basic principles underlie the provisions in Chapter 7 are similar to SOLAS Convention but with regard to the category of craft and the potential fire hazard involved:

1) maintenance of the main functions and safety systems of the craft, including propulsion and control, fire-detection, alarms and extinguishing capability of unaffected spaces, after fire in any one compartment on board;
2) division of the public spaces for category B craft, in such a way that the occupants of any compartment can escape to an alternative safe area or compartment in case of fire;
3) subdivision of the craft by fire-resisting boundaries;
4) restricted use of combustible materials and materials generating smoke and toxic gases in a fire;
5) detection, containment and extinction of any fire in the space of origin;
6) protection of means of escape and access for fire fighting; and
7) immediate availability of fire-extinguishing appliances.

Fire-resisting divisions and materials are classified by fire protection time. The fire-resisting materials are to have properties complying with IMO Fire Test Procedure Code.

The structural fire protection time for areas of major fire hazard is 60 min, from which the evacuation time is calculated. If any other lesser structural fire protection time is determined for category A craft and cargo craft by shorter evacuation time, then the times given in 7.4.2.2 and 7.4.2.3 may be amended pro rata. But in no case the structural fire protection time for major fire hazard be less than 30 min.

Considering low melting point of aluminum and combustible material such as FRP structures, the insulations are to be applied at both sides of the bulkheads/decks, in accordance with tables 7.4-1 and 7.4-2. If the hull structure is made of aluminum, their installation shall be such that the temperature of core material does not rise more than 200 ^{o}C above the ambient temperature in accordance with the times in 7.4.1.1 and 7.4.2.2.

If the structures are made of combustible material such as FRP, their insulation shall be such that their temperatures shall not rise to a level where deterioration of the construction will occur during exposure to fire for times in 7.4.1.1 and 7.4.2.2.

For category B passenger craft, the public spaces shall be divided into fire zones as detailed in 7.11.

Automatic smoke-detection system and manually operated call points are to be provided for areas of major and moderate fire hazard and other enclosed spaces not regularly occupied within public spaces and crew accommodation. Main propulsion machinery rooms shall, in addition, be installed with detectors sensing other than smoke and be supervised by TV camera monitored from the operating compartment.

Areas of major fire hazard shall be protected by an approved fixed fire-extinguishing system operable from the control position. Detail requirements of the fixed fire-extinguishing systems are given in 7.7.3.

Fixed sprinkler system is to be provided for public spaces and service spaces and accommodation area on category B passenger craft. This requirement may be waived for category A passenger craft provided they can meet the requirements as indicated in 7.13.3 of the HSC Code.

Fixed sprinkler system is also required for crew accommodation where sleeping berths are provided, having total deck area greater than 50 m^2.

Part D of Chapter 7 stipulates requirements for carriage of dangerous goods.

2.6 Life Saving Appliances and Arrangements

Life saving appliances and arrangement shall enable abandonment of the craft in accordance with the escape and evacuation requirements of 4.7 and 4.8. Life saving equipment shall comply with IMO LSA Code and be certified by the Administration.

All craft shall carry survival craft with sufficient capacity of at lease 110% of the total number of persons the craft is certified to carry. Liferaft meeting the requirements of LSA Code may be accepted as survival craft. At least one rescue boat is required for passenger craft certified to carry more than 450 passengers.

For craft of less than 30m in length, rescue boat may be exempted provided the craft meets all the requirements as in 8.10.1.5.1 to 8.10.1.5.3.

Craft operating on voyages having a duration of 2 hours or more between each port of call shall be provided with a helicopter pick-up area.

2.7 Machinery, Auxiliary System, Remote Control and Electrical Installations

In addition to class rules, the HSC Code has specific requirements concerning machinery, auxiliary system, remote control and electrical installations, stipulated in Chapter 9 to Chapter 12.

Unique to HSC Code is Chapter 9, which requires a failure mode and effect analysis for machinery systems and their associated controls and electrical system.

In view of the operational characteristics of HSCs, the main propulsion machinery and all auxiliary machinery essential to the propulsion and the safety of the craft shall be designed to operate in upright, heeled and trimmed conditions with heel and trim angles as in 9.1.12.

As stabilization is essential for a high speed craft in maintaining the main parameters of the craft attitude, such as heel, trim, course and height and main parameters of the craft motion, such as roll, pitch, yaw and heave within safe limits, Chapter 16 stipulates requirements for stabilization systems.

The limits of safe use of any of the stabilization control system devices shall be based on demonstration and a verification process in accordance with annex 9.

2.8 Navigation system, radio communications and bridge layout

It has been recognized that modern equipment, if properly integrated into operational procedures, can enhance the operator's senses, automate certain navigational functions, and effectively increase the time available for operational decision-making. Chapter 13 of the HSC Code has minimum requirements for compasses, speed and distance measurement devices, radar installations, echo-sounding device, electronic positioning systems, rate-of-turn indicator and rudder angle indicator, steering arrangement and propulsion indicators, and automatic steering aid (or automatic pilot) etc.

More specifically, there are some new navigational equipment requirements in 2000 HSC Code comparing with 1994 HSC Code. These are voyage data recorders (VDR), automatic identification system (AIS) and electronic chart display and information system (ECDIS).

The requirements on radio communications are in Chapter 14, such as global maritime distress and safety system identification, radio installations and equipment and maintenance requirements etc.

The design and arrangement of bridge is also of importance to facilitate operation of the high speed craft, which is specified in Chapter 15.

2.9 Operational requirements

As required in Chapter 17, operational limitations and procedures are to be well documented. Instructions to crew shall be provided in the craft operating manual regarding required actions and craft limitations subsequent to prescribed failures.

The operational safety and limitations of the craft in normal service conditions and equipment failure situations are to be verified by full scale tests, supplemented by model tests where appropriate. As a minimum, the system shall measure accelerations in three axes close to the craft longitudinal centre of gravity. Special attentions are to be paid to the following aspects:

1) yawing
2) turning
3) automatic pilot and steering performance
4) stopping in normal and emergency conditions
5) stability in the non-displacement mode about three axes and in heave
6) trim
7) roll
8) plough in
9) lift power limitations
10) broaching
11) slamming and
12) bow diving

As a craft operational control, the following documents are required for a high-speed craft:

- The High-Speed Craft Safety Certificate
- The Permit to Operate High-Speed Craft
- Craft operating Manual

The Permit to Operate High-Speed Craft can only be issued by the Administration when it is satisfied that the operator has made adequate provisions from the point of safety generally and specifically the matters as in 18.1.3.

Training and qualifications are also required for master and each crew member. These are normally under the jurisdiction of the Administration. Each country will have their own requirements in this regard. 18.3.3 listed minimum training required.

The craft and equipment shall be maintained and inspected in accordance with Chapter 1 and Chapter 19. Like SOLAS vessel, high-speed craft is also required for the following surveys:

- an initial survey
- renewal survey at intervals specified by the administration but not exceeding 5 years
- periodical survey (annual with window of ±3 months)
- additional survey as the occasion arises.

3. Probability Concept and Risk Assessment

As stated in several chapters of the IMO HSC Code , it encourages mathematical analyses to assess risk and determine validity of safety measures implemented. Specifically, Annex 3 of 2000 HSC Code gives guideline on use of probability concept in general, while Annex 4 of the Code specifies procedures for failure mode and effects analysis (FMEA).

The IMO Guidelines for Formal Assessment and the five steps of FSA may also be applied to High Speed Craft:
- Identification of hazards;
- Assessment of risks associated with those hazards;
- Ways of managing the risks identified;
- Cost benefit assessment of the options identified in the above step; and
- Decisions on which options to select.

FMEA shall be conducted for machinery systems as required under the provisions of 5.2, 9.1.10, 12.1.1 and 16.2.6 of 2000 HSC Code. Some Classification Societies extend the scope of the FMEA to cover more systems.

4. National Administrations As Flag Administrations and Port States

Most Administrations view the HSC Code and its philosophy as an effective instrument for regulating the design, outfitting, and operation of high speed craft, regardless of whether the craft is operated domestically or internationally.

4.1 USCG Regulations on High Speed Craft

To facilitate use of the HSC Code, the USCG issued a Navigation and Vessel Inspection Circular (NVIC 6-99) that provides guidance on plan review, inspection, and certification for vessels built to the Code. The NVIC serves as guidance for both Code and non-Code vessels.

The guidelines described in Enclosure (1) to NVIC 5-01 should be applied to domestically built, non-HSC Code, passenger vessels, inspected under 46 CFR, Subchapter T, K, or H, capable of loaded service speeds of 30 knots or more.

4.2 UK and EU Regulations on High Speed Craft

UK Statutory Instrument 1996 No.3188, i.e., The Merchant Shipping (High-Speed Craft) Regulation 1996, requires high speed craft to comply with the IMO HSC Code. MCGA has also drafted guidelines in the application and interpretation of each sections of the HSC Code.

EC Council Directive 1999/35/EC of 29 April 1999 on a system of mandatory surveys for the safe operation of regular ro-ro ferry and high-speed passenger craft services.

Directive 95/21/EC lays down all the general safety requirements to be applied to new and existing passenger ships and to high speed passenger crafts.

Directive 98/18/EC applies to:
new passenger ships;
existing passenger ships of 24 metres in length and above;
high-speed passenger craft, regardless of their flag, when engaged on domestic voyages.

4.3 Other Applicable Regulations

Apart from the IMO HSC Code, other IMO Conventions/Regulations such as MARPOL, STCW, ISM, IMDG and SOLAS together with Flag State interpretations and requirements, Port State requirements and Classification Society Rules are all regulations applicable to HSC.

Basically, all countries signatory to SOLAS are obliged to apply HSC Code 2000 since it was made mandatory by Chapter X of the SOLAS Convention. However, it is noted that the HSC Code applies to craft operating on international routes. For domestic voyages, each Administration may have their own existing regulations on passenger vessels or cargo vessels in place to comply with. These regulations may not be specific to the design and operational characteristics of high speed craft. One of the implications from applying these different rules to build domestic HSC with restrictions is that one craft operating in one country may not be able to operate in another country without additional modifications. The financial impediment for an owner and operator is obvious.

5. Concerns From Ship Owners and Builders

The 2000 IMO HSC Code is applicable to cargo craft over 500 GRT and passenger craft regardless of tonnage performing international voyages. However, some HSC, although operating between two countries, say, between Singapore and Thailand, are at all times operating close to refuge along the route. Since the Code will be applied in such situation in any case, some operators claim that strictly applying the IMO HSC Code may not be reasonable.

In recognizing the dilemmas faced by the owners and operators, some international safety guidelines for smaller high speed craft approved with restrictions are desirable so that all countries may have more consistent standards and interpretations to high speed craft of this category. Reference may be made to IACS Guideline No.2 for vessels less than GRT 500 and association may be made to the existing risk profile with the risk control measures required for this category of high speed craft. Requirements for small passenger craft may also be revised to reflect the design constraints in full compliance with IMO HSC Code.

Following factors may be considered:

1) Actual environment where the high speed craft is operating
2) Size of the craft and nature of the design
3) Operating function of the craft
4) Number of passengers

One of the issues for smaller high speed craft is damage stability. As we discussed in 2.2 above, the requirement of two-compartment damage stability will require closer spacing of watertight subdivision bulkheads. However, for a slender small high speed craft, the length of the engine room may be half of the vessel length, which makes the two compartment damage stability very difficult to comply with.

Similar scenario goes to the raking damage stability. To comply with the raking damage stability requirement, it will require the craft to be fitted with a very high double bottom in relation to the depth of the craft. Designers very often find the raking damage stability criteria not achievable, if not impracticable, for many small high speed craft of 20-50meters in length, since for HSC to operate efficiently at high speed their hulls need to be of lightweight construction and the double bottom occupied so much of the limited spaces. See Figure 2 for example the raking damage applying to a 50m high speed craft. A smaller craft will have more difficulty in this aspect.

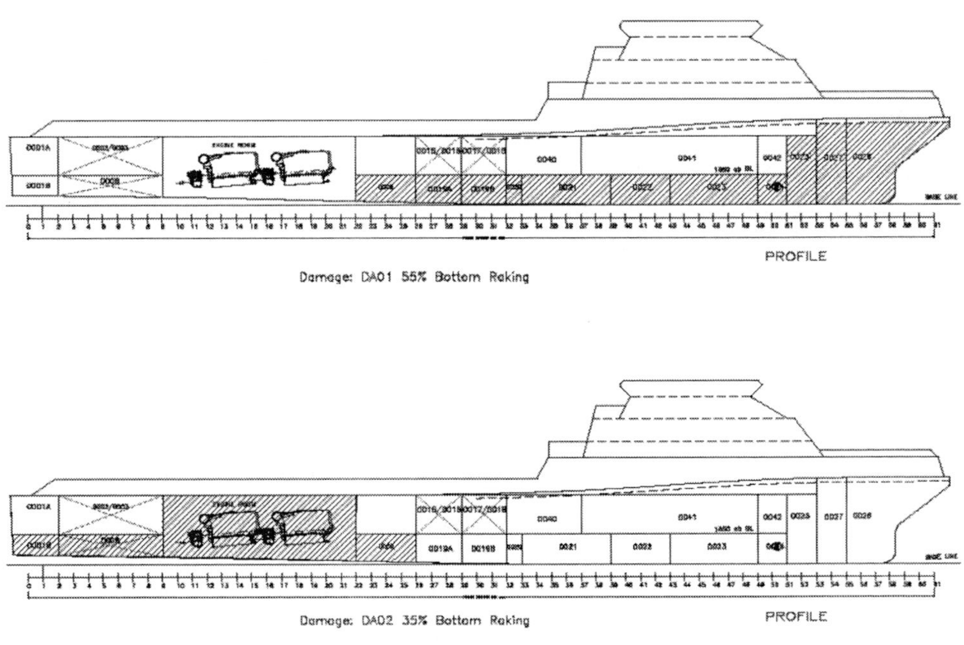

Figure 2 Raking Damage Example

6. More Discussions on Damage Stability Standards

6.1 Overview of Subdivision Standards

Vessels of different type, size and operation nature have different damage standards. These standards stipulate criteria for residual stability after damage and extent of damage based on the damage risk profile of each category. A list of existing international subdivision standards are shown in Table 4.

Table 4: Subdivision Standards

Standard	**Application**	**Required Subdivision**
Harmonized Probabilistic Subdivision Standard for Passenger and Cargo Ships	Will apply once completed and amended into SOLAS to cargo and passenger ships	Probabilistic Standard
SOLAS 1974 as amended by the November 1997 amendments adopted by resolution MSC.65(68).	Applies to all new passenger ships carrying more than 400 persons or more built on or after July 1, 1999.	**2** Compartment Deterministic Standard
SOLAS 1974 as amended by the November 1995 amendments, adopted by Res 1 of the Conference of Contracting Governments to the International Convention for the SOLAS, 1974.	Applies new Ro-Ros that carry more than 400 persons built on or after July 1, 1997 Applies to Ro-Ros that carry more than 400 persons built before July 1, 1997 on a phase in schedule noted in SOLAS.	**2** Compartment Deterministic Standard
SOLAS 1974 as amended by the 1990 amendments, which were adopted by resolution MSC.19(58) - Part B-1	Applies to cargo ships constructed on or after February 1, 1992.	**Probabilistic** Standard
SOLAS 1974 as amended by the October 1988 amendments adopted by resolution MSC.12(56).	Applies to all passenger ships built on or after April 29, 1990	**1** Compartment deterministic standard with addition of Damage Stability requirements of Regulation 8 paragraph 2.3.
SOLAS 1974	Applies to ships constructed on or after May 25, 1980	**1** Compartment Deterministic Standard
A.265(VIII)	Optional and considered equivalent to SOLAS 1960	**Probabilistic** Standard
SOLAS 1960	Applies to ships constructed on or after May 26, 1965	**1** Compartment Deterministic

		Standard
Reg 27 1966 Load Line	Applicable to Cargo ships (primarily bulk carriers and tankers) can be taken in lieu of SOLAS II-1 Part B-1.	**1** Compartment Deterministic Standard
Code of Safety for Special Purpose Ships	Carrying more than 50 special personnel	**1** Compartment Deterministic Standard
	Carrying more than 50 special personnel but less than 200	**1 Compartment Deterministic Standard**
	Carrying more than 50 special personnel but less than 200 and greater than 100m	**1 Compartment Deterministic Standard including breach of the collision bulkhead**
	Carrying more than 200 special personnel	Same as SOLAS Passenger Vessel Standard
International Gas Carrier Code	All types - with length considerations on 2 compartment requirements	**2** Compartment Deterministic Standard
International Bulk Chemical Code	All types - with length considerations on 2 compartment requirements	**2** Compartment Deterministic Standard
Annex I to MARPOL 73/78	Applicable to all new Oil Tankers greater than 225 meters in length	**2** compartment throughout
	Applicable to all new Oil Tankers greater than 150 meters but not more than 225 meters	**2** compartment throughout with exception of machinery space located aft (treated as one floodable compartment)
	Applicable to all new Oil Tankers not exceeding 150 meters	**1** compartment exclusive of the machinery space
	Applicable to all new Oil Tankers 100 meters and less in length	**Administrations may relax requirement**
High Speed Craft Code	Applicable to high-speed craft engaged in international voyages	**2** Compartment Deterministic Standard

As shown in the above table, the standards other than HSC Code are applying two compartment deterministic damage stability to the vessels over certain length or passengers over certain number, except the IBC Code and IGC Code which are applied to vessels of higher risk profile.

6.2 Damage Risk Profile of High Speed Craft

Damage of vessel usually is caused by collision or contact. Collision is defined as the vessel striking or being struck by another vessel, regardless of whether under way, anchored or moored. Contact is defined as the vessel striking or being struck by an external object (eg. navigation buoy or berth) but not another vessel or the sea bottom. Collisions may occur in either confined waters or open sea whilst contacts generally occur in confined waters.

The main differences between HSC and conventional ships are that HSC operate at high speed, are of lightweight construction using aluminium or composite materials for main hull and superstructure construction and have significantly higher power to weight ratio propulsion systems. By nature, HSC may pose higher damage risk profile than conventional ships.

In a proposal submitted to IMO by United Kingdom, it was recommended that the length of raking damage should be related to the displacement, speed, hull material and hull scantlings. The proposal introduced a concept of Grounding Damage Index (GDI), the longitudinal extent of raking damage to any part of the surface of the hull(s) defined in 2.6.8.1 of HSC Code 2000 shall be equal to 0.244*L*(*GDI*)0.7 but not greater than *L*1, where:

$$\text{GDI} = \frac{0.405MV^2}{LF_H}$$

Where M is the displacement (including added mass in surge), V is the impact speed and FH is the average horizontal force.

The proposal is based on their grounding damage research. The statistics of grounding accidents of high speed craft is scarce so that a recently assembled database for the grounding of conventional ships has been considered. It is assumed that the non-dimensional probability distributions for impact speed, location, width and height of damage also apply to HSC.

Based on their grounding damage research, the raking damage length for an aluminum monohull of 24m vessel length with 24.7 knots vessel speed may be as long as 0.7L. While an aluminum catamaran of 24 m with 32 knots vessel speed may have damage length of 1.27L, which means the whole length of the vessel is subject to raking damage, and therefore a double bottom is required for the whole vessel length.

However, there are arguments that since the accident database is too small for a reliable statistical analysis, the analysis was based on conventional big vessel database and the simplified probability distributions of relevant parameters are not appropriate to resemble the grounding of real-life high-speed craft.

The analysis performed by MCA research projects also indicated that maximum damage length was about 70% of the vessel length and it is possible for an effective damage length of 100% to be experienced. There is recommendation that the most severe scenario with100% effective raking damage length but with reduced residual stability be considered.

There is also recommendation that HSC criteria should be revised to take into account of sea state and to include some formula on the significant wave height at which the craft will capsize depending on the wave frequency, the size of the vessel relative to the wave, and its stability.

The formal safety assessment conducted by the British Maritime and Coast Guard Agency (MCA) points to alternative measures. In its studies of incident statistics, it concludes that the two highest accident categories are collision and contact, with significantly more injuries from collisions. Groundings are significantly less frequent, although groundings on submerged or visible rocks, rather than sandbanks, tend to be the highest consequence incidents affecting HSC. Majority of collision cases occurred in the harbor or in coastal waters where other vessel traffic is relatively dense. These accidents are more related to human errors rather than without the arrangement of double bottom. It suggests that crew training and qualification, safety of operational and navigational equipment and control/pilot of the sea traffic may be more important than requiring an impracticable structural arrangement. Perhaps, providing better guidance to the master, in the form of permissible operational sea states, capsizing wave heights based on stability range, permissible speeds at given displacements, and other safety boundaries should be given more emphasis.

8 Conclusions

While high-speed craft have come of age, they have, in their formative years, been treated as a novel ship type with their own operational characteristics and safety concerns. This warranted the development of a safety code by IMO for this specific ship type. In recognizing the lack of extended service experience, IMO has, in the HSC Code, encouraged the use of risk assessment and other formalized hazard assessment techniques to address some safety concerns. However, the code has also specified certain prescriptive requirements that may not be entirely practicable in some circumstances, some of which have been briefly mentioned in this paper. In comparison, some of these prescriptive requirements are actually more onerous than prevailing practice for conventional vessels in similar service. For example, current SOLAS requirements for passenger vessels specify two-compartment damage stability only for vessel length over 80m, while all HSC, regardless of length, are required to satisfy two-compartment damage stability.

As also briefly touched on in the paper, it may be interesting to examine the possibility of rebalancing the "management and reduction of risk" with "passive protection". The former may be further weighted with the intention of lightening the latter – particularly in the light of impracticability in some of the passive protection measures in the code. It appears that perhaps in the future development of the HSC Code, considerations can be given to limiting the two-compartment damage stability and raking damage criteria to HSC of over certain length and speed, and number of passengers, while extending risk mitigating measures to include reducing human errors.

Further in-depth application of FSA and pooling of operational experience may assist to more systematically identify the hazards and the safety golas of high speed craft. More emphasis should be given to differentiating hazards and safety goals based on sizes, types and service areas of these crafts.

References:

1. International Code of Safety for High Speed Craft, 2000, IMO
2. The Merchant Shipping (High-Speed Craft) Regulations 1996, UK
3. Navigation and Vessel Inspection Circular No. 5-01, CH-1, Guidance for Enhancing The Operational Safety of Domestic High-Speed Vessels, USCG
4. Maritime and Coastguard Agency, MCA Research Report – FSA of Shipping, UK

AUTHOR INDEX